Supercritical Fluid Chromatography

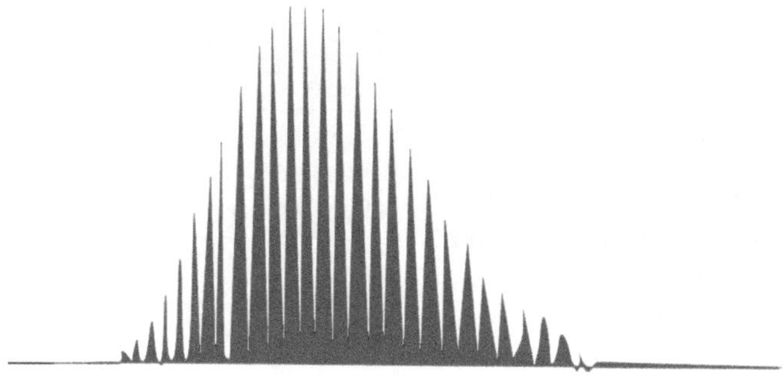

Supercritical Fluid Chromatography

Advances and Applications in Pharmaceutical Analysis

edited by
Gregory K. Webster

PAN STANFORD PUBLISHING

Published by

Pan Stanford Publishing Pte. Ltd.
Penthouse Level, Suntec Tower 3
8 Temasek Boulevard
Singapore 038988

Email: editorial@panstanford.com
Web: www.panstanford.com

British Library Cataloguing-in-Publication Data
A catalogue record for this book is available from the British Library.

Supercritical Fluid Chromatography: Advances and Applications in Pharmaceutical Analysis

ISBN 978-981-4463-00-3 (Hardcover)
ISBN 978-981-4463-01-0 (eBook)

Contents

Preface

After a Pittsburgh Conference symposium, Stanford Chong of Pan Sanford approached me to ask if I was interested in writing a book on supercritical fluid chromatography (SFC). I quickly replied, "No!" Thinking about it further, I saw a need to update the many applications of SFC in the pharmaceutical industry. I told Stanford that I would look to industry experts to help guide me on this endeavor. Thank you, Stanford, and my good friend, Laila, for your support and encouragement on this endeavor.

When I look at supercritical fluid chromatography (SFC), the first concept that comes to mind is that of an analytical toolbox. A past co-worker, or "Pope Van Deempter" as we sometimes called him, often liked to label people by their specialty. He was a "chromatographer." Because my graduate work was mostly in atomic spectrometry, he would call me a "spectroscopist." This would challenge me because (1) my academic days have been in the rear view mirror for a quite a while and (2) such labeling is opposite of what I was taught in regard to analytical chemistry. If I had to be labeled as a scientist, I would much prefer the title "analytical chemist" for a true analytical chemist tries not to marry oneself to a single technique, but uses the "right" technique for the task at hand. We value the generalist who can maintain a broad view as much as the specialist in a single technique. In truth, both are needed in the pharmaceutical industry. The success of SFC is along these same lines. For much of its existence, other techniques could perform in a similar manner, but SFC has several niches where it is the "right tool" for the task at hand. Conversely, whereas SFC can be investigated as a tool for several applications, it should not be used in areas where other techniques have technical, efficiency, or cost advantages.

SFC developed its place in the pharmaceutical industry because it has simply outperformed other techniques in preparative chromatography. From there, it developed other applications at the analytical, preparative, and production scale as scientists looked to broaden their use of this analytical tool. In some areas, such as in chiral analysis, SFC has become the primary technique. In others, such as in achiral analytical chromatography, the advances have been less apparent to date and researchers are still looking for improvements. This book is a mix. We have included areas where SFC is the dominant technique as well as areas where its application is still emerging in the pharmaceutical industry.

When I graduated from college, the pharmaceutical industry accounted for approximately seventy percent of the jobs for young chemists and accounted for most of the sales for analytical instruments. While we all know this number seems to be changing, pharma today still plays a dominant role in the chemical industry. The goal of this book was not to rewrite what others have written and pioneered in SFC but rather to add to this existing body of work as to how SFC is being used in the pharmaceutical industry today. In this last regard, I invited current pharmaceutical industry scientists who are currently using SFC in a specific role to write about their applications. I would like to thank the authors for their contributions resulting from their experiences, dedication, and work. In addition, I would like to thank Phillip Searle, Erin Jordan, Paul David, Cindy Pommerening, Ken Miller, and Christine Havrilla of AbbVie, as well as the authors, for their help in reviewing the content.

I would like to thank my dog, Murphy, who continually reminds me that life is more about the joy of chasing a tennis ball than the stress of meeting deadlines. Finally, I would like to thank my wife, Tammy, who shares my life and reviews my grammar on our wonderful journey together.

Gregory K. Webster
December 2013

Chapter 1

The SFC Market: "Yesterday, Today, and Tomorrow"

Gregory K. Webster
AbbVie, Global Research and Development, 1 N. Waukegan Rd., North Chicago, IL 60064 USA
gregory.webster@abbvie.com

1.1 Introduction

Chemistry majors in the 1980s were aware of the excitement of a new and "revolutionary" technique that was coming. The technique was called supercritical fluid chromatography (SFC). Capillary SFC was thought to be the next great innovation in column chromatography. As undergraduate students, we didn't quite know much about this technique; but the news at the time was that this technique was projected to take over gas and liquid chromatography (LC) and bring analytical separation science into a new dynamic in chromatographic analysis. The use of a supercritical fluid mobile phase had a potential advantage in not only chromatographic efficiencies but also cost and ease of use. However, as with capillary electrophoresis, the advanced marketing and hype of early capillary SFC was never achieved.

Supercritical Fluid Chromatography: Advances and Applications in Pharmaceutical Analysis
Edited by Gregory K. Webster

ISBN 978-981-4463-00-3 (Hardcover), 978-981-4463-01-0 (eBook)
www.panstanford.com

While SFC found a niche in the petrochemical industry where nonpolar aromatics are of interest, the relatively polar nature of pharmaceuticals limited their ability to be analyzed by SFC. The earlier excitement of SFC fizzled. Instead, as a capillary chromatography technique, SFC was challenged by its inability to solvate enough polar molecules to maintain the attention of the chemical industry as a whole, and the pharmaceutical industry in particular. Primarily through the efforts of Berger [1], SFC evolved into a packed column technique that found a niche in preparative and chiral analysis. Before SFC could make strides in achiral analysis, SFC was essentially sidetracked as a technique for effective impurity discrimination by the Ultra High Performance Liquid Chromatography (UHPLC) revolution.

Today's SFC has had a rocky road to get here. As we will see in this text, traversing this path has been worth it. Although still today SFC has yet to match the advanced levels projected in the 1980s, it has established itself as a valuable chromatographic separation tool in the pharmaceutical analytical chemist's toolbox. The introduction of traditional LC column format for packed column SFC eliminated the difficulties many laboratories had in running capillary SFC. The advent of chiral column chromatography created the demand for efficient normal phase separations.

SFC is now the stalwart technique in preparative-scale chiral chromatography and is rapidly becoming the technique of choice for routine analytical applications of chiral chromatography as well. Since nearly 40% of drugs in use are known to be chiral and approximately a quarter of these are administered as pure enantiomers, SFC is involved in a substantial analytical and preparative market.

Today's SFC instrumentation enables the analytical chemist to develop highly efficient chromatographic methods and fast re-equilibration. The dynamics of a supercritical fluid mobile phase enables chromatographic coupling and ease in interfacing with mass spectrometric (MS) detection. Preparative SFC has proven to significantly reduce development costs, minimize waste handling, and replace alkane solvents in many laboratories. Since SFC adds no additional carbon dioxide to the atmosphere, it is designated as a "green" analytical technique. It has several proven advantages over traditional HPLC (Table 1.1). The benefits of modern SFC have lead to the availability of a chromatographic tool that enables fast speeds and high resolution with low operating costs. SFC provides performance now only beginning to be achieved through the use of UHPLC.

Table 1.1 Advantages of SFC over HPLC

Advantages	Opportunities
The higher diffusivity/lower viscosity using supercritical CO_2 mobile phases leads to faster methods with higher efficiency than traditional liquid chromatography. (3-10×)	Analyte solubility in mobile phases
Method development screening systems are faster.	Equipment
Preparative SFC fractions are collected in small volumes of volatile organic solvent.	
SFC mobile phases are more compatible with mass spectrometer systems.	
Higher efficiency in SFC allows preparative injection stacking for rapid collection of fractions.	
CO_2 mobile phases improve operational costs thru reduced solvent consumption and solvent disposal.	
CO_2 mobile phases with alcoholic modifiers that are much greener than other solvents are generally used in NPLC and RPLC.	

With the advances of SFC, chromatography is returning to its true capability in orthogonal analysis. In recent years, differences in C18 stationary phases were deemed "orthogonal" due to the differences in selectivity seen with various phases [2, 3]. Traditionally, chromatographers challenged the purity of their chromatograps by analyzing their sample in both reversed phase and its "orthogonal" compliment of normal phase chromatography. Normal phase chromatography has become less popular for analytical applications over the last few decades, and mass spectrometry has become more routine for chromatographic detection. Thus, challenging method selectivity by changing chromatographic modes became less commonplace, allowing this new definition of orthogonality to creep in. Perhaps with the return of normal-phase chromatography with SFC, the original notion of orthogonality for purity confirmation will re-establish itself in some regard.

1.2 Development of SFC

The first reported use of SFC was in 1962 when Klesper demonstrated the technique [4]. The first commercial system came two decades later in a kit available from Hewlett Packard. The kit could upgrade their HP1084 HPLC. This packed column system could use normal-phase packed columns with ultraviolet (UV), MS, and flame ionization detection (FID). Jasco commercialized the ability to couple a supercritical fluid extraction (SFE) and SFC system in a design HP1084-based SFC system. Jasco's system introduced the mechanical back pressure regulator too. Unfortunately with the introduction of the HP 1090, the SFC modification was no longer available.

Chemists who worked with early SFC fondly remember constructing early laboratory versions of the technique with an old gas chromatograph, syringe pump, and pressure transducer setup. Fixed restrictors were either pulled from silica tubing, packed tubing, or similar designs. While these restrictors were quite capable in maintaining column pressure, analysts recall the struggle of replacing them as they broke with handling or coupling.

Soon companies such as Isco (headquartered in Lincoln, Nebraska, USA) made syringe pumps, pressure programming, restrictors, and detector systems commercially available. SFC began to generate academic and industrial interest. However, the lack of pressure management and fraction collection technologies made early SFC products impractical in commercial operations.

1.2.1 Capillary SFC

Capillary SFC was introduced in 1984 [5]. The first commercially available systems were available from Lee Scientific two years later. The advent of capillary SFC led to great interest in the theory and necessary instrumentation in SFC. Selectivity in the capillary technique was through both column stationary phase selection and the variety of compressed gases available, such as CO_2, N_2O, SF_6, and xenon.

Because of its efficiency in nonpolar alkyl and aromatic compound resolution and simulated distillation capabilities, SFC was successfully implemented in the petrochemical industry. Yet, because of its limited ability to test polar analytes, capillary SFC could not attract interest in the largest segment of analytical chromatographic equipment buyers, the pharmaceutical market. As such, capillary SFC diminished rapidly after 1990.

1.2.2 Packed-Column SFC

In the mid-1990s, Berger Instruments introduced the first commercial modular packed-column SFC system. The development in moving to packed columns and alcohol co-solvents reestablished the analytical viability of normal-phase chromatography. SFC in this format could efficiently chromatograph any compound that could be dissolved in methanol. Like with capillary SFC, this version of SFC produced high chromatographic efficiencies but became more viable to several industries through its new capability in testing significantly more polar compounds. Multiple detectors such as ultraviolet, mass spectrometry, evaporative light scattering, and even atomic emission and infrared spectrometry detectors could be coupled.

Both Gilson and HP introduced commercial analytical-scale SFC systems. The introduction of semipreparative SFC led to library purification capabilities, and the stacked injection algorithm was implemented. Today commercial systems are available from major chromatography vendors such as Agilent, Jasco, and Waters.

1.2.3 Preparative SFC

While not with the flair of capillary SFC, work on preparative SFC continued in parallel to the HP 1084 in the 1980s. Preparative SFC began using CO_2 in petroleum applications [2]. Commercial systems that were manufactured by Prochrom and Novasep led to large-scale SFC for industrial applications. The innovation that enabled preparative SFC to collect fractions was controlling the decompression of the supercritical fluid phase without aerosol formation to allow the eluted compounds remain soluble in the modifier liquid phase. Preparative SFC enabled high-speed separations and low operational costs in compound isolation and recovery. Pharmaceutical industry's new demand for enantiomeric purity and isolations were met with preparative chiral chromatography. Preparative SFC was the method of choice for isolating pure enantiomers from analytical to kilogram scale.

1.3 The SFC Market

The combined market for analytical, preparative, and process SFC is $25–30 million worldwide [6]. Preparative and analytical instrumentation typically cost $60,000–120,000 depending on configuration. Analytical SFC instruments use flow rates up to 20 mL/min, semipreparative instruments from 20–200 mL/min, and preparative instruments capable of purifying up to 100 kg amounts of product have flow rates up to several L/min with columns. The primary market application is high-purity preparative separations of enantiomeric and achiral drugs routinely in use by pharmaceutical companies. SFC has replaced traditional normal-phase HPLC and flash chromatography methods of isolation. Both preparative and analytical SFC are likely to grow in double digits, while process SFC, now accounting for as much as a third of this market, could grow more than 30% in the near term. The largest markets for SFC are the pharmaceutical, agriculture, and the food and beverage industries. Interest is also growing in biofuels and environmental applications.

SFC is considered a niche market, with product demand primarily focused in the U.S., Canadian, and European markets. Market growth is expected to be seen in the developing regions due to the increasing demand for laboratory scale research and development-related applications. The SFC column and aftermarket segment accounts for nearly two-thirds of the purification SFC market demand.

1.4 SFC in the Pharmaceutical Industry

This book builds on the work that began in capillary and packed-column SFC. The goal of this book is to focus on the application of SFC as it is used in today's pharmaceutical industry. Details on the development and theory of capillary SFC [7], packed column [1], and industrial SFC [8] can be found in the literature.

1.4.1 The Use of SFC in Discovery

Today's pharmaceutical industry is tasked with processing new chemical entities in a rapid and efficient fashion. The chromatographic efficiency of SFC has enabled Discovery groups to use rapid chromatography methods to test and isolate thousands of candidates a year. Supercritical fluid mobile phases have enabled rapid and efficient isolation of these candidates upon synthesis. Qifeng Xue and coauthors of Theravance present the current use of SFC in modern pharmaceutical discovery operations.

1.4.2 IQ/OQ/PQ for SFC Instrumentation

Qualifying analytical instrumentation for GxP use is of great importance in the pharmaceutical industry. Ludwig Huber of LabCompliance presents a strategy and the requirements for qualifying SFC instrumentation for compendial and regulated-testing applications.

1.4.3 Method Development for Achiral SFC

With the advent of SFC, normal-phase chromatography is being used in impurity analysis as never before. While reversed-phase method development is commonplace in the pharmaceutical laboratory, most analytical chemists today have not had the opportunity or experience in developing achiral methods by normal-phase chromatography. Jeff Caldwell of Princeton Chromatography, Gregory Webster of AbbVie, and Zhenyu Wang of Merck Research Laboratories present current SFC column technologies and strategy for method development of achiral pharmaceuticals.

1.4.4 Achiral Preparative SFC

Preparative chromatography was developed as a primary tool for the isolation of impurities and small-scale active pharmaceutical ingredients. The cost advantages of using SFC over reversed-phase and normal-phased preparative HPLC have transitioned the technique to primarily use supercritical fluids. In the past five years, with advances in the instrumentation and the operational adoption of similar hardware and software as that widely used in reversed-phase HPLC, SFC has experienced steady growth in acceptance, particularly in pharmaceutical and chemical laboratories for achiral purifications. The incentive of employing green, sustainable SFC technologies in industrial processes as well as in analyses has also increased. To embrace this renaissance, in this chapter, Vivi Lazarescu, Mark Mulvihill, and Lifu Ma of OSI Pharmaceuticals provide an overview of the development of the achiral preparative SFC field since 2000 and its evolution from a standard technique of chiral separation to a viable tool for high throughput purification of challenging isomeric mixtures and crude library compounds. After a historical review of commercial SFC instrumentation and stationary phases, we will focus on the integration of SFC technique as a routine purification tool to pharmaceutical discovery laboratories.

1.4.5 SFC for Chiral Method Development Screening and Analysis

With many of the new chemical entities under development in the pharmaceutical industry being chiral, the specific stereochemistry of these substances affects their biological activity and commercial viability of the potential new drug. Thus, enantioselective separation techniques play a vital role in the development of these entities into commercial product.

In an attempt to improve upon the efficiency of chiral method development, column manufacturers and industry scientists have developed screening procedures to efficiently evaluate various chiral separation conditions in an unattended mode. Gregory Webster of AbbVie and Ted Szczerba of Regis Technologies present the focus of current method development strategy using SFC.

1.4.6 Chiral Preparative SFC

The demand for chiral separations has existed for decades and increased markedly since it was discovered that certain pharmaceutical molecules consist of enantiomers with highly divergent activities. Chiral preparative SFC is the leading technology now for small molecule enantiomer separations. Manuel Ventura of Amgen reflects on the development of the technologies utilized in chiral preparative SFC. It describes the stationary phases and mobile phases utilized as well as the basic instrument components necessary for the application.

The general standard of practice currently employed in the pharmaceutical industry for chiral purification of unknown racemates is given. The description of this process is made from analytical method development to preparative separation and fraction collection. The way the preparative SFC equipment functions to facilitate this process is also explained. Also described are strategies to achieve specific purification outcomes with various applications of SFC stationary phases and mobile phases. This includes techniques to enhance solubility and selectivity through modification of standard solvents used in the preparative SFC to a brief description of supercritical fluid recycling separation techniques.

1.4.7 SFC in Process Analytical Chemistry

Process analytical chemistry serves as a critical function in the drug development process. A variety of analytical techniques are needed to support analytical research and development activities. SFC has been used in process analytical chemistry to meet the increasing demand for better efficiency. In this chapter, the typical applications of SFC in process analytical chemistry are summarized with examples as the highlight.

In-process testing of chemical synthesis schemes often requires testing samples in synthesis matrices that are not compatible with reversed-phase chromatography. These matrices, however, are compatible with SFC mobile phases. As such, SFC has become a primary technique for pharmaceutical in-process testing. Yanqun Zhao of AbbVie presents how SFC is being used in modern process analytical chemistry and chemical in-process testing for pharmaceutical synthesis.

1.4.8 Analytical SFC for Impurities

Supercritical fluid chromatography has been used primarily for chiral purification and enantiomeric purity assessment for non-GMP activities during drug discovery and development due to its historically low sensitivity as well as not meeting the insufficient accuracy and precision required for GMP testing. In recent years, with the constant improvement in the sensitivity and the advancement of the instrumentation, SFC can meet the method validation requirements to be used for GMP API release and stability testing.

Yun Huang of Pfizer presents the advances SFC has made in GMP assay and related substances applications with the pharmaceutical industry. This chapter presents how to implement SFC method in GMP testing. First, the chapter provides the pioneer experience on how to qualify the SFC system for GMP use, including the system qualification protocol write up and acceptance criteria definition. Second, the chapter also demonstrates the feasibility of applying SFC method in GMP testing by validating chiral SFC methods to compare with the corresponding HPLC methods. The successful validation of the SFC methods proves that SFC can meet the acceptance criteria for chromatographic purity method, thus can be the alternative to replace HPLC for API clinical release and stability testing in a GMP-regulated laboratory. In addition, the method transferability among different SFC systems was studied and confirmed the suitability of the chiral SFC as a regulatory filing method.

Alone with continuous improvement on SFC instrumentation and further exploration on how to better apply the technology in the pharmaceutical analysis, SFC will be widely implemented as one of the major fast and efficient separation tools for pharmaceutical industry. In the near future, SFC will not be limited only to the chiral analysis; SFC in achiral impurity analysis is on the way in the regulated analytical QC laboratories.

1.4.9 SFC-MS

Although SFC-MS has been around since the late 1980s, it has not become as prevalent as liquid chromatography–mass spectrometry , since the bulk of the SFC work, until recently, was mostly done employing preparation chromatography. With more sensitive instruments coming online, the analytical interest in SFC has increased, and along with that there has been an increase in SFC-MS activities.

This chapter describes the progression of the technology, the types of interfaces, back-pressure regulators, and analyzers that have been coupled to SFC. Laila Kott of Millennium Pharmaceuticals presents an overview of chiral SFC-MS and prep SFC-MS in the pharmaceutical industry.

1.4.10 SFC of Natural Products

As the pharmaceutical industry has moved to incorporate biological drugs into its portfolio along with small molecules drugs, the need has developed to analyze these higher molecular weight compounds. SFC is ideally suited to solvate and chromatograph these compounds at low temperatures. Y. Wang of Novartis presents how SFC is being used in natural product testing of pharmaceuticals.

Natural products have historically been the most important source of drugs for the treatment of various diseases. As their response to needs and challenges of the natural environment, natural products have been apparently optimized through the evolution of many centuries. Therefore, they cover an unsurpassed chemical diversity with unique breadth of biological functions. The separation and purification of natural products are quite challenging, but it also stimulates the development and integration of innovative technologies, which have higher performance, selectivity, and throughput in both analytical and preparative scales. As a continuum between high-performance LC and gas chromatography, SFC has been becoming an important complementary technique to other chromatographic methods in the field of natural products. Numerous biologically active natural products have been separated by SFC, including various secondary metabolites from plant, such as sesquiterpenes, diterpenes, triterpenes, alkaloids, and flavonoids, and some macrolides, cyclic, or linear peptides and polyethers from microorganisms. Different optimization strategies, methodologies, and processes for both capillary SFC and packed-column SFC techniques were reviewed with selected representative examples. Established methods and optimized parameters should be very useful for the analysis of each major metabolite types. Concise technique features in each reported case study, as well as some technical failures in the early feasibility studies, may also support the method development for new applications. In our systematic method development with SFC for microbial metabolites purification, the results also showed clearly that SFC is a versatile and powerful technique in natural product research. As a complimentary method to other techniques, SFC will make its unique contributions to high-throughput purification and play an important role to further increase the chemical diversity of natural products for new drug discovery.

1.4.11 Polarimetry Detection in SFC

Chiral chromatography has made significant advances in the pharmaceutical industry, mostly with traditional ultraviolet-visible (UV-Vis) detection. Yet, for the most part, UV-Vis detection provides little ai in distinguishing is a chiral pair is present within a particular chromatographic profile. Thus, the process of chiral method development depends upon not only resolution but identification of the optically active peaks detected. Gary Yanik of PDR Chiral discusses the advantages of incorporating polarimetry for chiral-specific detection in SFC.

1.4.12 New Frontiers in SFC–USFC

SFC has been well adopted for pharmaceutical analysis for its efficiency and economical benefits. Its higher mobile-phase diffusivity and low viscosity have been well utilized to overcome some outstanding issues and complement in some important application fields in which conventional LC has inherent limitations. The concept to integrate and develop ultra-performance particles (less than 2–3 μm in diameter) has long been a desired direction to further improve the performance and to offer new growth area.

Both theoretical and practical operational aspects with ultra-performance particles are discussed; considerations of the benefits vs. potential negative impacts have been carefully studied, and supporting evidences for arguments have been presented. It demonstrated that the overall impact of ultra-performance particles has improved the technique from current state of art level, and the implementation of instrumentation shall promote this technology.

Ziqiang Wang of Waters Corporation discusses recent advances on application and development across the fields, and instrumentation developments with ultra-performance particles have been demonstrated and discussed.

Since the 1990s, SFC has remained a packed-column technique using 3 to 5 μm particle stationary phases. The revolution of UHPLC has made sub-2 μm particles for chromatographic stationary phases a commercial reality and available for USFC application as well. Zi Wang of Waters presents the theory, advantages, and instrumentation of USFC for pharmaceutical analysis.

1.4.13 Pilot-Scale SFC

Since its inception, SFC has been a pilot- and plant-scale technique and not simply an analytical-scale tool. Scale SFC from laboratory to commercial operations has been the focus of the pharmaceutical industry as it searches for economical tools to bring new drugs to market. Geoff Cox of PIC Solution presents the current technology that is being used to enable SFC to be used in pilot-scale operations.

Despite the acceptance of SFC as a green chemistry technique and its rapid adoption as a preparative technique over the past few years, it so far has not been used to a significant extent in production processes. The chapter discusses the potential benefits in terms of production rate and costs as well as some of the perceived obstacles to incorporation of the technique into production processes. The available techniques for large-scale processing, including simulated moving bed–supercritical fluid chromatography (SMB-SFC), are described. Finally a short review of the separations that have been scaled up to pilot and production is presented.

References

1. Berger, T.A. (1995). *Packed Column SFC* (The Royal Society of Chemistry, Cambridge, UK).
2. Pellett, J., Lukulay, P., Maob, Y., Bowen, W., Reed, R., Mac, M., Mungerc, R.C., Dolan, J.W., Wrisley, L., Medwide, K., Toltl, N.P., Chanf, C.C., Skibic, M., Biswas, K., Wells, K.A., Snyder, L.R. (2006). "Orthogonal" separations for reversed-phase liquid chromatography. *J. Chromatogr. A*, **1101**, 122–135.
3. West, C., Lesellier, E. (2006). Orthogonal screening system of columns for supercritical fluid chromatography. *J. Chromatogr. A*, **1203**, 105–113.
4. Klesper, E., Corwin, A.H., Turner, D.A (1962) High pressure gas chromatography above critical temperature. *J. Org. Chem.* **27**, 700–701.
5. Novotny, M., Springston, S.R., Peaden, P.A., Fjeldsted, J.C., Lee, M.L. (1981) Capillary supercritical-fluid chromatography. *Anal. Chem.*, **53**, 407A–411A.
6. 2007 Strategic Directions International Inc, Los Angeles, CA 90045.

7. Lee, M.L. , Markides, K.E (1990) *Analytical Supercritical Fluid Chromatography and Extraction* (Chromatography Conferences, Inc., Provo, UT).
8. Dean, J.R. (1993) *Applications of Supercritical Fluids in Industrial Analysis* (Blackie Academic, London).

Chapter 2

The Use of SFC in Discovery Sciences

Kanaka Hettiarachchi, Andersen Yun, May Kong, John R. Jacobsen, and Qifeng Xue
Research Analytical, Department of Medicinal Chemistry,
Theravance Inc., 901 Gateway Boulevard, South San Francisco, CA 94080, USA
QXue@Theravance.com, KHettiarachchi@Theravance.com

2.1 Introduction

In recent years, the aggressive pace of drug discovery programs has placed increasing emphasis on implementing high-throughput analytical and preparative technologies to support medicinal chemistry in delivering novel high-purity molecules for testing. Conventional liquid chromatography coupled with mass spectrometry (LC-MS) is used extensively in drug discovery for compound identification, purification, purity assessment, as well as in pharmacokinetic studies [1–3]. Many discovery analytical laboratories have automated the initial sample preparation and screening process [4,5]. High-throughput sample screening and purification has become an indispensable tool to enable the rapid exploration of chemical space [6–8]. Mass-directed collection of

Supercritical Fluid Chromatography: Advances and Applications in Pharmaceutical Analysis
Edited by Gregory K. Webster

ISBN 978-981-4463-00-3 (Hardcover), 978-981-4463-01-0 (eBook)
www.panstanford.com

fractions, as well as highly automated software systems to track samples and associated chromatographic data, increases both the efficiency and quality of purification and is now commonplace [9,10]. The need for rapid turnaround places limitations on the time available for method development and drives the need for systems that can reliably provide acceptable separations with minimal development time, often for sets of chemically diverse compounds. Screening and purifying isomers (enantiomers, diastereomers, etc.) at the early stages of discovery are also growing in importance [11,12]. Supercritical fluids with lower viscosities and better diffusion properties as mobile phases in packed-column chromatography have improved the analysis and handling of these pharmaceuticals, as well as achiral ones, with a fast purification turnaround time and decreased solvent consumption [13,14].

2.2 High-Throughput Screening and Purification

High-throughput screening and purification are critical in the strategy of a company for finding drug-like compounds [15,16]. During the exploratory stage, pharmaceutical companies can screen a million or more compounds by employing combinatorial and parallel synthesis. Successful isolation is a priority during all stages of a project, and instrument reliability is essential. As the chemistry becomes more focused, fewer compounds are produced and high purity becomes critical for cell biology and pharmacokinetic studies.

2.2.1 Chromatographic Technologies

Utilizing complementary chromatographic screening and purification technologies is necessary in high-throughput environments to maximize the output of pure compounds [17,18]. Reversed-phase liquid chromatography (RPLC) is often the technique of choice as many drug-like compounds can be effectively screened and purified with appropriate reversed-phase methods. More recently, ultra-high-performance liquid chromatography (UHPLC) and the use of small particle (<2 μm) columns have allowed for rapid 1–2 min analytical-scale separations at higher efficiencies and a wider range of flow rates but at much higher ~1200 bar back pressures [19]. Supercritical fluid chromatography (SFC) has also experienced a growing acceptance in pharmaceutical laboratories this past decade because of significant savings in operation costs and improved efficiency at low ~120 bar back pressures [20]. Implementing these high-throughput technologies is essential in the drug discovery sciences to maximize the rate of success in finding lead compounds from diverse sets of projects with short turnaround time.

2.2.2 Laboratory Workflow

The Research Analytical Group at Theravance operates a dedicated, highly automated facility with a multi-step process (Fig. 2.1) set up for rapid compound library screening and purification support for Medicinal Chemistry. Implementing a specific workflow has significantly improved efficiency and reduced the turnaround time of sample purification. By employing rational drug design by using structural information of validated targets, potential leads across several chemical classes can be quickly identified from focused libraries on the scale of hundreds to thousands of compounds.

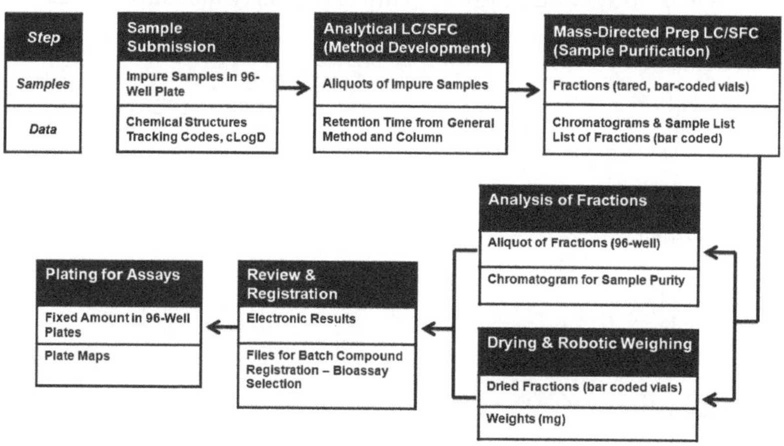

Figure 2.1 Workflow in a high-throughput discovery environment.

The laboratory has several commercial mass-directed preparative RPLC systems as well as a mass-directed preparative SFC system with the ability to handle challenging separations on high-value compounds as well as arrays containing dozens of 10–100 mg-scale samples. Prior to purification, elution times and presence of impurities are quickly determined on the basis of the results of scouting runs with analytical RPLC-MS or SFC. After a suitable column has been selected, a focused gradient approach can be utilized to save time and maximize resolution in order to isolate the compound of interest by mass. Figure 2.2 shows typical results of high-throughput screening and purification of a compound with RPLC-MS and SFC-MS. For most achiral separations, RPLC or SFC may be employed successfully. Selection is often determined by instrument availability and quantity of material submitted. However, with SFC we can typically expect a factor of 3× improvement in purification time. Same day delivery of data after purity analysis, evaporation, and weighing is a reality when performing SFC. Batch turnaround time for RPLC is typically two to three business days. Sample purity criteria are established on the basis of individual project chemistry and biology; from either technique, 95% is an achievable target in many cases. These instruments can handle several hundred samples per week. Because of prior sample screening, the purification success rate is >90%. Purified compounds are delivered in formats ready for biological assays and physicochemical property measurements.

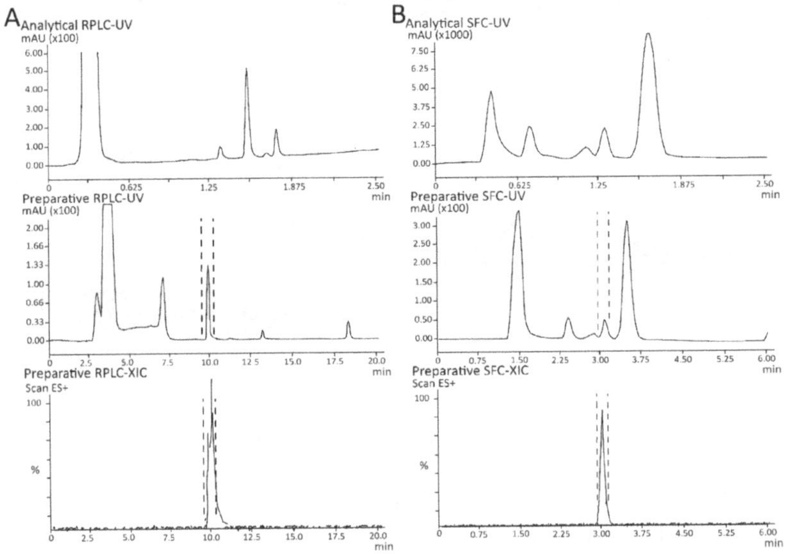

Figure 2.2 Chromatograms illustrating the different selectivity between (a) RPLC-MS and (b) SFC-MS in mass-directed achiral purification of a compound.

An automated electronic data-flow path from sample submission to registration of final compound allows for a fast and error-free process. To date, the group has purified greater than 150,000 compounds from rationally designed arrays. Working in conjunction with other departments has enabled Research to deliver 23 development candidates since 1998, which has translated into a robust pipeline of small molecules that have demonstrated clinical proof of concept.

2.3 Implementation of SFC

Supercritical fluid chromatography is increasingly used for method development, analytical screening, and improved compound (chiral and achiral) purification efficiency with reduced solvent consumption [21,22]. The greatly increased facility of chiral separations is an added benefit that has been a significant driver for the development of SFC purification. Continued advances in the technology have made SFC an attractive alternative to RPLC for general purification of reaction products [23,24].

2.3.1 SFC Fundamentals

The low viscosity of supercritical CO_2 and higher diffusivity of a solute in SFC allow for higher separation speeds and/or the use of longer columns [25]. An empirical formula for this principle is the van Deemter equation (Eq. 2.1), which describes the relationship between linear velocity (flow rate) and plate height H (HETP or column efficiency) [26]. Term A is the coefficient of eddy diffusion, B the coefficient for longitudinal diffusion, C the coefficient for analyte mass transfer kinetics between the mobile and stationary phases, and μ the mobile phase linear velocity.

$$H = A + \frac{B}{\mu} + C\mu \tag{2.1}$$

As shown in Fig. 2.3, the SFC and UHPLC curves are much flatter in comparison to HPLC in the high-flow region due to the reduced C-term. The C-term correlates directly with peak broadening.

The resolution (R_s) equation (Eq. 2.2) describes the quality of chromatographic separations of adjacent bands. Parameters t_1 and

t_2 are the retention times of the two peaks, and W_1 and W_2 are the baseline widths obtained from the intersection of the tangents with the baseline. Adequate separation can be defined as the condition when $R_s \geq 1.0$ and is useful in determining parameters for scale-up.

$$R_s = \frac{2(t_2 - t_2)}{W_1 + W_2} \tag{2.2}$$

In SFC, one can expect higher peak resolution and increased loading capacity without compromising efficiency or running at the high-pressure regime of UHPLC. In practice, this can enable higher throughput, reduced need for method development, and increased utility for difficult applications such as chiral separations.

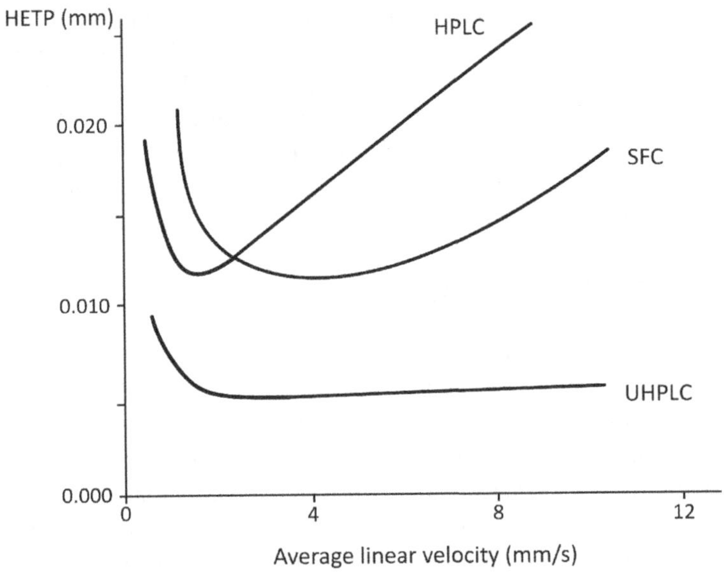

Figure 2.3 Comparison between HPLC, SFC, and UHPLC on the effect of flow rate on plate height.

2.3.2 Benefits of SFC

SFC is a versatile chromatographic tool that complements RPLC well in improving selectivity [27,28]. Analytical SFC has some unique advantages over conventional analytical HPLC and UHPLC such as the ability to use higher flow rates, improved separation efficiency, lower solvent usage, and lower operational costs [29]. Analytical SFC screening is useful to determine the retention time and method selection for preparative systems. Trace-level pharmaceutical impurity analysis with SFC is also possible [30]. The implementation of an MS-directed SFC preparative instrument enables us to significantly increase the efficiency in purifying compound arrays by taking advantage of high flow rates (80–100 mL/min) and short (5–10 min) run times with faster post-purification compound recovery [31–33]. Addition of the mass spectrometer increases the likelihood of collecting the desired compound by triggering on detected mass. These new systems can have similar hardware components and software as those present in conventional MS-directed RPLC systems.

The use of CO_2 as a mobile phase in SFC reduces waste generation and leaves only a small volume of methanol modifier to evaporate. This volume can be dried in a few hours, allowing for same day reporting of results in some cases. In comparison, RPLC purification of complex samples can be time-consuming to achieve high-purity requirements. This can constrain the number of compounds that can be purified during a typical workday such that overnight operation is generally required to meet throughput objectives. RPLC mobile-phase systems also generate significant volumes of chemical waste, and evaporation of collected aqueous fractions is slow, requiring run times that can last 12–24 hours or more depending on volume. Acetonitrile, commonly employed as a mobile phase in RPLC, is a major contributor to the cost of purification. Shortages experienced in the past have led to extreme price fluctuations and uncertainty about the availability of sufficient quantities of solvent to maintain operations [34].

Table 2.1 summarizes typical experimental conditions for preparative RPLC and SFC, and compares waste generation and cost savings for compound purification in a given year. Our RPLC-MS purification systems produce on average 4800 L of solvent (ACN + H_2O) waste per year and cost the company >$80,000 in disposal fees. With SFC-MS, only a fraction (35% of RPLC) of waste as methanol would be produced per year for the same number of injections, dramatically reducing costs.

Table 2.1 Comparison of key experimental parameters in preparative RPLC and SFC, including waste generation for a typical year

	RPLC-MS	SFC-MS
Conditions		
Method	20–60% ACN/H_2O 95%/10% : 1/19 min	5–50% MeOH/CO_2 No 2nd column regeneration
Column	21 × 250 mm^2 C18	21 × 250 mm^2 Pyridine
Flow rate	20 mL/min	100 mL/min
Run time	20 min/inj	7 min/inj + 3 min regeneration
Flow rate	20 mL/min	100 mL/min
Injections	6000	6000
Results (in 1 year)		
Waste	960 L ACN | 342 L ACN 1440 L H_2O | 2058 H_2O	1800 L of MeOH 0 L CO_2
Total waste	4800 L	1800 L
Total time	2000 hours	1000 hours

Normal-phase liquid chromatography (NPLC) has its own unique separation benefits but is not suited for mass-directed fraction collection. The use of volatile and flammable solvents with NPLC reduces the practicality of the technique for routine chiral separations. Chiral separations, however, are well suited for SFC. Most drug discovery programs involve synthetic chemical compounds, either a pure enantiomer or diastereomer that was required for biological screening and late-stage drug candidate development. Testing the correct stereoisomer is mandated by the FDA, and stereoselective synthesis is quite challenging [35,36]. SFC can often separate enantiomers without extensive method development or stationary phase screening, making this a practical approach to obtaining single stereoisomers when they are not readily accessible through stereoselective synthesis [37–39]. This can avoid the testing of mixtures of stereoisomers at an early stage of discovery and can postpone the need for stereoselective syntheses and optimization until later when the value of a lead chemical series is well established.

2.4 Instrumentation

2.4.1 Analytical Instrumentation

The analytical reversed-phase UHPLC system is a Nexera series system from Shimadzu Corporation (Columbia, MD), consisting of a SIL-30AC autosampler, LC-30A pumps, CTO-30A column oven, and a SPD-M20A photodiode array (PDA) detector. The system is connected to a Shimadzu 2020 mass spectrometer containing an electrospray ionization (ESI) source and operated with Shimadzu LabSolutions software. This is our primary system for comparing RPLC analytical screening results with SFC data. The mobile phases are LC-MS grade water (A) and acetonitrile (B) from Honeywell Burdick & Jackson, both containing 0.1% TFA. For UHPLC, the 2.1 mm × 50 mm and 2.1 mm × 100 mm analytical columns used are Zorbax RRHD Bonus RP from Agilent (Santa Clara, CA), each with a particle size of 1.8 μm and capable of withstanding 1200 bar. The flow rate is 1 mL/min (4 min methods) for the longer column and 1.7 mL/min (2 min methods) for the shorter column.

Our current analytical SFC instrument, a Thar (a Waters Company, Pittsburgh, PA) Method Station SFC System, serves as an orthogonal tool for impurity analysis, and provides column and modifier selection allowing us to quickly screen various achiral and chiral columns and modifiers. The system consists of a Thar fluid delivery module, Alias autosampler, Analytical-2-Prep 10 column oven (set at 40°C), automated back pressure regulator (set at 120 bar), Waters 2996 PDA, and a Thermo DigitalOne coolant circulator. The mobile phases are LC-MS grade methanol (without additives) from Honeywell Burdick & Jackson and liquid CO_2, provided by Praxair Incorporated (Danbury, CT). The system is operated with Waters MassLynx software.

There are many achiral and chiral SFC columns available with different packing materials and surface properties that show different performance in separating various compounds. For achiral separation studies with this analytical SFC system, the 4.6 mm × 250 mm columns used are a 2-ethylpyridine and silica from Princeton Chromatography (Cranbury, NJ) with a particle size of 5 μm. For chiral separation studies, the 4.6 mm × 250 mm analytical columns used are Chiralcel OD-H, OJ-H, and Chiralpak AD-H from Chiral Technologies (West Chester, PA).

2.4.2 Preparative Instrumentation

Both the semi-preparative to preparative-scale LC-MS and SFC-MS systems are Waters Corporation instruments. Each system consists of a 2767 sample manager providing open bed fraction collection, a 2998 PDA coupled with a 3100 series mass spectrometer containing an ESI source, and full software control provided by MassLynx software. The preparative LC-MS contains high-pressure 2545 solvent pumps and a dual-column fluidics organizer. The SFC-MS Prep-100 system is additionally equipped with TharSFC high-pressure CO_2 and solvent pumps, 6-port preparative column-switching column oven (set at 40°C), preparative automated back pressure regulator (set at 120 bar), tunable splitter, and a gas–liquid separator (GLS). An Airgas Engineering Solutions (Oakwood Village, OH) dual gas booster pumping system provides the necessary 900–1100 psig delivery range from two 230 L liquid CO_2 cylinders with a 300 psig inlet pressure to the pump.

For reverse-phase purification, the mobile phases are LC-MS grade water (A) and acetonitrile (B) from Honeywell Burdick & Jackson, both containing 0.1% TFA. The flow rate is 20 mL/min. The 21.2 mm × 150 mm and 21.2 mm × 250 mm preparative columns used are Zorbax Bonus RP C14 from Agilent and Luna C18 from Phenomenex (Torrance, CA), respectively. For SFC purification, the mobile phases are LC-MS grade methanol (without additives) from Honeywell Burdick & Jackson and liquid CO_2 from Praxair Incorporated. The flow rate is 100 mL/min, density of CO_2 is 0.89–0.91 g/cm^3, and back pressure is 120 bar. For chiral separations, the 21.2 mm × 250 mm preparative SFC columns used are OD-H and AD-H from Chiral Technologies. For routine achiral separations, the 21.2 mm x 250 mm preparative SFC columns used are 2-ethylpyridine and silica columns from Princeton Chromatography.

Method gradients are determined on the basis of analytical scouting runs, and fractions are collected by MS triggering on the $(M+H)^+$ ion or UV at a specified wavelength (typically 214 nm). Collected fractions are dried on a GeneVac HT 24 evaporator (Genevac LTD, Gardiner, NY), weighed on a Bohdan automated robotic workstation by Mettler-Toledo AutoChem (Columbia MD), and analyzed by gradient analytical LC-MS to determine the purity (UV area percent at 214 nm).

2.5 Enantiomeric Separation in SFC

2.5.1 Chirality and Chromatographic Separation

Chirality has a large impact on drug properties such as potency, selectivity, metabolism, and toxicity, and performing enantioselective analysis with SFC can shorten the early lead optimization stage of drug discovery [40]. Adequate column screening is necessary to determine which chiral stationary phase (CSP) and modifier combination will provide the best separation.

Four types of polysaccharide-based CSPs (amylose AD, AS, and cellulose OJ, OD derivatives) were selected for screening because they have proved to be able to resolve more than 80% of the chiral drugs currently on the market [37]. The Chiralpak AD-H and Chiralcel OD-H chiral columns are a particularly good combination for initial enantiomeric separation testing [41].

2.5.1.1 RPLC and SFC Separation of Two Diastereomers

We examined the separation of stereoisomers from a crude sample of a compound with a mixture of configurations at two stereocenters, as a test case. For reverse-phase analytical screening, 4 min gradient or isocratic methods were used with a 100 mm column and a 2 min wash and re-equilibration time was added at the end of each method. C14 column chemistry (Agilent Zorbax RRHD Bonus RP) demonstrated partial separation of diastereomers (Fig. 2.4a) using a 55%B isocratic run over 4 min at 50°C and 1000 bar. Individual isomer separation was not possible. The best NPLC separation of stereoisomers was performed using a Chiralpak AD-H column with a 5%B isocratic method over 50 min, but complete isomer separation was not possible.

It is often not feasible to separate enantiomers with RPLC mode due to the lack of availability of suitable CSPs. In our experience, chiral compounds that have shown poor resolution in RPLC or NPLC can be baseline resolved with SFC at much shorter run times. For SFC, 4 min gradient or isocratic methanol in CO_2 methods were used with 250 mm columns at a flow rate of 8 mL/min and a system back pressure of 120 bar. A 1 min wash and re-equilibration step was added at the end of each method. By using the Chiralpak AD-H column, the separation between diastereoisomers was possible.

The mixture was resolved well into four individual stereoisomers (Fig. 2.4b) using a Chiralcel OD-H column with an isocratic 4 min method. In fact, we were able to obtain resolutions greater than 1.0 for the first two isomer peaks with all isocratic methods that ranged from 10–17%B from runs that lasted 2, 4, and 8 min using the OD-H column. The last two peaks had resolutions greater than 0.9. The analytical method that provided the largest resolution (R_s = 1.7 at 10%B isocratic) was selected for scaling up to preparative scale.

The 20 mg crude sample containing four stereoisomers (two diastereomeric pairs of enantiomers) was then purified by RPLC-MS using a 45–60%B method over 30 min on a 21.2 mm × 150 mm Agilent Zorbax Bonus RP C14 column (Fig. 2.4c). Two injections at 10 mg per injection were performed and compound recovery was 82 ± 2%, with 100% calculated purity by UV. While the diastereomers were efficiently separated and purified by RPLC, resolving the pairs of enantiomers was only possible by SFC. NPLC purification was not attempted on the basis of the analytical screening results described in the previous section. Instead, approximately 15 mg of the same crude sample was then purified by SFC-MS using an isocratic 10%B method over 8 min on a 21.2 mm × 250 mm Chiralcel OD-H column (Fig. 2.4d). For improving resolution, four injections at 3 mg per injection were performed and each of the four stereoisomers was baseline separated and successfully purified, with compound recovery at approximately 90%.

SFC was found to be ideal for fast small molecule chiral separations and analysis for our Medicinal Chemistry group. We are able to shorten our chiral run times by a factor of 4× and improve the separation while reducing the fraction collection volume (3 mL MeOH compared to 30 mL aqueous in RPLC). Implementation of a preparative SFC for chiral work can be beneficial, as RPLC is not ideally suited for high-throughput analysis or purification of chiral compounds due to separation inefficiencies. Chiral separation by NPLC has some advantages over chiral RPLC, but NPLC requires long run times (35–50 min methods on 250 mm columns) and is incompatible with some classes of drug-like compounds. An entire chiral method development study with SFC can be performed within an hour using 4 min methods with four different columns and three different modifier combinations, with the back pressure, temperature, and flow rate remaining the same.

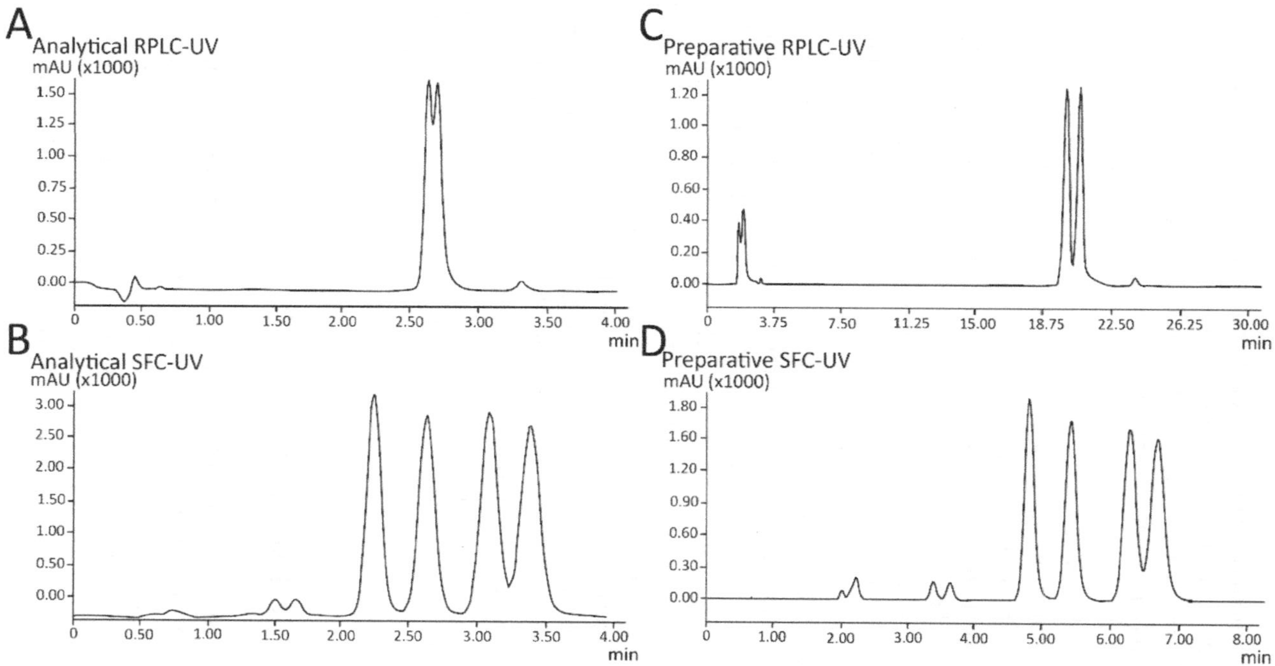

Figure 2.4 (a) Screening of a chiral compound with an Agilent Zorbax RRHD Bonus RP 2.1 mm × 100 mm UHPLC column. (b) Chiral screening with SFC using a 4.1 mm × 250 mm Chiralcel OD-H column. Chiral purification with (c) RPLC-MS using a 21.2 mm × 150 mm Agilent Zorbax Bonus RP column over 30 min and (d) SFC-MS using a 21.2 mm × 250 mm Chiralcel OD-H column over 8 min.

2.5.1.2 SFC Separation of a Lipophilic Prodrug

Analytical support of drug discovery programs often requires complex multi-component sample analysis, including those of prodrugs and metabolites. Prodrugs [42] increase the clinical usefulness of a drug when the parent molecule has pharmacokinetic based issues due to lower oral bioavailability, inefficient cellular uptake, and poor metabolism. Enhanced water solubility is important for parenteral or oral administration, while an increase in lipophilicity enhances cellular uptake. In this example, a phosphate prodrug variant was prepared as a mixture of diastereomers due to chirality, and a separation scheme was investigated.

Although a prodrug can be separated by traditional extraction, crystallization, or RPLC chromatographic techniques, SFC was chosen for the solvent system benefits (no trifluoroacetic acid and fast drying times). Analytical SFC methods using cellulose and amylose CSPs were developed for separation of the lipophilic prodrug. The separated diastereomers can then be tested for activity. The mobile phase organic modifier was also varied to achieve the best resolution. When moving from methanol to a less polar alcohol such as ethanol, an increase in retention time was observed. An optimal baseline separation ($R_s > 1.5$) was obtained with the amylose-based column by using an ethanol-hexane modifier combination. The two diastereomers were eventually separated with SFC on a preparative scale by using a 15–35%B method over 8 min on a 21.2 mm × 250 mm Chiralpak AD-H column with methanol as the eluent (Fig. 2.5). In this case, the shallow gradient was sufficient to separate the compounds from the mixture. Changing the nature of the organic modifier is often not desirable for preparative SFC.

2.6 Achiral Separation in SFC

SFC offers some significant advantages for preparative processes and can lead to higher throughput in achiral purification. In fact, the majority of compounds received in our laboratory in a given week are achiral mixtures, and purification is routinely performed on mass-directed RPLC systems using dual 21.2 mm × 150 mm C14 columns or 21.2 mm × 250 mm C18 columns. Each purification lasts 10, 20, or 30 min, and the duration is determined by the difficulty of separation on the basis of analytical scouting runs and purity requirements. SFC can complement RPLC well in terms of selectivity and can even be considered a normal-phase separation technique but without the issues of lengthy run times and MS compatibility. Achiral purification with SFC is 2–4× faster than RPLC, but this significant speed advantage may be sacrificed if better resolution is required. For testing the system, two achiral columns (2-ethylpyridine and silica) were selected and then used for purification of the basic compounds that we routinely receive.

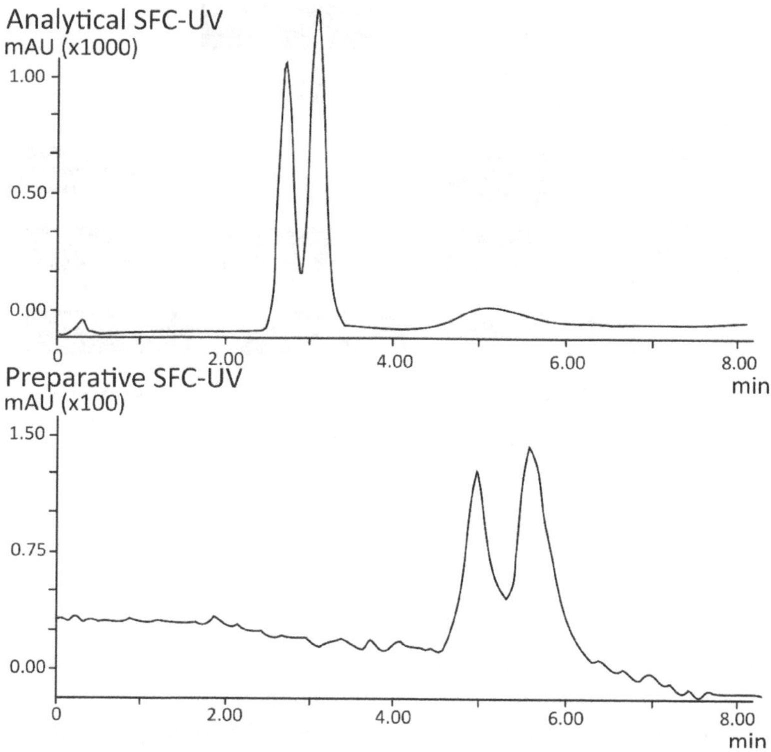

Figure 2.5 Separation of a chiral prodrug compound with a 4.1 mm × 250 mm analytical and 21.2 mm × 250 mm preparative ChiralPak AD-H column over 8 min.

2.6.1 Screening Samples

By pre-screening samples with SFC and defining a separation strategy, it is possible to quickly determine if purification is possible with the appropriate column or even necessary (Fig. 2.6). Our analytical instrument can select from 10 different columns and 6 solvent modifiers for rapid, automated method development. In SFC, column, flow rate, and modifier are the most common parameters adjusted for improving retention and selectivity. Temperature and pressure are secondary variables but can significantly impact selectivity in some cases (data not shown).

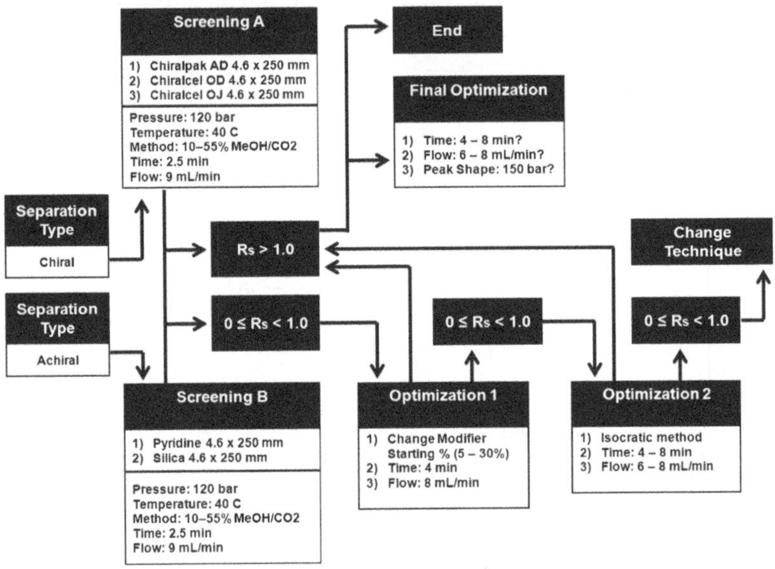

Figure 2.6 General scheme of a separation strategy in SFC.

2.6.1.1 Separation of Routine Compounds

Both analytical UHPLC and SFC techniques are successful in separating compounds from submitted arrays. However, the separations between UHPLC and SFC are different, primarily due to the different selectivity between the two techniques. The lower back pressure generated in SFC supports higher flow rates than are possible with RPLC. The regular use of 250 mm columns with SFC (compared to 50–150 mm columns typically employed in RPLC purification) offers more stationary phase to improve the separation and allows for higher sample loading. High flow rates (10 mL/min for analytical and 100 mL/min for preparative) allow for faster separations without generating excessive back pressure. An inversely linear relationship between flow rate and retention time is observed for all of the routine compounds tested with SFC (Fig. 2.7).

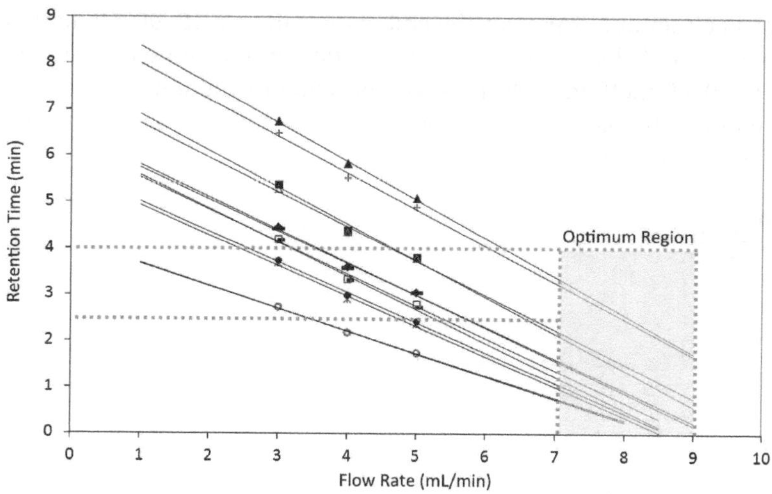

Figure 2.7 Optimal flow rate determination from elution order of multiple compounds at 3, 4, 5 mL/min with a 4.1 mm × 250 mm 2-ethylpyridine column. The gradient is set with modifier content from 10–55% MeOH in 6 min.

As outlined in the separation strategy, 2.5 min methods at 9 mL/min are often sufficient for initial screening. Run times can easily be varied from 2–12 min depending on separation difficulty. The reproducibility of the retention times after multiple injections is very good for analytical SFC, and we have observed a typical variation in retention time of only +/– 0.02 min.

2.6.1.2 Challenging Separations

Mixtures that contain strongly polar or highly hydrophobic compounds with close running impurities present some of the most significant separation challenges. The successful separation of hydrophilic compounds (Fig. 2.8a) and highly hydrophobic compounds (Fig. 2.8b) by both UHPLC and SFC is possible. However, the retention times and selectivity are vastly different for the same sample using either the UHPLC or SFC technique. For the example shown in Fig. 2.8a, the desired peak elutes at 35%B near the middle of the run on a 2–50%B method with UHPLC. With SFC, the compound elutes at 30%B near the middle of the run on a 5–45%B method and with improved baseline resolution. For the highly hydrophobic compound separation shown in Fig. 2.8b, the desired peak elutes at 82%B near the end of the run on a 40–90%B method with UHPLC, but comes out at the middle of the run with SFC on a 10–50%B method with improved baseline resolution. Six different compounds spanning two projects were tested with SFC, all having values of $R_s > 0.9$.

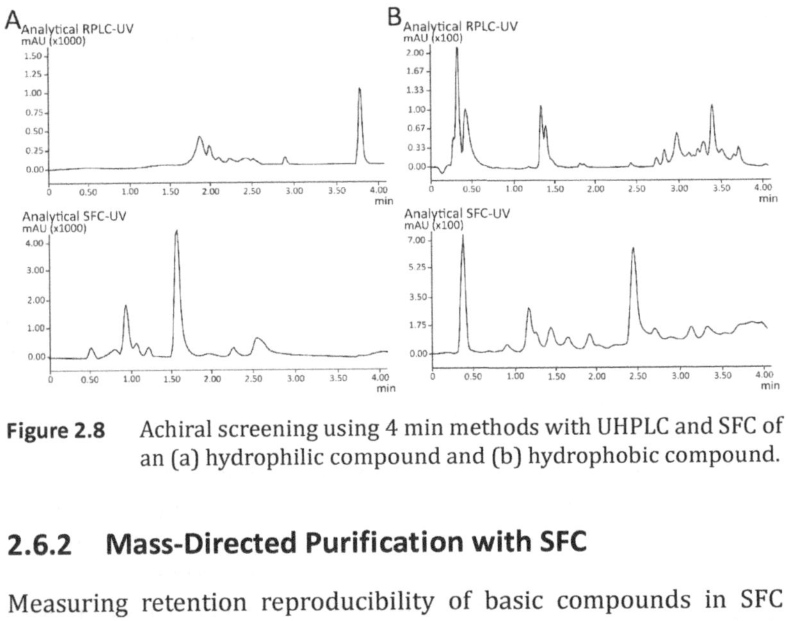

Figure 2.8 Achiral screening using 4 min methods with UHPLC and SFC of an (a) hydrophilic compound and (b) hydrophobic compound.

2.6.2 Mass-Directed Purification with SFC

Measuring retention reproducibility of basic compounds in SFC and observing peak shape are important for maintaining a high level of integrity among varying batches from multiple projects. Among the compounds we routinely test for system performance, one is a neuroleptic agent with anticholinergic, hypotensive, and sedative effects. A series of such compounds were separated and the neuroleptic agent was collected by mass detection using SFC-MS with an ESI source.

A 15 mg crude sample containing multiple compounds as well as the desired agent (approximately 2.7% in the crude) was purified by SFC-MS using a 5–50%B method over 7 min on a 21.2 mm × 250 mm Princeton Chromatography 2-ethylpyridine column (Fig. 2.9). Six injections at approximately 2.5 mg per injection were performed for low sample volume performance testing. With SFC, all peaks were sharp without significant fronting at 2.5 mg loading. Retention reproducibility was excellent for each compound in all 6 injections with a standard deviation of <4 s. Some peak tailing was observed due to the lack of additives in the MeOH modifier solution. Even with just 2.7% of the target compound ($m/z = 285$) in the crude mixture, a recovery of 70% was achieved with SFC and >95% purity between collected fractions.

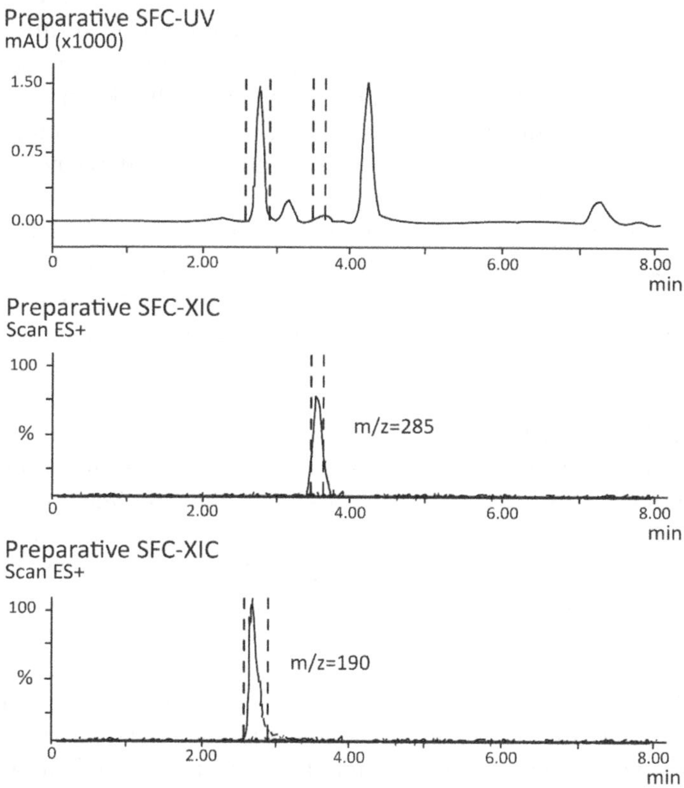

Figure 2.9 Mass-directed purification of multiple compounds with SFC.

2.6.3 Achiral Purification Comparison of RPLC-MS and SFC-MS

Purification of first lot pharmaceutical compounds from Medicinal Chemistry is an ongoing laboratory priority and has taken the bulk of instrument time on our RPLC-MS systems during the past decade. In order to increase efficiency and reduce turnaround time, standardized methods have now been adopted for our analytical and preparative SFC systems. These methods are compatible with our overall laboratory workflow.

This example involves an approach for difficult achiral separations and purifications on both RPLC and SFC. After scouting runs revealed the separation profiles on analytical instruments, the 15 mg crude sample containing approximately 6 mg of the desired agent was first purified by RPLC-MS using a 65–85%B method over 20 min on a 21.2 mm × 250 mm Phenomenex Luna C18 column. The compounds formed profiles with some peak tailing using RPLC-MS and a very close running impurity was observed (Fig. 2.10a). The compound was successfully separated from the impurity, and recovery was 70% (4.2 mg), with 100% purity by UV.

The same crude mixture was concurrently purified with SFC-MS on a 21.2 mm × 250 mm 2-ethylpyridine column from Princeton Chromatography. With SFC, a 5–50%B method was used over 5 min. One 15 mg injection was performed and some peak tailing was observed due to the lack of additives in the MeOH modifier solution (Fig. 2.10b). The compound recovery was excellent at 96% (5.8 mg), with 95% purity by UV.

With SFC we saw a factor of 4× improvement in purification time (5 min vs. 20 min) for this achiral separation, although 2× to 3× is typical. When the optimal fraction collection thresholds are applied, preparative mass-directed SFC can provide a similar success rate in terms of recovery when compared with mass-directed RPLC. The same day delivery of data after purity analysis, evaporation, and weighing was achievable.

RPLC-MS and SFC-MS systems provide complementary approaches to routine purifications, and the availability of both methods is helpful for meeting the throughput demands and isolations performed at early stages of drug discovery. Despite the differences in technology from RPLC-MS, we have found that SFC-MS is easily integrated into our existing automated laboratory workflow. Data from either purification system can be transferred to a custom built software tool on the network for management of chemical analytical data generated in our lab and remote viewing via a web browser. Figure 2.11 shows the results in a graphical format of the achiral sample purified with RPLC-MS and SFC-MS. While eliminating

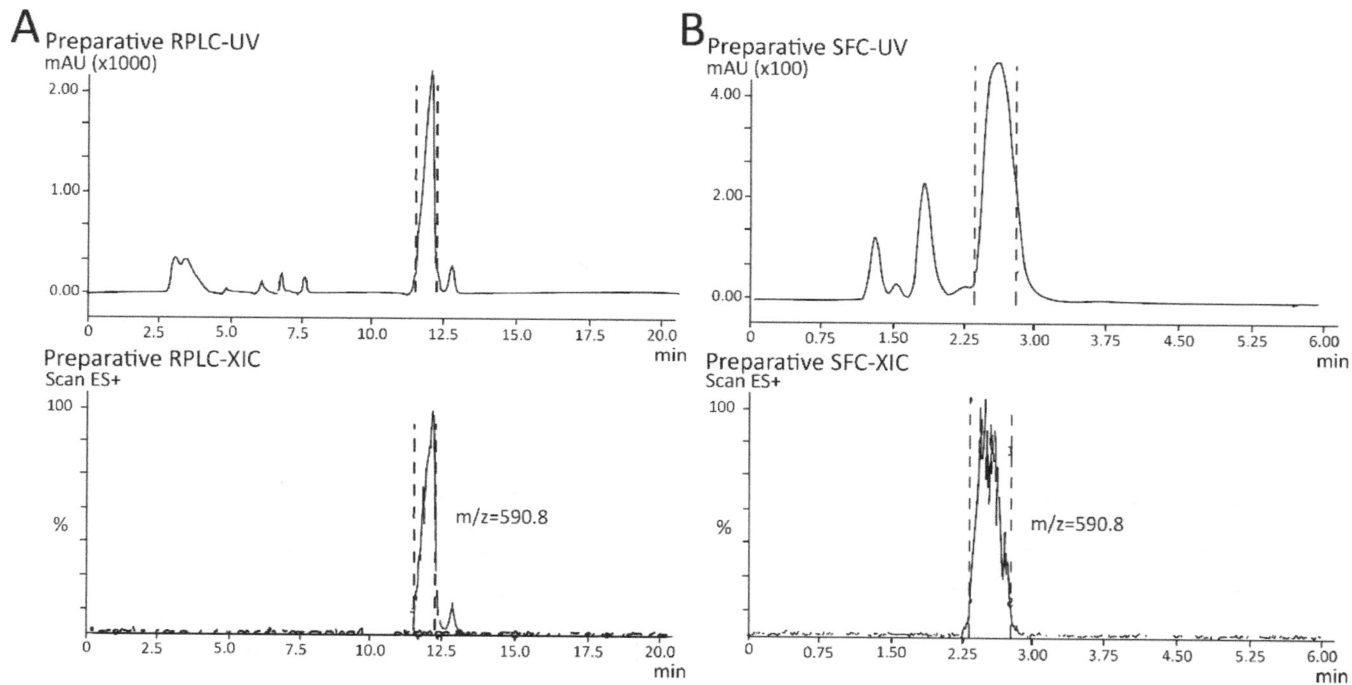

Figure 2.10 Achiral purification with (a) RPLC-MS using a 21.2 mm × 250 mm C18 column over 20 min and (b) SFC-MS using a 21.2 mm × 250 mm 2-ethylpyridine column over 5 min.

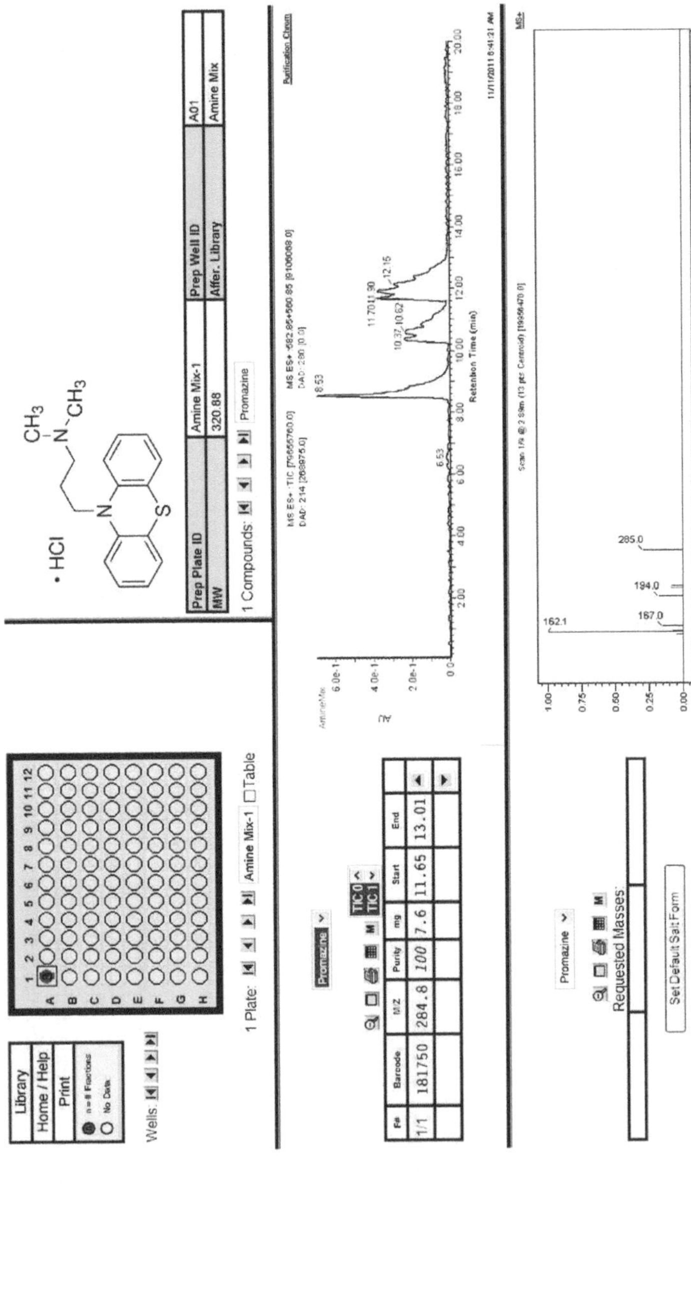

Figure 2.11 Data browser for internal compound review and registration.

paper waste, this tool also improves the efficiency in data review and registration of purified compounds into the corporate database.

2.7 Remarks of SFC in Drug Discovery

The current climate in pharmaceutical research calls for the continuous adoption of new technologies that provide increased speed with decreased cost. SFC technology is one efficient solution that can be leveraged to address the challenges faced in Medicinal Chemistry and drive future progress. One of the main attractions for implementing SFC is the fact that it is a green technology (using liquid CO_2, a by-product in generating other gases from air). SFC is an acetonitrile-free chromatography technique that reduces waste and operating costs.

Recent advances in sample collection have improved the reliability of SFC systems [43], but the technology is at a much earlier stage of development than RPLC and instrument reliability is likely to improve further in coming years. For high throughput SFC to be routinely used in drug discovery labs and allow for overnight library runs, interruptions in the automated purification of compounds must be minimized to levels of preparative RPLC systems.

Significant progress has been made in integrating SFC into our automated high-throughput research laboratory. SFC in conjunction with RPLC covers most of our chemical space. Analytical screening with SFC allows for expedited method development. SFC reduces our achiral purification time in half and eliminates the need for overnight runs, improving throughput and reducing costs. We are confident that our investment in broadening the purification infrastructure to include SFC will enable many separations that were difficult or impossible with RPLC and that our discovery programs will benefit from these added capabilities. It is clear that both technologies will continue to be valuable tools for the foreseeable future.

References

1. Zeng, L., and Kassel, D. B. (1998). Developments of a fully automated parallel HPLC/mass spectrometry system for the analytical characterization and preparative purification of combinatorial libraries. *Anal Chem*, **70**(20), 4380–4388.

2. Lee, M. S., and Kerns, E. H. (1999). LC/MS applications in drug development. *Mass Spectrom Rev*, **18**(3–4), 187–279.

3. Korfmacher, W. A. (2005). Foundation review: principles and applications of LC–MS in new drug discovery. *Drug Discov Today*, **10**(20), 1357–1367.

4. Villa, J. S., Cass, R. T., Karr, D. E., Adams, S. M., Shaw, J.-P., and Schmidt, D. E. (2004). Increasing the efficiency of pharmacokinetic sample procurement, preparation and analysis by liquid chromatography/tandem mass spectrometry. *Rapid Commun Mass Sp*, **18**(10), 1066–1072.

5. Isbell, J. J., Zhou, Y., Guintu, C., Rynd, M., Jiang, S., Petrov, D., et al. (2005). Purifying the masses: integrating prepurification quality control, high-throughput LC/MS purification, and compound plating to feed high-throughput screening. *J Comb Chem*, **7**(2), 210–217.

6. Weller, H. N., Young, M. G., Michalczyk, S. J., Reitnauer, G. H., Cooley, R. S., Rahn, P. C., et al. (1997). High throughput analysis and purification in support of automated parallel synthesis. *Mol Divers*, **3**(1), 61–70.

7. Yan, B., Collins, N., Wheatley, J., Irving, M., Leopold, K., Chan, C., et al. (2004). High-throughput purification of combinatorial libraries I: a high-throughput purification system using an accelerated retention window approach. *J Comb Chem*, **6**(2), 255–261.

8. Liu, M., Chen, K., Christian, D., Fatima, T., Pissarnitski, N., Streckfuss, E., et al. (2012). High-throughput purification platform in support of drug discovery. *ACS Comb Sci*, **14**(1), 51–59.

9. Xu, R., Wang, T., Isbell, J., Cai, Z., Sykes, C., Brailsford, A., et al. (2002). High-throughput mass-directed parallel purification incorporating a multiplexed single quadrupole mass spectrometer. *Anal Chem*, **74**(13), 3055–3062.

10. Goetzinger, W., Zhang, X., Bi, G., Towle, M., Cherrak, D., and Kyranos, J. N. (2004). High throughput HPLC/MS purification in support of drug discovery. *Int J Mass Spectrom*, **238**(2), 153–162.

11. Caner, H., Groner, E., Levy, L., and Agranat, I. (2004). Trends in the development of chiral drugs. *Drug Discov Today*, **9**(3), 105–110.

12. Piras, P., and Roussel, C. (2008). A screening study of ChirBase molecular database to explore the expanded chiral pool derived from the application of chiral chromatography. *J Pharmaceut Biomed*, **46**(5), 839–847.

13. Crowther, J. B., and Henion, J. D. (1985). Supercritical fluid chromatography of polar drugs using small-particle packed columns with mass spectrometric detection. *Anal Chem*, **57**(13), 2711–2716.

14. Taylor, L. T. (2009). Supercritical fluid chromatography for the 21st century. *J Supercrit Fluid*, **47**(3), 566–573.

15. Ripka, W. C., Barker, G., and Krakover, J. (2001). High-throughput purification of compound libraries. *Drug Discov Today*, **6**(9), 471–477.

16. Isbell, J. (2008). Changing requirements of purification as drug discovery programs evolve from hit discovery. *J Comb Chem*, **10**(2), 150–157.

17. Ventura, M., Farrell, W., Aurigemma, C., Tivel, K., Greig, M., Wheatley, J., et al. (2004). High-throughput preparative process utilizing three complementary chromatographic purification technologies. *J Chromatogr A*, **1036**(1), 7–13.

18. Miller, L., and Potter, M. (2008). Preparative chromatographic resolution of racemates using HPLC and SFC in a pharmaceutical discovery environment. *J Chromatogr B*, **875**(1), 230–236.

19. Jerkovich, A. D., Mellors, J. S., Jorgenson, J. W., and Majors, R. E. (2003). The use of micrometer-sized particles in ultrahigh pressure liquid chromatography. *LC GC Asia Pac*, **6**(4), 8–12.

20. White, C. (2005). Integration of supercritical fluid chromatography into drug discovery as a routine support tool Part I. Fast chiral screening and purification. *J Chromatogr A*, **1074**(1–2), 163–173.

21. McClain, R. T., Dudkina, A., Barrow, J., Hartman, G., and Welch, C. J. (2009). Evaluation and implementation of a commercially available mass-guided SFC purification platform in a high throughput purification laboratory in drug discovery. *J Liq Chromatogr R T*, **32**(4), 483–499.

22. Searle, P. A, Glass, K. A, and Hochlowski, J. E. (2004). Comparison of preparative HPLC/MS and preparative SFC techniques for the high-throughput purification of compound libraries. *J Comb Chem*, **6**(2), 175–180.

23. Pinkston, J. D., Wen, D., Morand, K. L., Tirey, D. A., and Stanton, D. T. (2006). Comparison of LC/MS and SFC/MS for screening of a large and diverse library of pharmaceutically relevant compounds. *Anal Chem*, **78**(21), 7467–7472.

24. Li, F., and Hsieh, Y. (2008). Supercritical fluid chromatography-mass spectrometry for chemical analysis. *J Sep Sci*, **31**(8), 1231–1237.

25. Brunelli, C., Zhao, Y., Brown, M.-H., and Sandra, P. (2008). Development of a supercritical fluid chromatography high-resolution separation method suitable for pharmaceuticals using cyanopropyl silica. *J Chromatogr A*, **1185**(2), 263–272.

26. Van Deemter, J. J., Zuiderweg, F. J., and Klinkenberg, A. (1956). Longitudinal diffusion and resistance to mass transfer as causes of nonideality in chromatography. *Chem Eng Sci*, **5**, 271–289.

27. Gere, D. R., Board, R., and McManigill, D. (1982). Supercritical fluid chromatography with small particle diameter packed columns. *Anal Chem*, **54**(4), 736–740.

28. Berger, T. A, Fogleman, K., Staats, T., Bente, P., Crocket, I., Farrell, W., et al. (2000). The development of a semi-preparatory scale supercritical-fluid chromatograph for high-throughput purification of "combi-chem" libraries. *J Biochem Bioph Meth*, **43**(1–3), 87–111.

29. White, C., and Burnett, J. (2005). Integration of supercritical fluid chromatography into drug discovery as a routine support tool Part II. Investigation and evaluation of supercritical fluid chromatography for achiral batch purification. *J Chromatogr A*, **1074**(1–2), 175–185.

30. Wang, Z., Zhang, H., Liu, O., and Donovan, B. (2011). Development of an orthogonal method for mometasone furoate impurity analysis using supercritical fluid chromatography. *J Chromatogr A*, **1218**(16), 2311–2319.

31. Ventura, M. C., Farrell, W. P., Aurigemma, C. M., and Greig, M. J. (1999). Packed column supercritical fluid chromatography/mass spectrometry for high-throughput analysis. Part 2. *Anal Chem*, **71**(19), 4223–4231.

32. Wang, T., Barber, M., Hardt, I., and Kassel, D. B. (2001). Mass-directed fractionation and isolation of pharmaceutical compounds by packed-column supercritical fluid chromatography/mass spectrometry. *Rapid Commun Mass Sp*, **15**(22), 2067–2075.

33. Aurigemma, C., and Farrell, W. (2010). FastTrack to supercritical fluid chromatographic purification: implementation of a walk-up analytical supercritical fluid chromatography/mass spectrometry screening system in the medicinal chemistry laboratory. *J Chromatogr A*, **1217**(39), 6110–6114.

34. Welch, C. J., Brkovic, T., Schafer, W., and Gong, X. (2009). Performance to burn? Re-evaluating the choice of acetonitrile as the platform solvent for analytical HPLC. *Green Chem*, **11**(8), 1232–1238.

35. Ariëns, E. J. (1986). Chirality in bioactive agents and its pitfalls. *Trends Pharmacol Sci*, **7**, 200–205.

36. Shinkai, I. (1997). Design and development of practical asymmetric syntheses of drug candidates. *Pure Appl Chem*, **69**(3), 453–458.

37. Zhao, Y., Woo, G., Thomas, S., Semin, D., and Sandra, P. (2003). Rapid method development for chiral separation in drug discovery using

sample pooling and supercritical fluid chromatography–mass spectrometry. *J Chromatogr A*, **1003**(1–2), 157–166.

38. Zeng, L., Xu, R., Zhang, Y., and Kassel, D. B. (2011). Two-dimensional supercritical fluid chromatography/mass spectrometry for the enantiomeric analysis and purification of pharmaceutical samples. *J Chromatogr A*, **1218**(20), 3080–3088.

39. Hamman, C., Wong, M., Hayes, M., and Gibbons, P. (2011). A high throughput approach to purifying chiral molecules using 3 μm analytical chiral stationary phases via supercritical fluid chromatography. *J Chromatogr A*, **1218**, 3529–3536.

40. Helmy, R. O. Y., Biba, M., Zang, J. I. A., Mao, B., Fogelman, K., Vlachos, V., et al. (2007). Improving sensitivity in chiral supercritical fluid chromatography for analysis of active pharmaceutical ingredients. *Chirality*, **19**, 787–792.

41. Toribio, L., Bernal, J. L., del Nozal, M. J., Jiménez, J. J., and Nieto, E. M. (2001). Applications of the Chiralpak AD and Chiralcel OD chiral columns in the enantiomeric separation of several dioxolane compounds by supercritical fluid chromatography. *J Chromatogr A*, **921**(2), 305–313.

42. Rautio, J., Kumpulainen, H., Heimbach, T., Oliyai, R., Oh, D., Jarvinen, T., et al. (2008). Prodrugs: design and clinical applications. *Nat Rev Drug Discov*, **7**, 255–270.

43. Ebinger, K., Weller, H. N., Kiplinger, J., and Lefebvre, P. (2011). Evaluation of a new preparative supercritical fluid chromatography system for compound library purification: the TharSFC SFC-MS Prep-100 system. *J Lab Auto*, **16**(3), 241–249.

Chapter 3

Qualification of SFC Hardware and Validation of Systems

Ludwig Huber
Labcompliance, Obere Alm 27, Oberkirch 77704, Germany
Ludwig_huber@labcompliace.com

3.1 Introduction

GxPs require that each analytical instrument used in the generation, measurement, and the evaluation of analytical data is suitable for its intended use. This means, instruments should be well designed and qualified to ensure compliance with pre-determined specifications.

Equipment qualification and validation of computerized systems cover the entire life of a product. It starts when someone has a need for a specific product and ends when the equipment is retired. Computer system validation ends when all records on the computer system have been migrated and validated for accuracy and completeness to a new system. Because of the length of time

Supercritical Fluid Chromatography: Advances and Applications in Pharmaceutical Analysis
Edited by Gregory K. Webster

ISBN 978-981-4463-00-3 (Hardcover), 978-981-4463-01-0 (eBook)
www.panstanford.com

and complexity, the process has been broken down into shorter phases, so called lifecycle phases. Several lifecycle models have been described for qualification and validation.

This chapter will utilize the USP terminology and the 4Q lifecycle model steps as recommended by chapter <1058>, "Analytical Instrument Qualification" [1]. Since its release in 2008, this chapter has become the global standard for analytical instrument qualification for the pharmaceutical industry. It is the preferred qualification procedure for commercial supercritical fluid chromatography (SFC) instruments, and with some modifications it also can be used for the validation of configurable computerized SFC systems.

This chapter provides an overview on the USP process for qualification, validation, and maintenance of SFC hardware and systems. Detailed information on the process is available in the Agilent Technologies *Analytical Instrument Qualification and System Validation* primer [2].

3.2 Analytical Instrument Qualification According to USP <1058>

USP <1058> adopts the 4Q lifecycle using these four phases: design qualification (DQ), installation qualification (IQ), operational qualification (OQ), and performance qualification (PQ). The process is illustrated in Fig. 3.1. In the DQ phase, user requirements are compared with the vendor's specifications. In addition, users conduct an assessment of the vendor. In the IQ phase the user's environment is checked if it meets the vendor's environmental specifications. The instrument is installed according to the vendor's recommendations and correct installation is verified and documented. OQ checks if the instrument conforms to the functional specifications, as defined in the DQ phase. PQ verifies that the complete system works for selected applications. Preventive maintenance activities and controlled changes also are part of this phase.

All activities are defined in a validation or qualification plan and results are documented in a summary report. Configurations are defined in the requirement specifications document for configurable computerized SFC systems. They are implemented during the installation and tested as part of OQ.

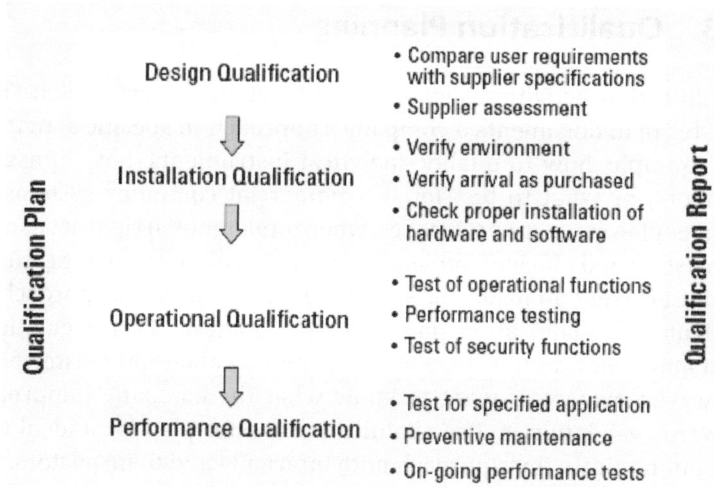

Figure 3.1 Analytical instrument qualification according to the 4Q model.

Figure 3.2 illustrates which questions should be answered during the four phases along with tasks and who usually performs the tasks. It also has information on the test objectives, the type of sample and the type of column used for testing.

	DQ	IQ	OQ	PQ
Question to be answered	Is the instrument's design suitable for the intended application?	Is the lab suitable for the instrument? Is the instrument correctly installed?	Does the instrument work in the the user's environment according to specifications?	Is the instrument performance suitable for the intended application?
Tasks	Compare user requirements with instrument specifications	Prepare lab environment. Install and verify correct installation	Qualify equipment operational specifications in the user's environment	Conduct system suitability tests
Who performs the tasks	User(s) assisted by vendor(s)	User(s) and vendor(s)	User(s) or vendor(s)	User(s)
Test sample	N/A	Generic	Generic	Application specific
Column	N/A	None or generic	Generic	Application specific

Figure 3.2 4Q phase test diagram.

3.3 Qualification Planning

Qualification activities should be laid out in a master plan. The master plan documents a company's approach to specific activities, for example, how to qualify analytical instruments, how to assess vendors, or what to test for in commercial computer systems. A master plan serves two purposes: when implemented right, it ensures consistent and efficient implementation of equipment qualifications, and it answers an inspector's question of a company's approach to instrument qualification and system validation. FDA regulations and guidelines do not specifically require a validation master plan. However, inspectors want to know what the company's approach towards validation is. The qualification master plan is an ideal tool to communicate this approach both internally and to inspectors.

In case there are any questions as to why things have been done or not done, the master plan should provide the answers.

Within an organization a validation master plan can be developed for

- the entire company at a corporate level;
- multiple or single sites;
- departments;
- system categories.

The master plan is a framework for individual project plans and should be written at the highest level possible. This ensures consistent implementation across an organization.

Equipment and computer validation master plans should include

1. introduction including the plan's scope, e.g., sites, systems, processes;
2. responsibilities, for example, user departments, QA, IT;
3. related documents, for example, risk management master plan;
4. products/processes to be validated and/or qualified;
5. qualification/validation approach;
6. risk assessment;
7. steps for equipment qualification and computer system validation with examples on type and extent of testing;
8. vendor assessment;

9. handling existing systems;
10. change control procedures and templates;
11. instrument obsolescence and removal;
12. training plans (system operation, GMP);
13. glossary;
14. attachments with templates and examples.

For each individual project, a validation project plan should be developed. This plan is derived from the validation master plan.

The project plan outlines what is to be done in order to get a specific system into compliance. For inspectors, it is a first indication of the control a laboratory has over a specific instrument or system and it also gives a first impression of the qualification quality.

For simple equipment qualification, a template in table form can be used to outline planned activities. An example template is shown in Fig. 3.3. The left column can be the same for all instruments in the same category, which makes the whole qualification process very efficient.

Scope	This plan outlines qualification activities for SFC equipment hardware.
Product description	An SFC system consisting of a CO_2 pump, modifier pump, back pressure regulator, automated liquid sampler, column heater, and UV/Vis diode array detector. Instrument control and data acquisition is performed through a chromatographic data system.
Validation strategy	The 4Q lifecycle model is used. Individual modules are tested as a part of the SFC system. This holistic test approach should be used, whenever reasonable.
Responsibilities	The end-user department has the ultimate responsibility that the system is qualified. The department leader will decide on a case-by-case basis who will be responsible for individual qualification activities. This person is known as the system owner.
Supplier assessment	ISO 9001 or equivalent is sufficient for supplier assessment.
Risk assessment	Risk assessment will be performed by the system owner according to ICH Q9. Risk assessment includes defining a risk category for the SFC system, major risk factors and mitigation steps for high risk factors.
Testing strategies	Performance functions as specified in the requirement specification list will be tested in the end-user department. Exceptions: functions that are not impacted by the user's environment.

DQ	Includes: 1. A comparison of requirement specification, vendor specifications and confirmation that the vendor's product design meets the user's specification. 2. A statement that the vendor designs, develops and manufactures the analytical equipment in a controlled quality environment. 3. A vendor statement that support requirements will be met.
IQ	IQ is performed by the vendor and reviewed and signed by the system owner.
OQ	OQ is performed before initial use. Requalification is performed annually and after significant system changes. The system owner decides who will perform OQ: the end user department or the vendor.
PQ	PQ tests will be performed by the end user on an ongoing basis: daily or whenever the system is used.
Traceability matrix	A traceability matrix links product requirement specifications to OQ tests. It is created by the system owner.
Procedures	Procedures are required for: DQ, IQ, OQ, PQ, change control, maintenance, operation, training and problem reporting.
Approval	Qualification documents will be developed under supervision of the system owner, approved by the user department's supervisor and QA.
Documentation control	Qualification documents will be created, reviewed, approved and archived following the laboratory's SOP for document control.

Figure 3.3 Elements and content examples for the SFC hardware qualification project plan.

3.4 Design Qualification

Activities in the DQ phase should ensure that the design of the instrument is suitable for the user's applications and that the instrument is developed, manufactured, tested, and supported by vendors with a certified quality system. Design qualification is a shared responsibility between the vendor and the user of an instrument.

The user writes requirement specifications for the instrument. This includes all functions the instrument should have and the performance specifications the equipment should meet as required for the intended application. Next, the user compares his/her specifications with the vendor's specification sheet. If the vendor's specifications are equal or better than what is required, the design is qualified for the intended use. Also included in the DQ phase is a formal vendor assessment. This can be made based on experience with the vendor, through a mail audit or through a direct audit.

Specific vendor activities

- Design, develop and manufacture instruments in a quality control environment.
- Provide functional and operational product specifications.
- Provide information on how software and instruments are validated during development and supported during the entire life of the products.
- Allow user audits, if required, and share approaches for development and testing.

Specific user activities

- Describe the analysis problem and selection of the technique.
- Describe the intended use of the equipment.
- Describe the intended environment (including computer environment).
- Select and document the functional and performance specifications (technical, environmental, safety).
- Verify that the vendor's functional and operational product specifications conform to the user requirements.
- Select and assess the vendor.
- Document the rationale for selecting the specific SFC equipment and the vendor.

3.4.1 The Importance of Requirement Specifications

DQ should ensure that instruments have all the necessary functions and performance criteria that will enable them to be successfully implemented for the intended application and to meet business requirements. Errors in DQ can have a tremendous technical and business impact, and therefore a sufficient amount of time and resources should be invested in the DQ phase. For example, setting wrong operational specifications for an SFC system can substantially increase the workload for OQ testing, and selecting a vendor with insufficient support capability can decrease instrument up-time with a negative business impact.

Figure 3.4 shows a template for documenting user requirements and vendor specifications of an SFC system. The user defines his/

Functions and performance	User requirements	Vendor specification	Passed failed
Autosampler			
Capacity			
Injection volume range			
Injection volume precision			
Sample carry over			
CO_2 Pump			
Pressure range			
Flow rate range			
CO_2 Pump flow rate precision			
CO_2 Pump flow rate accuracy			
Modifier Pump			
Modifier pump flow rate precision			
Modifier pump flow rate accuracy			
Step gradient composition accuracy at 10, 50 and 90%			
Back pressure regulator range			
Back pressure regulator precision			
Column Heater			
Column heater temperature range			
Column heater temperature precision			
Column heater temperature accuracy			
Column capacity			
Column length			
Diode Array Detector			
Baseline noise (peak to peak at 254 nm)			
Baseline drift			
Linearity			
Wavelength range			
Wavelength accuracy			
Software			

Figure 3.4 Template for user requirements and vendor specifications of an SFC hardware.

her requirement specifications and compares them with the vendor's specifications. The exact user requirement specifications depend on the intended use of the system. To set the functional and performance specifications, the vendor's specification sheets can be used as guidelines. However, it is not recommended to simply copy the vendor's specifications, because compliance to the functional and performance specifications must be verified later in the process during operational qualification and also when requalifying the instrument at a later time. Specifying too many functions and setting the values too stringently will significantly increase the workload for OQ. For example, if a company has a need for an isocratic SFC system but plans to purchase a gradient system for future use, only an isocratic system should be formally specified for regulatory purposes. This means, as long as the instrument is not used for gradient runs, no gradient test needs to be conducted. Later on, when the system is used for gradient analysis, the specifications should be changed through a change control procedure.

The specifications should be set so that there is a high likelihood that the instrument conforms to them, not only during initial OQ but also during requalification, for example, a year later. Otherwise users may be expected to initiate an investigation to determine if the non-qualified instrument could have had a negative impact on the quality of the product. For example, these possibilities are expressed in ICH Q7 [3]: "Deviations from approved standards of calibration on critical instruments should be investigated to determine if these could have had an impact on the quality of the intermediate(s) or API(s) manufactured using this equipment since the last successful calibration."

3.4.2 Vendor Assessment

Vendors of analytical instruments should be qualified through a formal process. The objective is to ensure that vendors provide the quality of products and can give adequate support. For basic equipment, such as pH-meters or a balance, this can be a single-page statement describing why the vendor XY has been selected. Certification for a recognized quality system is sufficient for simple instruments. The formal assessment statement should be supported by the quality system's certificate. Figure 3.5 shows a template with examples to document vendor assessment criteria for analytical instruments.

For more complex systems, especially for critical computer systems, such as chromatographic data systems, a more detailed assessment is recommended. Depending on the complexity and criticality of the system, this can be a mail audit, third-party audit, and a direct audit through the user firm.

Assessment Item	Requirement	Findings
Establishment in the market place	Vendor must be a leading supplier of SFC systems.	Vendor is a leading supplier of SFC systems. Vendor is also a leading supplier of equipment for pharmaceutical quality control laboratories.
Expertise with FDA regulated industries	Vendor must be recognized as a leading supplier of products and compliance services for FDA-regulated industry.	Vendor has several times been ranked as the #1 supplier of products and compliance services for FDA regulated industry.
Experience with the vendor	Our company must have good experience with the vendor.	Our laboratory already has 2 SFC systems with different configurations and 10 HPLC systems from the vendor. Experience with the vendor regarding instrument and support quality is very good.
Quality system	Vendor must have ISO 9001:2008 certification.	Vendor is ISO 9001:2008 certified.
Training	Vendor must provide onsite operator training.	Vendor provides onsite operator training.
Installation qualification	Vendor must install the product and perform installation qualification.	Vendor installs the product and performs installation qualification.
Operational qualification	Vendor must provide operational qualification services.	Vendor provides operational qualification services for SFC systems.
Product support	Vendor must provide phone and on-site support in case of defects and other problems.	Vendor provides phone and on-site support in case of defects and other problems.
Upgrade support	Vendor must have a user-accessible website with information on availability of new firmware upgrades.	Vendor has a useraccessible website with information on availability of new firmware upgrades.
Defect reporting	Vendor must notify users about known critical hardware, firmware and software defects.	Vendor has a procedure to notify users about known critical hardware, firmware and software defects.

Figure 3.5 Documenting vendor assessment.

The purpose of the vendor assessment is to ensure that products are designed, developed, and manufactured in a documented quality environment. The assessment should also verify that the vendor provides the right services and can maintain the instrument through phone and on-site support.

3.5 Installation Qualification

During the IQ phase, the delivery of equipment is compared with the purchase order for completeness and the vendor's installation instructions are executed. This should also include checking if the laboratory conforms to the vendor's environmental specifications, for example, humidity and room temperature. Finally, the IQ protocols are completed inserting the vendor's name, model number, serial number, and other relevant product information.

Tasks performed for IQ

- Prepare the laboratory facility according to the vendor's environmental specifications.
- Control and record environmental conditions, if critical. For example, temperature and humidity.
- Compare equipment received with the purchase order (including, accessories and spare parts).
- Check equipment for any damage. Verify that the instrument conforms to physical and construction requirements, as specified by the user.
- Check documentation for completeness (operating manuals, maintenance instructions, standard operating procedures for testing, safety and validation certificates).
- Install hardware (instrument, fittings and tubing for fluid connections, columns, power cables, data flow and instrument control cables).
- Switch on the instruments and ensure that all modules power up and perform an electronic self-test.
- List equipment manuals and SOPs.
- Record firmware revision.
- Prepare an installation report.
- Enter instrument data into an inventory data base.
- Prepare, review and sign formal IQ documentation.

Figure 3.6 shows a template with selected examples that can be used to document completeness of delivery.

Purchase order item	Vendor PN	Complete Yes	No
CO_2 pump		☐	☐
Modifier pump		☐	☐
Backpressure regulator		☐	☐
Autosampler		☐	☐
UV/Vis diode array detector		☐	☐
Column heater		☐	☐
10 µL flow cell		☐	☐

Figure 3.6 Examples for documenting completeness of delivery for IQ.

All instruments should be documented in the IQ protocol and entered into a database. The IQ documents should be updated whenever there is a change made to any entry in the IQ documents. Examples of changes are a firmware revision and the location of the instrument within a building or site.

Installation should verify that the instrument hardware and software are properly installed. It does not verify that the instrument conforms to the functional and performance specifications. This is done later in the OQ phase. For individual modules, testing is limited to perform and document the instrument's self-diagnostics when it is switched on.

For SFC systems comprised of multiple modules, correct connection between the modules should be verified. This can be easily achieved by running a well-characterized test sample and comparing the output with a reference plot. An example is shown in Fig. 3.7.

3.6 Operational Qualification

Operational qualification (OQ) should demonstrate that the SFC equipment will function according to its operational specifications in the selected environment as defined by the user. OQ tests can be performed by a vendor representative or by the user. In any case, ultimate responsibility lies with the user and a user representative should sign the OQ document. Test engineers should be formally trained. When the tests are conducted by vendor representatives, user firms should get a copy of their training certificate. When the test is conducted by user representatives, documented evidence that the test engineer has been trained on the principles and requirements for equipment qualification and on specific SFC OQ testing should be available.

Actions	Expected result	Pass/fail
1) Set instrument conditions according to the installation manual for analyzing the installation verification sample	A chromatogram similar to the one in the installation manual is obtained.	1) ☐ ☐
2) Inject the installation verification sample	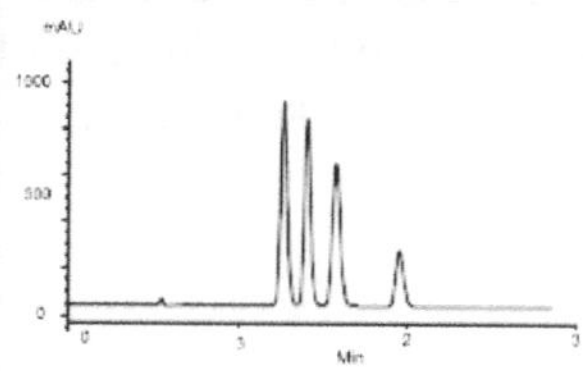	2) ☐ ☐
3) Verify if actual results are as expected	1. Chromatogram must include four peaks. 2. Peaks 1 and 2 are higher than 3 and 4 3. Retention time of 4th peak should be between 1.7 and 2.1 minutes	3) ☐ ☐

Figure 3.7 Testing an SFC system for correct installation.

Tools used for testing such as digital flow meter, thermo meters, and standards used for wavelength calibration should be formally calibrated and/or traceable back to national standards. Calibration certificates are initially provided by the vendor of the tools and may have to be recalibrated every year. Inspectors expect OQ tests to be quantitative. This means that the test protocol should include expected results and actual results.

The instrument's OQ is repeated at regular intervals. SFC systems classified as high risk equipment should be requalified every 6 to 12 months. Less frequent requalification should be justified based on documented risk assessment. In general, the time intervals should be selected so that the probability is high that all parameters are still within the operational specifications. Otherwise, analytical results obtained between the last and actual OQ with that particular instrument are questionable. Here the importance of proper selection of the procedures and acceptance limits becomes very apparent.

3.7 Tests for Operational Qualification

As a part of OQ, SFC equipment hardware should be tested for all specifications after it is installed in the user's laboratory even if it has been tested at the vendor's site. Equipment hardware, such as UV/Vis detector gratings, can be impacted by mechanical vibration as a result of shipping. The purpose of OQ testing is to verify the specifications as defined by the user for all individual modules or system components in the user's environment. The recommendation is to use the holistic approach for testing, which means to perform the test using a complete SFC system as shown in Fig. 3.8.

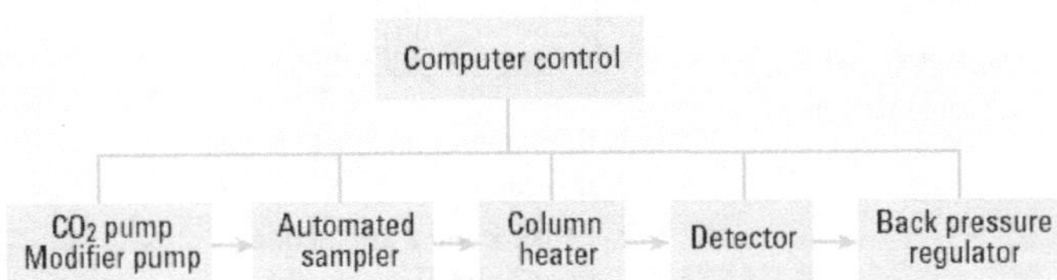

Figure 3.8 Block diagram of an SFC system.

Typical tests include precision of injection volume, detector linearity, precision, and accuracy of flow rates and temperature of the column compartment.

Examples for SFC hardware test items are shown in Fig. 3.9. The template can be used to document test items, and expected and actual results.

3.8 Performance Qualification

Performance qualification (PQ) should demonstrate that an instrument consistently performs according to the specifications as defined by the user and is appropriate for the intended use.

Here, emphasis is placed on the word "consistently." Important for consistent instrument performance are regular preventive maintenance checks, making changes to a system in a controlled manner and regular testing. The PQ test frequency is much higher than for OQ. Another difference is that PQ should always be performed under conditions that are similar to routine sample analysis. For an SFC system this means using the same column, the same analysis conditions and the same or similar test compounds.

Test	Expected result	Actual result	Passed/ failed
Precision of injection volume: Inject a standard solution 6 times and calculate the precision of peak areas.			
Carry over: Inject a high standard concentration followed by a blank solvent injection.			
Flow rate accuracy: Measure the flow rate at 1 and 5 mL/min with a digital flow meter over six oneminute fractions. Calculate the average and compare with the setpoint. (Important: The intial words "Flow rate accuracy" should be bold, nor new paragraph before "Calculate"			
Flow rate precision: Calculate the standard deviation of the six readings from the flow accuracy experiment.			
Colum heater temperature accuracy: Measure the temperature inside the column heater at two different locations and compare temperatures with the setpoint.			
Column heater temperature precision: Repeat the column heater accuracy experiment six times and calculate the relative standard deviation of the temperature readings.			
Baseline noise: Measure the peak-to-peak baseline noise over six one-minute fractions according to ASTM 1657 and calculate the average.			
Baseline drift: Measure the slope of the baseline noise according to ASTM 1657 over 10 minutes.			
Injection linearity: Inject 6 standard samples with increasing concentrations. Calculate the correlation coefficient of peak area versus concentration.			
Wavelength accuracy: Scan caffeine from 200 to 300 nm. Compare the actual WL at 205, 245 and 273 nm with the setpoints.			

Figure 3.9 Template and examples for SFC hardware testing.

PQ should be performed on a daily basis or whenever the instrument is used. The test frequency depends on the criticality of the tests, on the robustness of the instrument and on anything that may contribute to the reliability of analysis results. For an SFC instrument, this may be the analytical column or a detector's UV lamp.

In addition to tests during routine analysis, PQ tests should be conducted

- after developing and validating a new method;
- in the receiving laboratory when analytical methods are transferred between laboratories;
- when compendial methods are introduced into a laboratory;
- whenever a method is changed;
- whenever SFC hardware changes;
- after instrument maintenance;
- after instrument repair;
- after software upgrades and other software and computer system changes;
- before a new column is routinely used;
- as part of a failure investigation in an OOS situation.

In practice, PQ testing can mean system suitability testing or the analysis of quality control samples. This is supported by USP chapter <1058>: "Some system suitability tests (SST) or quality control checks that are performed concurrently with the test samples also imply that the instrument is performing suitably." For system suitability testing, critical system performance characteristics are measured and compared with documented preset limits. For example, a well-characterized standard can be injected five or six times and the standard deviation of amounts is then compared with a predefined acceptance criteria. If the limit of detection and/or quantitation is critical, the lamp's intensity profile or the baseline noise should be tested.

For chromatographic equipment, the following SST tests are recommended in USP chapter <621> [4]:

- precision of the amounts
- resolution between two peaks
- peak tailing factor
- theoretical plates (N) for column performance

- capacity factor (k′)

USP <621> makes it clear that these tests should be performed frequently: "System suitability must be demonstrated throughout the run by injection of an appropriate control preparation at appropriate intervals." The results of SST should be evaluated before the sample run. If SST does not pass, the system must not be used for any sample analysis: "No sample analysis is acceptable unless the requirements of system suitability have been met. Sample analyses obtained while the system fails requirements are unacceptable." Figure 3.10 shows an example an SFC system suitability test run.

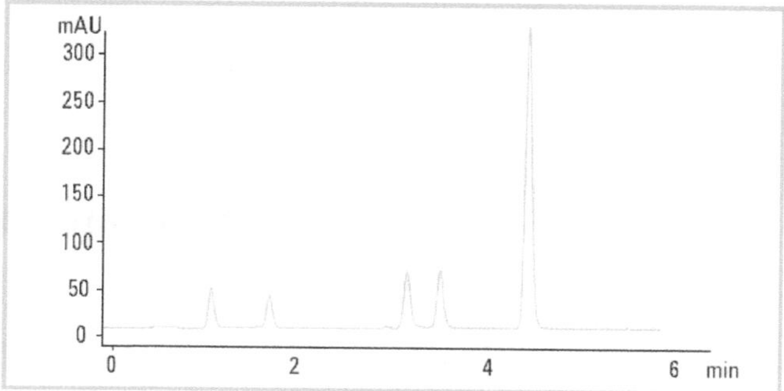

Figure 3.10 Example of an SFC system suitability test run.

The type and frequency of system suitability tests should be defined in an SOP. Conformance to the SOP should be checked by QA in regular internal audits. For routine analysis a possible scenario for test frequency is

- whenever samples are analyzed;
- for a series of 1 to 9 consecutive sample analyses: before starting the series;
- for a series of 10 to 20 consecutive sample analyses: before starting the series and after finishing the last sample analysis;
- for a series of >20 consecutive sample analyses: before starting the series, after finishing the last sample analyses and after each tenth sample analyses.

Figure 3.11 shows a template with examples for ongoing system suitability test protocol.

Test #	Date	Time	T21	T22	T23	T24	T25	Comment
			Baseline noise	Tailing factor	Peak resol.	Precis. #1	Precis. #2	
	mm/ day	a=am p=pm	Accept limit $<1\text{x}10^{-4}$	Accept limit <1.3	Accept limit >2.0	Accept limit <1%	Accept limit <1%	Passed/ failed

Test #	Date	Time	T21	T22	T23	T24	T25	Comment
001	08/04	06.20 p	$4.5\text{x}10^{-5}$	1.1	2.3	0.61	0.50	passed
002								
003								
004								

Figure 3.11 Template with examples for an ongoing system suitability test protocol.

3.9 Specific Considerations for Software and Computer Systems

Computers with associated software are used to control the SFC hardware, to acquire SFC signals and for data evaluation, printing and storage. Frequently complete computerized systems are purchased from a single vendor to ensure smooth system setup and operation and to get full system support from a single vendor. Sometimes computer hardware is purchased from a different vendor because of economic reasons. The final decision should be based on risk assessment. In any case, complete computerized systems should be validated. The approach is similar to the one for equipment. The major differences are as follows:

- More focus should be put on the qualification of vendors. Vendors should provide documented evidence that development followed a documented process and that the software has been validated as part of this process.
- Whereas for hardware equipment qualification all user specifications are verified in the user's environment, this is not required for software. It is sufficient to verify a selection of key software functions and to perform a complete system test. Examples of functions that should be tested are security access and electronic audit trail.

- Many times users customize computer systems, for example, through report generators or when setting network configurations. Users should include these configurations in the requirement specifications document. Configurations are implemented during installation and tested as part of OQ.

More information on validation and examples for software and laboratory computer system validation are included in references [2,6].

3.10 (Preventive) Maintenance and Repair

SFC instruments should be well maintained to ensure proper ongoing performance. Procedures should be in place for regular preventive maintenance of hardware to detect and fix problems before they can have a negative impact on analytical data. These procedures should describe

- the maintenance to be done;
- when it is to be done;
- what should be requalified after maintenance is done; for example, a PQ test should always be performed after instrument maintenance;
- how to document maintenance activities.

Instruments should be labeled with the dates of the last and next scheduled maintenance.

Planned maintenance activities should follow a documented instrument maintenance plan. Some vendors offer maintenance contracts with services for preventive maintenance at scheduled time intervals. A set of diagnostic procedures is performed and critical parts are replaced to ensure ongoing reliable system uptime.

Unplanned activities that are necessary in addition to the planned activities should be formally requested by the user of the instrument or by the person who is responsible for the instrument. The reason for the requested maintenance should be entered as well as priority. All maintenance activities should be documented in the instrument's logbook.

Defective SFC systems must not be used. Procedures should be available on how to handle most common problems. Procedures should also include information if and what type of requalification is required after repair. Uncommon problems should be handled through a special procedure that guides users of instruments through the repair and reinstallation process, for example, if a CO_2 pump becomes defective. In this case, the impact of the failure on the previously generated data should be evaluated.

Figure 3.12 shows a flowchart on how defective instruments should be handled. The following steps are recommended:

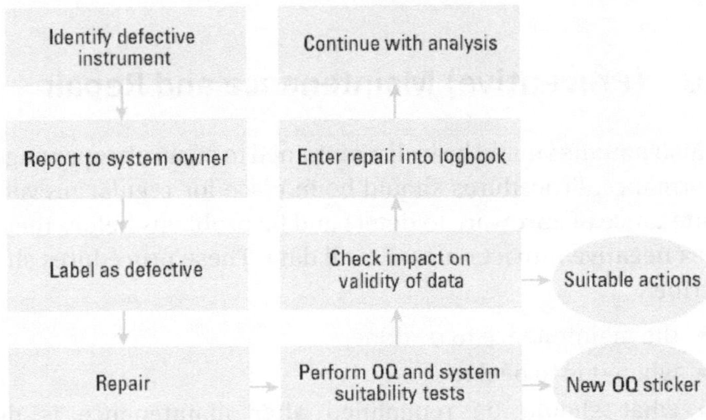

Figure 3.12 Flowchart for handling repair of SFC instruments.

- The problem should be reported to the laboratory supervisor, or to the person responsible for the instrument, who will decide on further action.
- The instrument should be clearly labeled as "Out of Service."
- After repair, correct functioning must be verified. The type and extent of testing depends on the failure and possible impact on the system. Depending on the failure, this may require partial or full requalification or only system suitability testing.
- The impact of the defect on previous test results should be examined.
- Suitable actions should be taken in case the defect instrument could have had an impact on the validity of data; for example, samples may have to be re-analyzed.
- An entry on the defect, repair, and performance verification should be made in the instrument's logbook

3.11 Change Control

Analytical instruments and systems go through many changes during their lifetime. New hardware modules may be added to enhance functionality; for example, an automated sampling system replaces a manual one for unattended operation. Vendors may change the firmware to a new revision to remove software errors or application software may be upgraded to be compatible with a new operating systems. Or a complete system is moved to a newly designed laboratory.

Any changes to instrument hardware, firmware, and software should follow written procedures and should be documented. Requests for changes should be submitted by users and authorized by the user's supervisor or department manager and by QA.

Before any change request is approved, business benefits should be compared with risks a change may bring. USP chapter <1058> states, "Implementing changes may not always benefit users. Users should therefore adopt changes they deem useful or necessary and should also assess the effects of changes to determine what, if any, requalification is required."

USP also recommends following the same 4Q model for changes as for initial qualifications. This requires the following:

- Specifications should be updated, for example, in case a new automated sampling system replaces a manual one.
- IQ documents should be updated, if a new firmware revision is installed. Installation documents should also be updated when a system is moved to a new laboratory.
- OQ documents with new test cases and test protocols should be added if the software is upgraded with new functionality.
- PQ tests need to be performed to verify correct functioning of the complete system, for example, when updating firmware of any SFC module.

Before any change is approved and implemented, a thorough evaluation should be made if OQ tests should be repeated. Depending on what the change is, an instrument may need no, partial, or full testing of a system.

3.12 Validation Reports

At the end of validation, a summary report should be developed. This should be a mirror of the validation project plan. It should be organized so that it contains all the elements and follows the outline of the validation plan. This makes it easy to check if all plan items have been completed successfully. Deviations should be documented if there are any, together with corrective actions and/or work around solutions. The report should include a statement that the instrument or system is qualified or validated. When the statement has been signed by laboratory and QA management, the product can be released for operation.

Typically, the validation plan and the report are the first documents inspectors want to see when they inspect a validation project. If everything is well organized and documented, it may well be that after looking at both documents inspectors get such a good impression about the validation work that they will focus on other inspection areas.

References

1. Unites States Pharmacopeia, Chapter <1058>, Analytical Instrument Qualification, Rockville, USA, 2008 Agilent Technologies, *Analytical Instrument qualification and System Validation*, Primer, Publication Number: 5990-3288EN, 2009.
2. Agilent Technologies, *Analytical Instrument qualification and System Validation*, Primer, Publication Number: 5990-3288EN, 2009.
3. ICH Q7: *Good Manufacturing Practice Guide for Active Pharmaceutical Ingredients*, update 2000.
4. 4-23 General Chapter 621, United States Pharmacopeia 30, *National Formulary 25*, Rockville, Md., USA, The United States Pharmacopeial Convention, Inc., 2007.
5. GAMP, Good Automated Manufacturing Practice, A Risk-based Approach for Compliant GxP Computerized Systems, Version 5: 2008.

Chapter 4

Method Development for Achiral SFC

Jeffrey W. Caldwell,[a] Walton B. Caldwell,[a] Gregory K. Webster,[b] and Zhenyu Wang[c]

[a]*Princeton Chromatography, Inc., 1206 Cranbury South River Rd, Cranbury Township, NJ 08512 USA*

[b]*AbbVie, Inc., Global Research and Development, 1 North Waukegan Road, North Chicago, IL 60064 USA*

[c]*Merck Research Laboratories, Merck & Co., Inc., 556 Morris Avenue, Summit, NJ 07901 USA*

jcaldwell@pci-hplc.com, gregory.webster@abbvie.com, zhenyu.wang@merck.com

4.1 Introduction

It may seem strange at first to deem a chromatography chapter for "achiral" chromatography. After all, isn't almost all chromatography "achiral"? Indeed it is. Yet, with supercritical chromatography (SFC), the technique regained its niche in analytical chromatography as being primarily a chiral chromatography technique.

Supercritical chromatography began as an achiral technique using capillary columns. The technology had expected to make a significant compromise between the efficiency of GC and the solvation properties of liquid chromatography (LC). Much of the early development and theory are summarized by Lee and Markides [1]. The marketplace

Supercritical Fluid Chromatography: Advances and Applications in Pharmaceutical Analysis
Edited by Gregory K. Webster

ISBN 978-981-4463-00-3 (Hardcover), 978-981-4463-01-0 (eBook)
www.panstanford.com

drove this technique to address more polar compounds, but SFC remained a relatively minor analytical technique until it was found to operate on HPLC platforms using mechanical restrictor technology. The use of packed column technology with SFC further expanded its applications [2].

The real change for industrial SFC came with its use in pharmaceutical preparative applications. Preparative SFC offered significant cost and recovery advantages to traditional preparative HPLC. Zelesky and DaSilva have reported that by using SFC, the purification turnaround time reduction was greater than 50%, and solvent cost reduction was over 70%, compared with prep-HPLC [3,4]. Pharmaceutical laboratories further expanded the technique with chiral applications. The technique was successful in preparative chiral separations, which drove the desire for analytical scale separations [5,6]. SFC emerged as a mainstream chromatographic technique again.

The transition to analytical SFC drove further instrument capabilities for the analytical methods needed to be held to stringent separation and precision requirements. Instrument manufacturers responded with instruments with improved analytical and qualification capabilities necessary for SFC to fulfill the analytical needs in the regulated laboratories of the pharmaceutical industry. With its newfound success in chiral applications, chemists are turning their attention back to looking at SFC separation capabilities in the achiral mode—first in discovery laboratories and more gradually to in-process and quality testing for assay, impurities and ultimately, pharmaceutical stability testing [7]. Previous pharmaceutical separations focused on traditional normal-phase stationary phases [2]. Now that pharmaceutical chromatography has returned attention to achiral SFC, applications are expanding as new phases and impurity method approaches are now being targeted for SFC [8–11]. Recently, SFC was successfully applied to cell lysates and peptides [12,13]. As with chiral SFC, several achiral column screening strategies can now be found in the literature [14–17].

4.2 Overview of Achiral SFC Separations

4.2.1 Achiral SFC vs. Achiral HPLC

Packed column supercritical fluid chromatography in the 21st century is probably better described as "carbon dioxide-based

HPLC" [18]. To new users, the current SFC instrument on the market "looks and feels" like traditional liquid chromatography and, even for certain models, SFC can be easily converted to traditional liquid chromatography and vice versa. SFC is a normal phase-like separation technique without the problems usually associated with normal-phase liquid chromatography. Practical problems associated with the latter include (a) slow equilibration times, (b) difficulty in running gradients, (c) traces of water in the solvents, and (d) flammable waste. SFC offers important advantage over normal-phase LC in that achieving robust and reproducible separations is more facile and routine. For pharmaceutical analyses, due to the (1) normal-phase mechanisms, (2) high speed, and (3) resolution in SFC, it is unsurprising that enantioselective separations are the dominant application area wherein this technique has had most impact [19]. However, SFC is not limited to chiral separations. It is getting more and more attention in the area of achiral separations, which has been dominated by reversed-phase liquid chromatography (RPLC). Compared with well-established RPLC, SFC provides comparable or even higher separation efficiency and faster analyses while offering low operation cost and facile tunable selectivity for method development. More attractively, SFC offers complementary selectivity to RPLC. This makes SFC a valuable tool in the pharmaceutical arena analyses of complex mixtures comprising analytes that may be both structurally related and ranging in polarity.

4.2.2 Commercially Available Achiral Stationary Phases for SFC

As a normal-phase chromatographic technique, achiral SFC uses polar stationary phases along with a less polar mobile phase. Table 4.1 illustrates many of the stationary phases in use today.

As with any mode of chromatography, the choice of stationary phase can significantly affect the resolution of the chromatographic profile. In normal-phase liquid chromatography (NPLC), the general elution order in SFC is typically opposite from that observed in reversed-phase liquid chromatography. The elution order generally follows the increasing polarity of the analytes in the matrix. Poole [20] notes that internal hydrogen bonding between functional groups and bulky alkyl groups adjacent to polar functional groups diminish retention. In addition, significant differences in steric fitting of isomers and diastereoisomers with fixed-position adsorption sites provide enhanced separations of these otherwise difficult to separate compounds.

Table 4.1 Common achiral SFC stationary phases

Silica	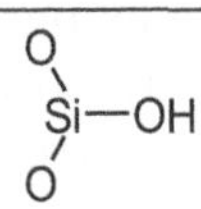
Diol	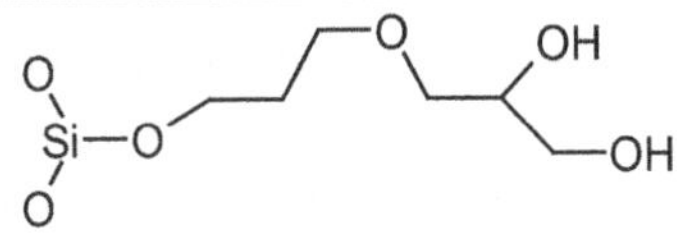
Amino-propyl	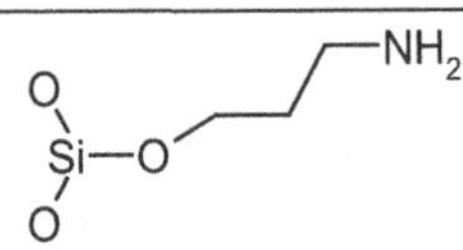
Cyano-propyl	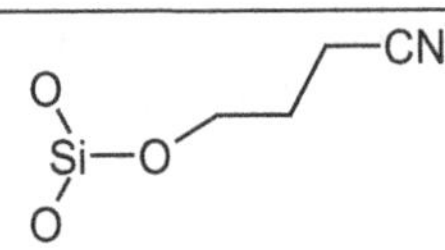
2-Ethylpyridine	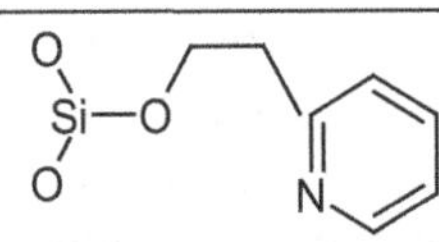
4-Ethylpyridine	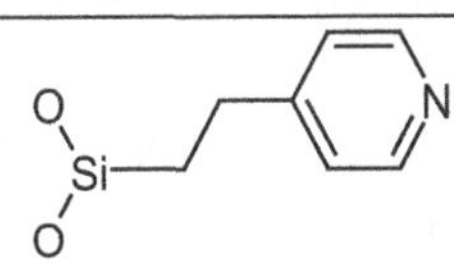

SFC bonded stationary phases such as the ones found in Table 4.2 have surfaces of the bonded ligand with unreacted silanol groups. The retention characteristics of 3-cyanopropylsiloxane-bonded silica are similar to deactivated (weak) porous silica. Complementary selectivity can be found with the strong hydrogen-bonding properties of 3-aminopropylsiloxane-bonded silica. Using bonded propanediol phases generally illustrate intermediate properties between the 3-aminopropylsiloxane-bonded and 3-cyanopropylsiloxane-bonded stationary phases [20]. For example, in Fig. 4.1, the selectivity behavior of 10 common pharmaceutical compounds on eight stationary phases using identical SFC conditions is illustrated.

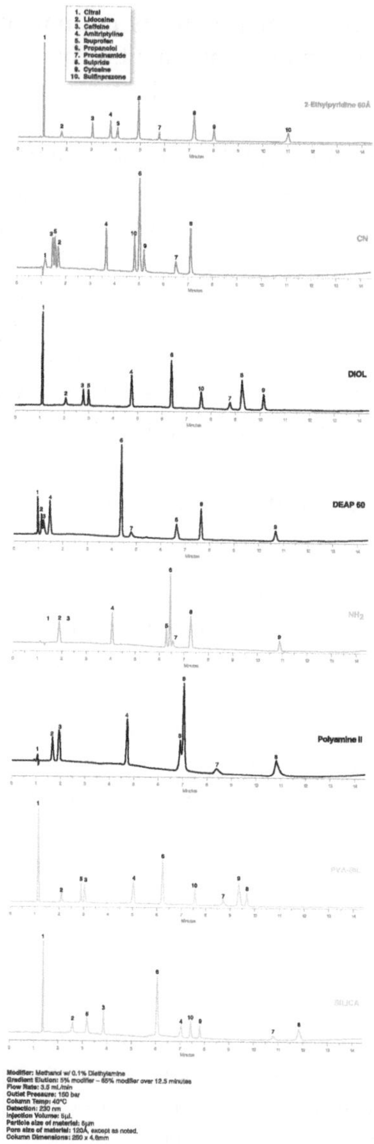

Figure 4.1 Selectivity behavior of SFC stationary phases. (1) Citral, (2) lidocaine, (3) caffeine, (4) amitriptyline, (5) ibuprofen, (6) propanolol, (7) procainamide, (8) sulpride, (9) cytosine, (10) sulfinprazone. CO_2 150 bar, 3.5 mL/min, 40°C, modifier: methanol with 0.1% diethylamine, 5% to 65% in 12.5 min. All columns were 250 × 4.6 mm, 5 μm particles. Graphics from YMC America, Inc. Used with permission.

The choice of stationary phase affects not only resolution but the elution order of the sample components. The ability to manipulate the chromatographic profile through stationary phase choice is a powerful tool for the chromatographer.

The choice of achiral SFC stationary phases can be selected through screening approaches as well as quantitative approaches using solvation parameter modeling. Generally, a screening approach is used in the pharmaceutical industry rather than the more academic minded modeling approaches. The goal in industry is to quickly produce a separation for the analytical testing requirements of drug development. The separation is to move a molecule along the critical path of developing products. A detailed review of SFC stationary phase modeling can be found in the literature [20].

4.2.3 Novel Stationary Phases for SFC

While traditional normal-phase stationary phase chemistries have served achiral SFC well, chromatographers are placing more selectivity demands upon SFC for pharmaceutical analysis. Specifically, to be a viable achiral technique in the industry, SFC needs to exhibit better selectivity for pharmaceutical impurities. Achiral SFC competes with Fast LC and ultrahigh performance liquid chromatography (UHPLC) for routine use and thus, the lower selectivity traditionally found in normal-phase chromatography is often not good enough for pharmaceutical testing. To address this gap, developers are looking at new SFC stationary phases with the goal of improving selectivity for pharmaceutical impurity testing as well as more routine analysis for compounds coming directly from the chemists.

4.2.3.1 "Amino" stationary phases

Pharmaceutical impurity testing has had success in achieving suitable selectivity using amino phases [21]. Further optimization of these phases continues and, as with any stationary phase development, work is ongoing. The basic structure of the stationary phases is shown in Fig. 4.2 and the phase constituents and abbreviations are detailed in Table 4.2. Using a 12-component test mixture (Table 4.3), the selectivity of a novel amino stationary phase series was investigated. The primary goal with the amino phase work was to highlight the chromatographic effect of increasing alkyl chain length in the stationary phase. The changes in alkyl chain length were made at both R positions as illustrated by the Fig. 4.3. An aromatic group was substituted at the R position as well for the purpose of comparison with the alkyl groups, but also to investigate the effect (if any) of π electrons present in the aromatic functionality. Meaning, is there any major change in selectivity from the alkyl substituents? Clearly, the aromatic group shows greatly diminished retention times in comparison, while selectivity changes for this test compound set are inconclusive. We would not expect large changes in the selectivity if the method were to be optimized and reasonable separation of the test components achieved. The representative chromatograms for these various amino phases are illustrated in Fig. 4.3. We can make several general observations from the chromatographic data presented here. As the length of the alkyl chain increases, overall retention times tend to decrease. We only see slight variations in selectivity (elution order), which is to be expected. As a general rule, dramatic changes in stationary phase chemistries are needed to effect large changes in the elution profile for a given mixture. What we see here is the ability to fine-tune a separation using only the column with no need to change the chromatographic conditions. It is possible the most likely explanation for the behavior illustrated in the study of these amino-based phases has to do with the ability of the analytes to access surface silanols and the nitrogen component of the phase. As the alkyl chain length is increased, there is decreased accessibility due to steric effects, both shielding the surface of the silica and blocking access to the nitrogen. The benzyl-substituted phase shows the least retention and presumably offers the greatest shielding of the silica surface, thus decreasing access to the polar silanols partially (but very importantly) responsible for chromatographic separation in SFC. If we take a "face value" look at the chromatography, the DMAP and DBAP columns separated all 12 components of the mixture to the baseline. Of these two columns, DBAP may be the phase of choice in terms of total analysis time, with 21 min vs. 36 min for the DMAP. The DBzAP column was much different from the others in terms of retention and selectivity as previously noted. The authors are particularly interested in studying the DBzAP further and attempting to optimize a method for its employ. All six stationary phases studied show interesting properties and, depending on the application at hand, could be classified as the "column of choice." It is important to note that neither traditional amino nor the very commercially popular DEAP stationary phase outperformed the other amino-based phases presented. This serves to further highlight the unique nature of SFC separations and their reliance on varied stationary phase chemistry more than probably any other technique (NPLC, RPLC).

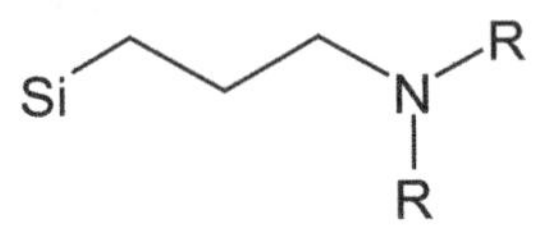

Figure 4.2 Amino stationary phases.

Table 4.2 Amino stationary phase constituents

Phase	Abbreviation	-R
Amino (NH_2)	NH_2	-H
Dimethylaminopropyl	DMAP	$-CH_3$
Diethylaminopropyl	DEAP	$-CH_2CH_3$
Dipropylamniopropyl	DPAP	$-CH_2CH_2CH_3$
Dibutylaminopropyl	DBAP	$-CH_2CH_2CH_2CH_3$
Dibenzylaminopropyl	DBzAP	$-CH_2-C_6H_5-H_5$

Table 4.3 SFC analytes tested on amino and hydroxy phases

1. Caffeine	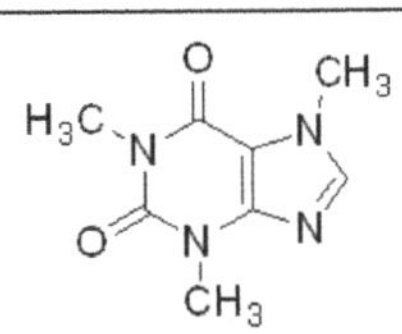
2. Theophylline	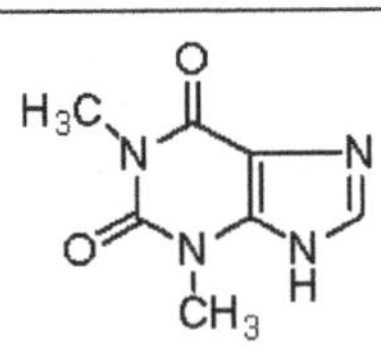
3. Thymine	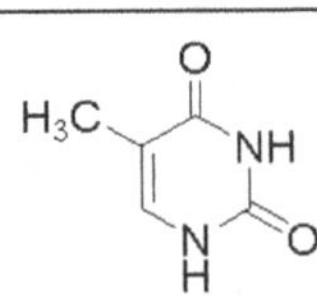

4. Uracil	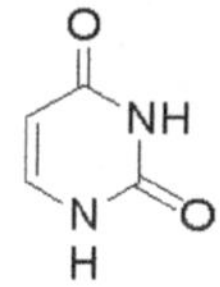
5. Fenoprofen	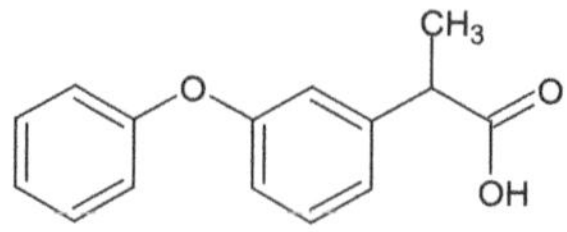
6. Flurbiprofen	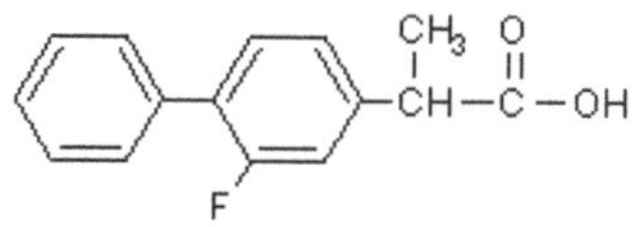
7. Prednisone	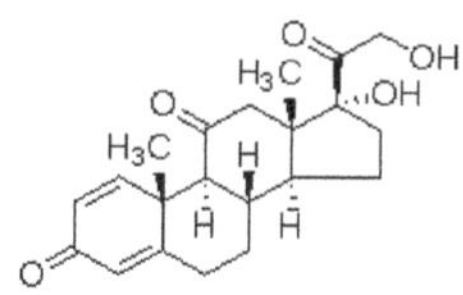
8. Hypoxanthine	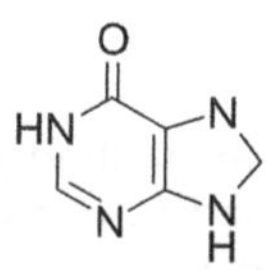
9. Hydrocortisone	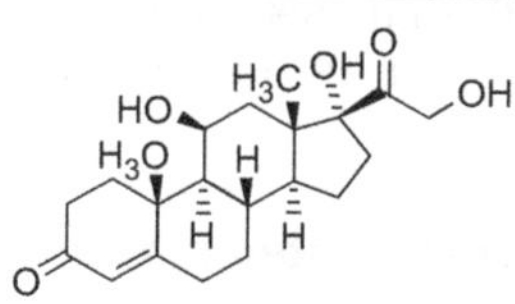
10. Sulfamerazine	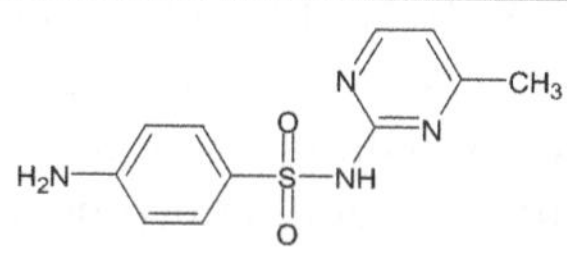
11. Sulfamethoxazole	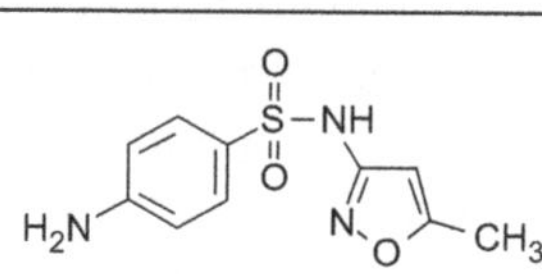
12. Sulfaguanidine	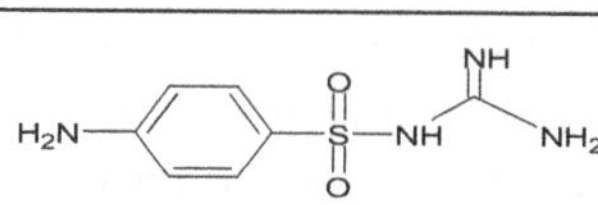

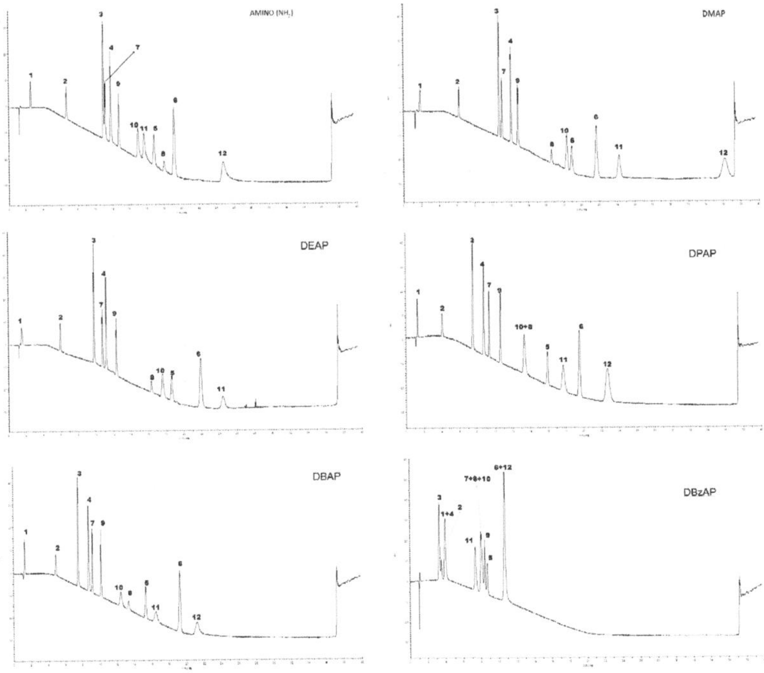

Figure 4.3 Novel SFC amino phases. Compounds listed in Table 4.3. SFC conditions: flow: 2.0 mL/min, outlet: 150 bar, column temp: 40°C, detection: 254 nm, column(s): 150 × 4.6 mm, gradient: 0–1 min 5% MeOH, 1–18 min 5–40% MeOH, 18–35 min 40% MeOH, 35–40 min—return to initial conditions.

4.2.3.2 Hydroxylated stationary phases

Silica has long been a popular SFC column due to its highly active surface silanols and extremely polar characteristics. It can be very good at accomplishing some separations in SFC and certainly has its place within screening protocols. Silica, however, has limitations owing to the very characteristics that originally made it so popular for SFC use. Silanols tend to be more acidic than, for example, the hydroxyl group(s) in a diol stationary phase. This simple fact can tend to exclude silica from use when working with very basic compound libraries. Without going into too much detail, it is widely accepted that silanols and hydroxyls are very different with respect to their chemical and chromatographic properties. As we know from our work with SFC , for a stationary phase to be effective for use in SFC it must have a polar component to it. The polar component can arise from two primary sources: from unbonded surface silanols or from the bonded stationary phase itself. The more we can control and manipulate the polar component of a stationary phase, the greater our ability to not only fine-tune separations but increase reproducibility with respect to the chromatography. This manipulation is achieved through innovative phase design, proper bonding technique, and above all, a willingness to try what has not yet been done. To address this notion in regard to hydroxylated stationary phases, a series of phases were investigated, which, while mostly related to traditional diol column, was fairly diverse in its own right. The series contains the following: diol, diol-HL (high carbon loaded), 2CN:DIOL (two parts CN, one part diol), DIOL:MONOL (one

Table 4.4 Hydroxylated SFC phases

DIOL DIOL-HL	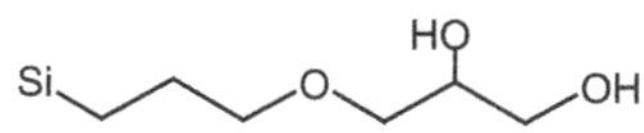
2CN:DIOL	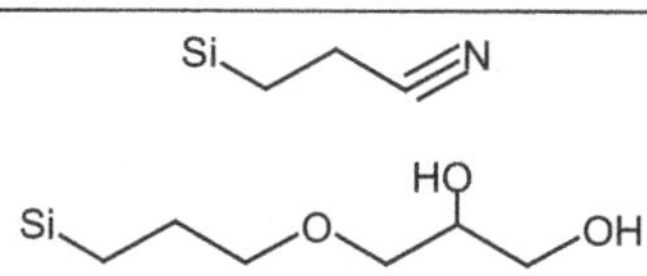
DIOL:MONOL	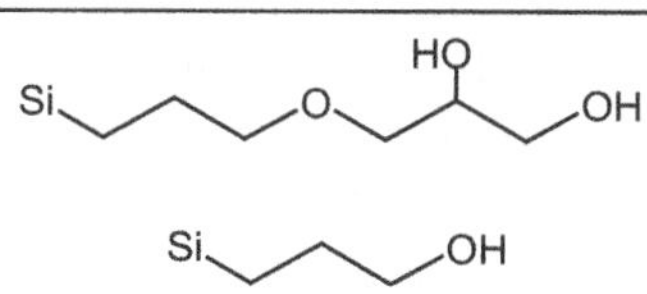
HA-DHP	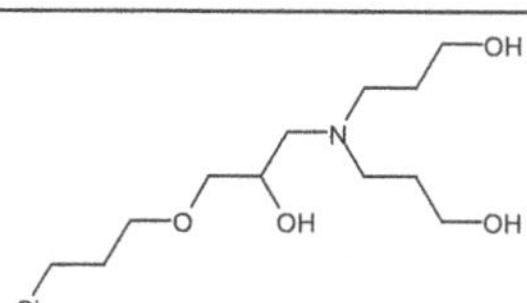
POLY-OH	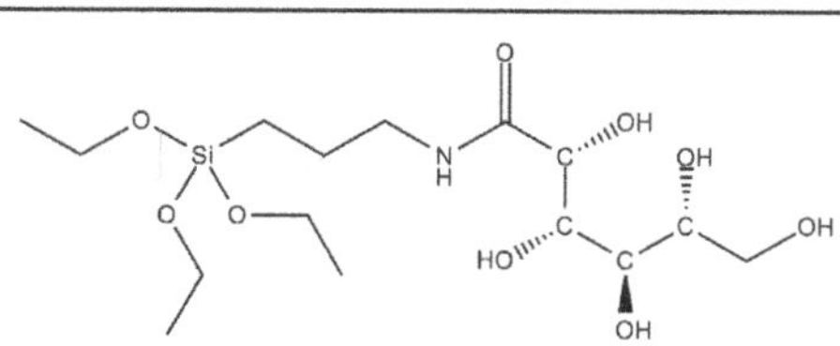
MONOL	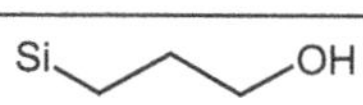

part DIOL, one part MONOL), HA-DHP (Dihydroxypropyl), MONOL, and a phase commercially named POLY-OH. The representative chromatograms for hydroxyl phases are illustrated in Fig. 4.4. Of the six columns evaluated, only 2CN:DIOL and DIOL:MONOL columns were able to resolve completely all 12 compounds of the test mixture. Diol was not successful, nor was Diol-HL (high load), and these two are most representative of commercially available diol columns. Selectivity remained very consistent for the DIOL and closely related phases. HA-DHP and POLY-OH exhibited selectivity differences as highlighted in Table 4.5. As with the amino-based columns in Section 4.2.3.1, it is noteworthy that the best performing stationary phase is not the most common in the marketplace.

4.2.3.3 Stationary phase pore size

Historically, many separation scientists have believed that for achiral SFC separations the most effective columns are those with stationary phases bonded on high-surface-area silica gel. It is our experience that the higher surface area material (>500 m^2/g) provides superior retention, selectivity and loadability. The purpose of this study was to investigate a group of sulfur-containing compounds (sulfanilamides; see Table 4.6) we had difficulty separating when using high-surface-area materials. The representative chromatograms for the pore sizes studied are illustrated in Fig. 4.5.

Silica and 2-ethylpyridine offer interesting selectivity differences over the range of nine sulfanilamide compounds tested. Lower surface area materials provided shorter retention times and in some cases enhanced separation resolution over higher surface area counterparts; 200 and 300 Å 2-ethylpyridine produce baseline separation of all nine components in a reasonable time frame, with the 300 Å material allowing for complete separation in just less than 14 min. As surface area decreased, so did retention times, but selectivity remained consistent across the entire pore size range for each respective material. This allows for easier prediction of chromatography as columns are changed to lower surface area materials. The 200 and 300 Å materials should be considered viable options for difficult separations, especially with compounds that are strongly retained in the SFC mode.

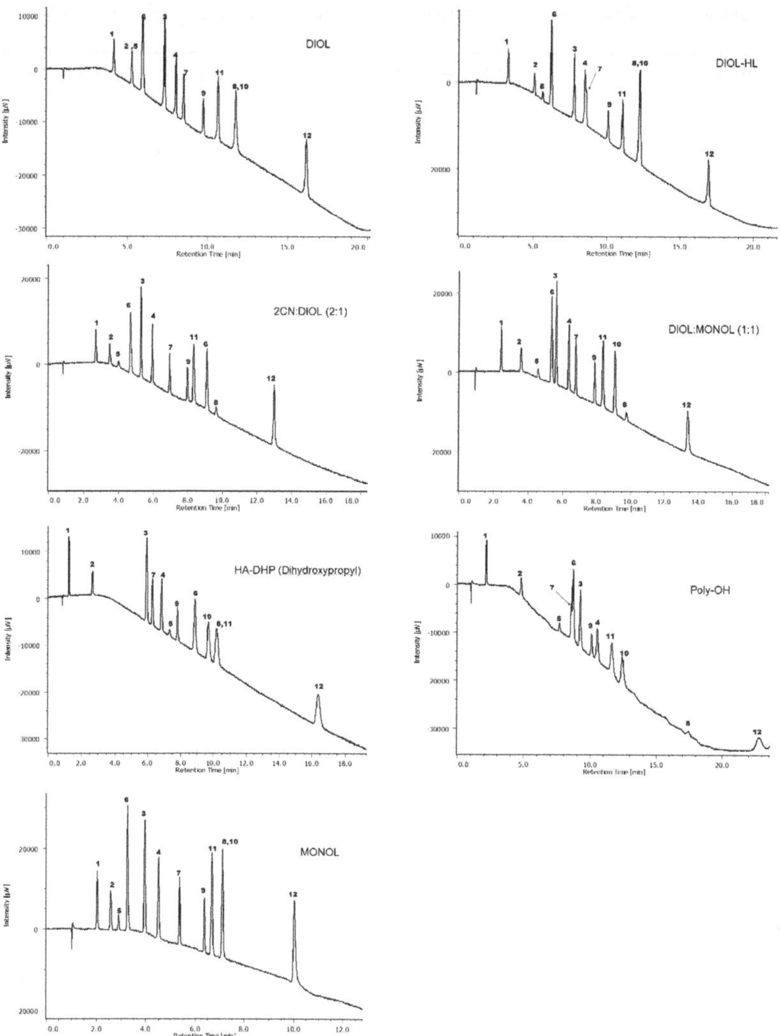

Figure 4.4 Novel SFC hydroxy phases. Compounds listed in Table 4.3. SFC conditions: flow: 3.0 mL/min, outlet: 150 bar, column temp: 40°C, detection: 254 nm, column(s): 150 × 4.6 mm; gradient: 0–1 min 5% MeOH (w/0.1% diethylamine), 1–18 min 5–40% MeOH, 18–22 min—return to initial conditions.

Table 4.5 Test compound elution order for the hydroxy phase study

DIOL	1		2,5	6	3	4	7	9	11		10,8	12
DIOL-HL	1	2	5	6	3		4,7	9	11		8,10	12
2CN:DIOL	1	2	5	6	3	4	7	9	11	10	8	12
DIOL:MONOL	1	2	5	6	3	4	7	9	11	10	8	12
HA-DHP	1	2	3	7	4	5	9	6	10		8,11	12
Poly-OH	1	2	5	7	6	3	9	4	11	10	8	12
MONOL	1	2	5	6	3	4	7	9	11		10,8	12

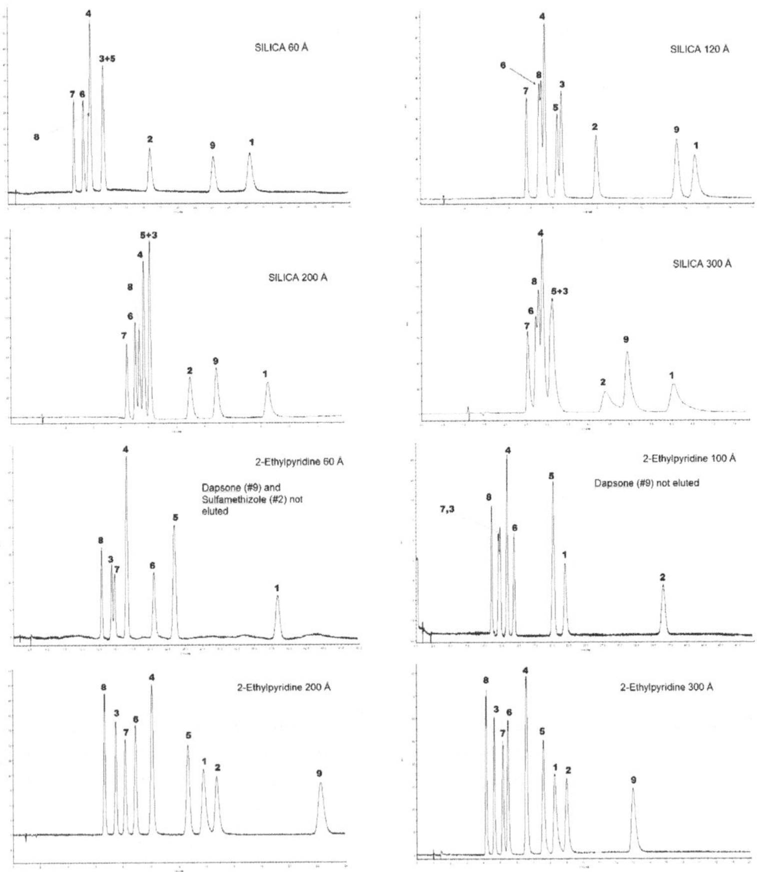

Figure 4.5 SFC pore size study phases. Compounds listed in Table 4.5. SFC conditions: flow: 3.0 mL/min; outlet: 100 bar, column temp: 40°C, detection: 254 nm, column(s): 250 × 4.6 mm; isocratic: 40% MeOH w/0.1% diethylamine).

Table 4.6 SFC analytes tested on pore size study

1. Sulfaguanidine	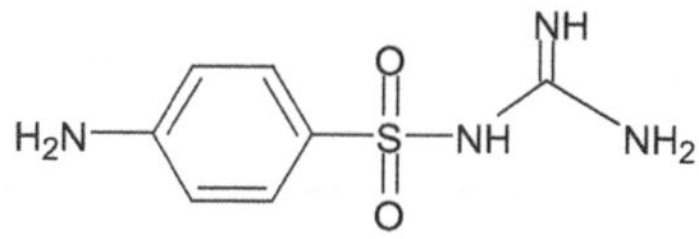
2. Sulfamethizole	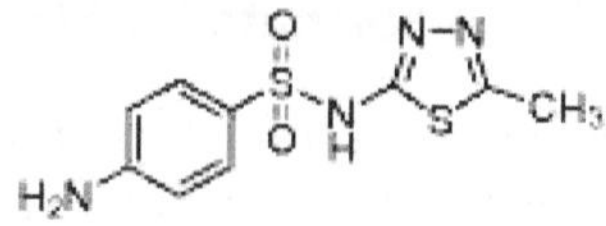
3. Sulfamerazine	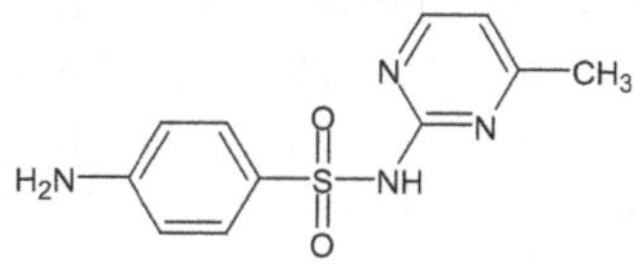
4. Sulfanilamide	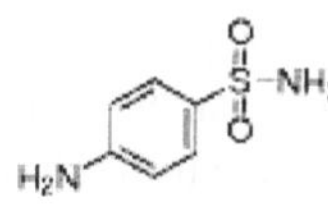
5. Sulfaquinoxaline	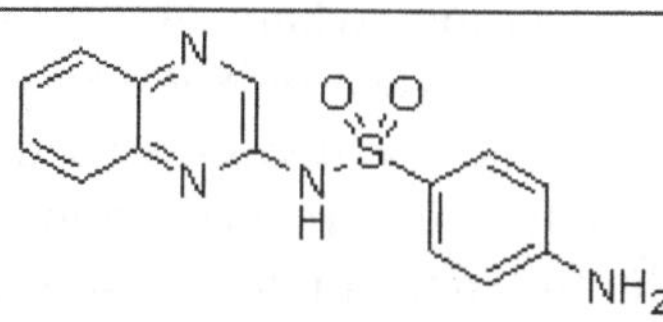
6. Sulfadimethoxine	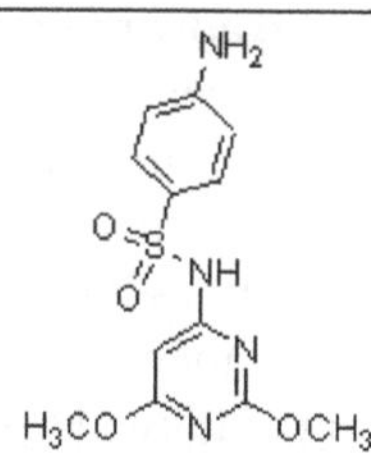
7. Sulfamethoxazole	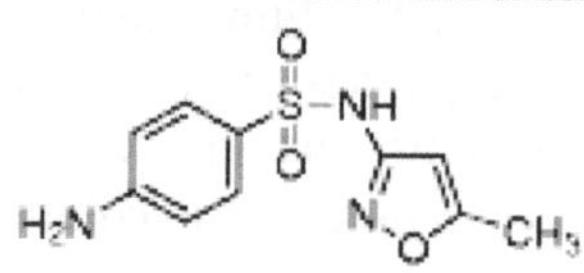
8. Sulfamethazine	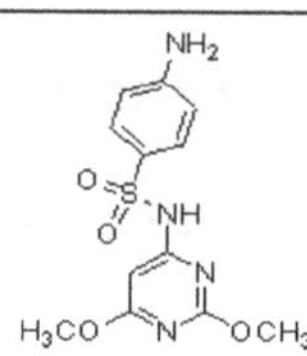

(*Continued*)

Table 4.6 (*Continued*)

9. Dapsone	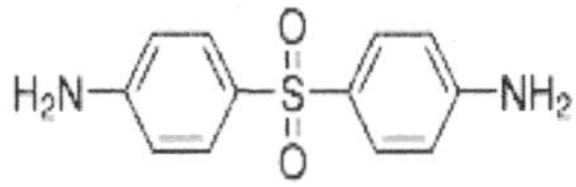

4.2.4 Column Formats for SFC

With the advent of ultrahigh performance liquid chromatography (UHPLC), as well as the development of superficially porous particle (SPP) stationary phases, the industry has been focused on fast separations and column formats. For reverse-phase chromatography, analytical laboratories are quickly moving to either UHPLC, SPP columns or a combination of these formats for the improved chromatographic efficiency advantages they can obtain. For SFC, the argument for smaller particles is less straightforward.

For typical mixtures, UHPLC is expected to yield greater chromatographic efficiencies than those found with supercritical fluid chromatography when conventional particle sizes and column lengths are used. However, SFC can yield higher plate counts than UHPLC when longer columns are used [20]. In reversed-phase liquid chromatography, researchers often look to monolithic columns to add column length under reasonable system pressures. In SFC, significant resolution advantages using monolithic columns have not been reported. The lower viscosity of SFC mobile phases enables the couple of traditional column formats. Berger [23] and Lesellier [24] noted an increase in performance for a superficially porous particle column but other authors have reported fronting issues [22]. At this time, the current advances in column technologies have not played a pivotal role in SFC. Many SFC separations are still viable on 150 to 250 mm, 4.6 mm i.d. columns using 5 um particles. This is attributed to the efficiency gains seen in using supercritical (or subcritical) fluid mobile phases. With further investigations, the market will settle on a predominant format for the column used with routine SFC testing.

4.3 Achiral Method Development

4.3.1 Role of Modifier and Additive

Supercritical fluid CO_2 is no more polar than hexane and its elution strength is too weak to elute most polar compounds. Its solvent strength can be increased by adding small volumes of polar organic solvents (typically 2% to 60%, v/v). In fact, modern packed column SFC almost invariably requires modified mobile phases to cover a wide range of solute polarity [18]. The typical organics are alcohols, such as methanol, ethanol, isopropanol, and acetonitrile. Acetonitrile differs significantly from alcohols, i.e., relatively weak eluent toward hydrogen-bond donor analytes while strong eluent for other types of analytes. Methanol is by far the most used modifier offering higher efficiencies and shorter retention time than ethanol and isopropanol. Methanol is probably the most polar, common modifier completely miscible with supercritical fluid CO_2 over a wide range of temperatures and pressures. The critical temperature and critical pressure of the mixture increase with the increase of the modifier concentration. For a CO_2–MeOH 82:12 (v/v) mobile phase, the critical temperature is increased to 75°C [21]. If the operating temperature is below the critical temperature of the new mixture, the definition of the fluid changes to subcritical. Fortunately, there is a continuum of the state of matter and properties when going between "subcritical" and "supercritical" region [24]. Also, in almost all practical circumstances, it is irrelevant whether these fluids are defined as subcritical or supercritical. No significant changes in either physical or chemical mobile phase characteristics occur when the definition of the fluid changes [18,25].

Modifiers can influence the SFC separation by changing analytes, mobile phase, and stationary phase, described as follows:

- *For the analytes*, the modifier may selectively solvate polar compounds in the mobile phase form clusters with different distribution properties [26].
- *For the mobile phase*, the modifier can alter the density and solvating power of the mobile phase.
- *For the stationary phase*, adding a modifier has multiple impacts [20]:
  (1) The modifier can block active sites on the stationary phase.
  (2) The adsorbed modifier can increase the volume of the stationary phase leading to a change in the column phase ratio.
  (3) The adsorbed modifier can act as a component of the stationary phase.

The proportion of the adsorbed modifier onto the stationary phase depends on the mobile phase composition. Practically in all situations, the amount of adsorbed modifier is greater than the modifier percentage in the bulk mobile phase. For instance, working with methanol/CO_2 (2/98), the percentage of methanol in the stationary phase was equal to 40% [27].

Even with polar organic solvent as modifier, SFC is still not always sufficient to facilitate elution in a reasonable time or with acceptable peak shapes for some ionic compounds and polar compounds. A third more polar component (i.e., additive) added into the mobile phase can help to mitigate this problem. Suggested roles for additives in the separation process include the following [28–32]:

(1) Enhance mobile phase solvating power.
(2) Suppress sample ionization.
(3) Ion pair with ionic analytes.
(4) Modify the stationary phase.

Typical additives are strong acids (e.g., formic acid, trifluoroacetic acid, citric acid), bases (e.g., isopropylamine, triethylamine), or salts (e.g., ammonium acetate). Water is also used as additive to help elute highly polar compounds, such as nucleobases and polypeptides where water introduces HILIC-like analyte partitioning [18]. There are no general guidelines for the choice of additives beyond that they should be a stronger acid or base than the sample components as well as compatible with the choice of detector [20].

4.3.2 Primary Screening on Mobile Phase and Stationary Phase

The role of stationary phase, modifier, additives, and their interactions makes the SFC achiral separation unpredictable. As a result, the initial choice of the mobile phase and stationary phase in SFC is a complex problem and could be quite time consuming. Most modern SFC systems come with a column selection unit and a mobile phase selection unit, which provide a practical work around for the lack of a suitable theoretical framework for SFC retention mechanism. With such systems, columns with different stationary phases are rapidly scouted using automatic column switching with a single or different mobile phase. The column and mobile phase screening sequence can be easily programmed using the instrument control and operation software. A wide gradient (e.g., 5–50%) is recommended during the initial screening. Luckily, SFC's rapid column equilibration makes it easy to use fast gradient analysis or to change the chromatographic parameters as well as the stationary phase for rapid method development.

Owing to lack of knowledge of the interactions established between the analytes and the SFC system, there are few clear guidelines on column selection for SFC method development [33]. In practice, multiple column need to be examined in order to obtain a suitable resolution. However, any stationary phases available for HPLC, and any solvent that is miscible with supercritical CO_2 can be used for SFC. This wide diversity is practically impossible for a chromatographer to conduct a thorough screening on the column. One way to simplify the column selection is to classify the stationary phases based on their potential retention mechanism and then choose one or two representative columns from each category for screening. Once the hit is identified, more columns in that particular class can further scrutinized.

West and Lesellier [33–38] have constructed a system constant database for commercially available SFC columns using a quantitative structure-retention relationship (QSRR) approach. As shown in Fig. 4.6, they developed a five-dimensional classification diagram based on a linear solvation energy relationship, representing the column selectivity similarity and difference. Three major groups of stationary phases can be defined: nonpolar, moderately polar and very polar [33]. The first group is constituted by alkyl-bonded stationary phases that possess no polar groups. With the non-polar phase, SFC behaves more like reversed-phase retention mode. The second group is constituted by stationary phases of intermediate polarity and comprises all ODS phases containing a polar-embedded group or a hydrophilic end-capping group, along with a wide variety of aromatic stationary phases. With the stationary phase in this group, SFC behaves like a mixed mode that is in between reversed-phase retention and normal-phase retention. The third group is constituted by polar stationary phases. This group comprises a fluoroalkyl phase (FD), polymer-coated phases (PEG and PVA), the ethylpyridine bonded silica (EP) along with more classical phases as bare silica gel (Si), 3-aminopropyl bonded silica (NH_2), 3-cyanopropyl-bonded silica (CN), and propanediol-bonded silica (DIOL). SFC behaves in a normal-phase retention mode with the polar phase. It is recommended that the initial column screening is started with picking one or two columns from each group. These stationary phase combinations would provide a high probability to find appropriate starting conditions for method optimization.

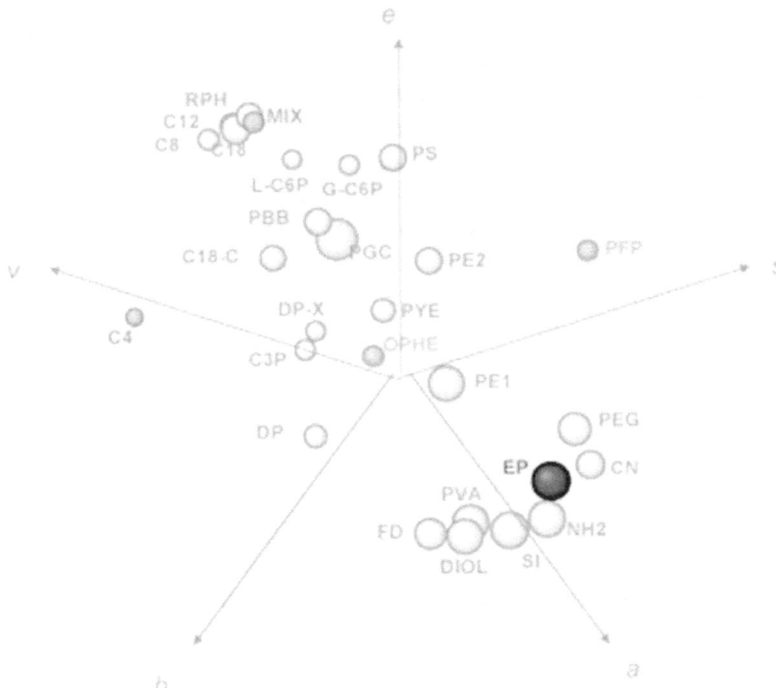

Figure 4.6 Spider diagram for a five-dimensional representation of the SFC column selectivities, developed by West and Lesellier [38].

4.3.3 Fine-Tuning on SFC Separation

Like HPLC, the SFC mobile phase strength is the function of the mobile phase composition, i.e., the nature of modifier and its concentration. What is unique to SFC is that the mobile phase strength is also the function of the density. This feature gives SFC additional means to adjust retention and selectivity. In fact, SFC method development can be divided into two parts: "primary" screening by searching the stationary phase and mobile phase, and "fine" tuning by changing the separation temperature, pressure, and even flow rate. The fine-tuning is best performed after selection of the best stationary phase and mobile phase composition for a given application.

Changing the pressure of the SFC separation changes the density of the fluid and, as a result, changes the relative retention of the analytes. Increasing pressure will increase the fluid density and the solvation strength, therefore resulting in shorter retention for the analytes. However, the magnitude of the changes due to variation in pressure is small compared with the changes due to modifier concentration. Pressure changes tend to produce larger changes in selectivity than do modifier concentration adjustments, although the differences are subtle [2]. Brunelli et al. pointed out that the change in mobile phase density with pressure had some influence on compound retention for less polar compounds, while such a difference in density did not alter the polarity of the mobile phase significantly enough to influence more polar compounds [19]. In other words, for the SFC separation with a high percentage modifier, the effect of pressure on retention is greatly reduced. It also means that the pressure variation normally has more influence on the early eluting (less polar) compounds than on those late eluting (more polar) compounds.

Changing the temperature of the SFC mobile phase, at constant pressure, also changes the density of the fluid. However, the impact of separation temperature on retention is hard to predict. Increased temperature causes a decrease in fluid density, and therefore increases the retention. On the other hand, increased temperature can also cause desorption of both CO_2 and modifier from the stationary phase, which tends to reduce the retention [2]. In fact, higher temperature can increase, decrease, or have no effect on the retention, and therefore, such parameter should be evaluated case by case.

One thing should be mentioned: Flow rate is not a typical parameter for selectivity tuning as it has no direct influence on selectivity. Owing to the low viscosity of supercritical fluid and a "flat" HETP-flow rate curve, flow rate is a useful tool in SFC to reduce the analysis time without compromising the efficiency. However, in SFC, the internal pressure also changes when adjusting the flow rate. This causes a change in the pressure drop and solvation strength, which can subtly shift the relative retention and selectivity.

Practically, in SFC analytical method development, pressure and temperature are two very valuable parameters to fine-tune the separation. Pressure should be varied while holding the modifier concentration and temperature constant. The largest changes in selectivity generally occur as pressure approaches the point where the fluid breaks down into two phases. However, working in such a region should be cautious since the separation robustness of the pressure needs to be studied. Similarly, temperature adjustment can be explored while holding pressure and modifier concentration constant. Generally, collecting two more chromatograms at +/- 5 to 10°C from the original set point gives an indication of the system susceptibility to temperature. If some significant selectivity changes occur over this temperature range, more extensive temperature changes are likely to produce even larger shifts [2]. One thing chromatographer should keep in mind when playing with temperature and pressure is to avoid forming two-phases. A two-phase system makes SFC separation almost impossible. Greater caution is required to avoid two-phase formation when operates at high temperature and low pressure. For instance, with CO_2 /MeOH mixture, when the temperature is set 40°C, the pressure should be kept at least 80 bar or higher.

4.4 Develop SFC Method for Mometasone Furoate Impurity Analysis

Mometasone furoate is a highly potent glucocorticoid anti-inflammatory agent and the active ingredient of several pharmaceuticals including Elocon® cream/ointment, Nasonex® nasal spray, Asmanex Twisthaler®, and Dulera® inhalation aerosol. The structures of mometasone furoate and its process impurities and degradation products are shown in Fig. 4.7. The current compendial method for mometasone furoate impurity analysis is based on reversed-phase LC that has a total run time of ~60 min. In this case study, following the achiral SFC method development principle elucidated in previous section, a fast packed column SFC method is developed for the quantitative analysis of mometasone furoate and its trace-level impurities [39].

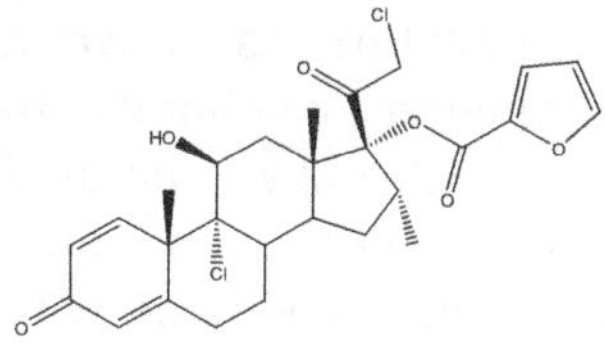

Mometasone Furoate

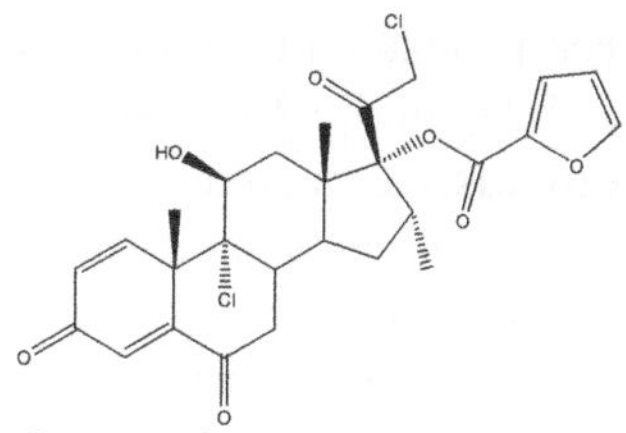

Compound 1

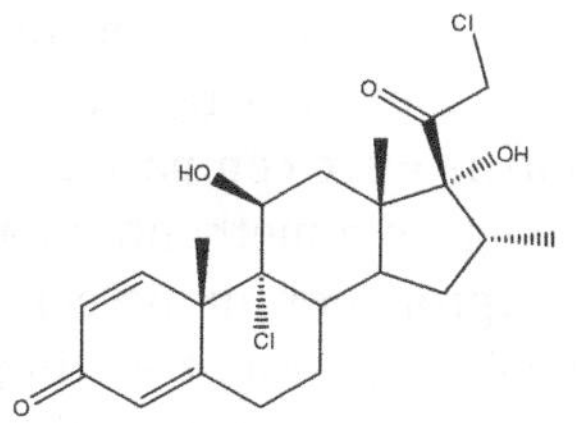

Compound 3

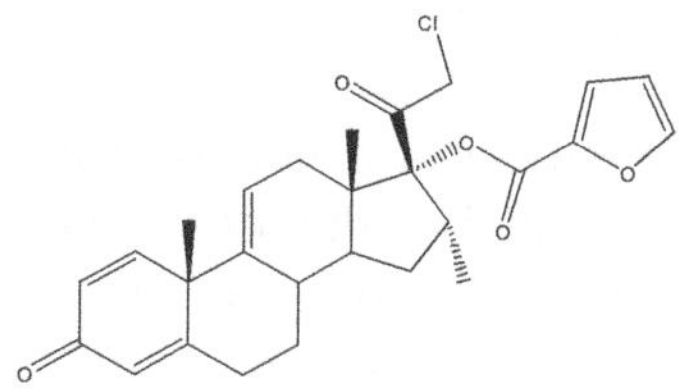

Compound 5

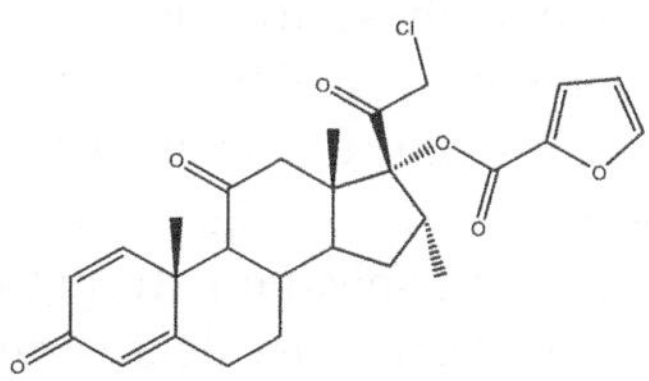

Compound 7

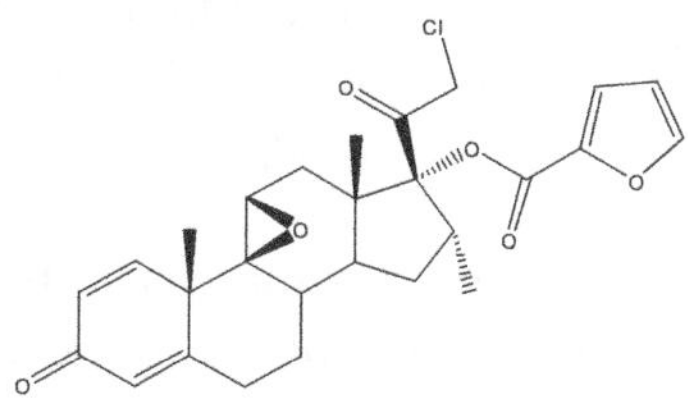

Compound 8

Figure 4.7 Chemical structure of mometasone furoate and compounds 1, 3, 5, 7, and 8.

As described in Section 4.3.2, a primary screening is conducted first to identify the best stationary and mobile phase. Since Mometasone and its impurities are relatively low polar compounds, three columns from the group of polar stationary phases were selected, including 2-ethylpyridine, cyano, and silica. Both cyano and silica columns have traditionally been used on normal-phase LC. The 2-ethylpyridine column is the first column that was developed specifically for SFC applications and has achieved wide success in the separation of various pharmaceutical compounds [40]. For the initial screening, neat methanol was used as the modifier. Considering the analytes are neutral and non-ionic compounds, no additive was used. A shallow gradient (5% to 15% methanol) was used to screen these columns to separate the spiked mixture. As shown in Fig. 4.8a, on the 2-ethylpyridine column, compound 2 co-eluted with mometasone furoate. Compound 3 was only partially separated from the main peak. On the cyano column (Fig. 4.8b), compound 5 and compound 8 were co-eluted, and compound 2 co-eluted with the main peak. Compound 3 eluted at the tail of the main peak with partial resolution. The best separation was achieved on the silica column (Fig. 4.8c) with all nine components baseline separated within 12 min. Further method development is then focused on the silica column and several modifiers are evaluated. Other than methanol, ethanol, isopropanol, and acetonitrile were compared as potential modifiers in this study. When moving from methanol to a less polar alcohol, in this case ethanol and isopropanol, an increase in the retention for all the components as well as several co-eluting peaks was observed. When acetonitrile was used as the modifier, more peaks co-eluted with mometasone furoate. Therefore, it was determined that methanol be used as the modifier in this method.

After the best stationary phase and mobile phase are identified, the second step of the method development is to fine-tune the separation by adjusting pressure and temperature. System back pressures at 100 and 150 bar were both evaluated (Fig. 4.9). As the pressure is increased, both the density and solvation strength of the mobile phase are increased. This explained the decreased retention at higher system pressure, as shown in Fig. 4.9b. However, the resolution was compromised at higher system pressure and compound 1 was only partially separated from compound 6. Changing the temperature (at constant pressure) will change the density of the SFC mobile phase and the kinetic energy of the solute. The combination of these effects can be used for retention/selectivity adjustment. Temperatures of 30, 35, and 40°C were evaluated while the pressure was kept constant at 100 bar to reduce the baseline noise. The column temperature control component can provide robust temperature control at ~5°C above the room temperature. For this reason, the lowest temperature evaluated was 30°C. As shown in Fig. 4.9c, when column temperature was increased from 30 to 35°C, the overall retention time increased. This may be due to reduced density of the mobile phase at higher temperature (when pressure remained constant). There is also a slight decrease in selectivity. Compound 2 and compound 6 were partially separated at 35°C. When temperature was further increased to 40°C (Fig. 4.9d), the trend of increased retention and deteriorated selectivity became

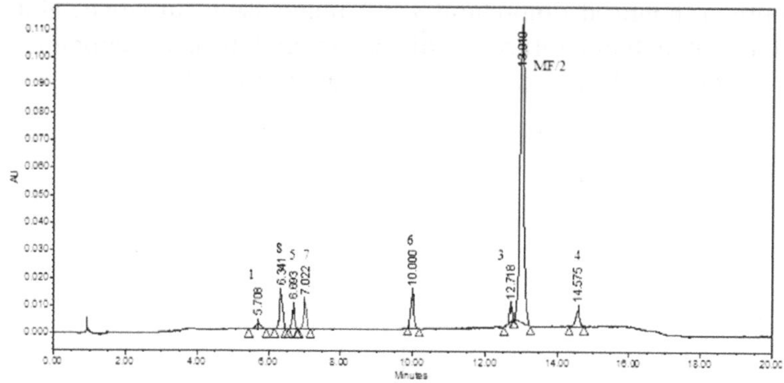

(a)

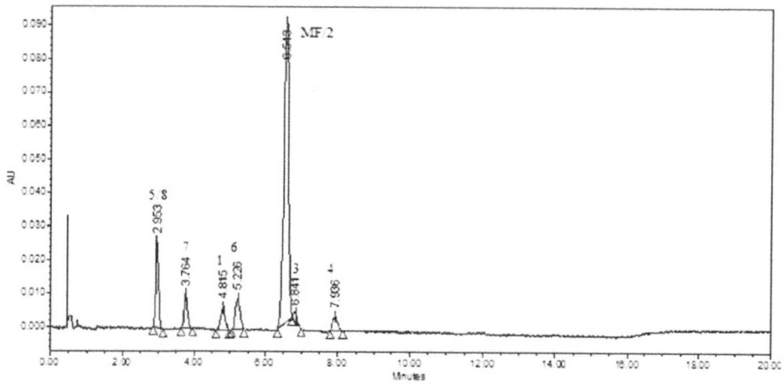

(b)

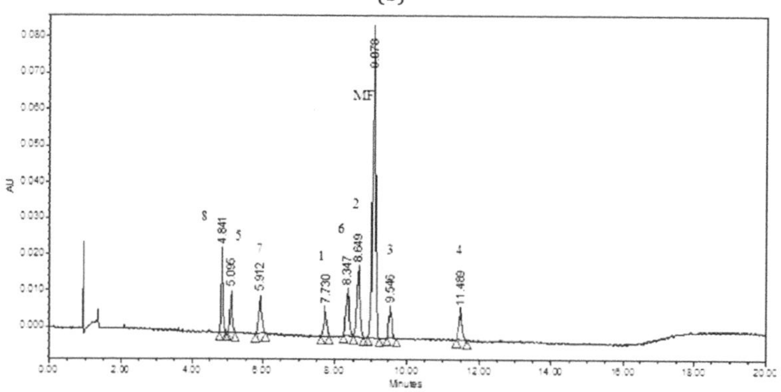

(c)

Figure 4.8 Column screening on (a) 2-ethylpyridine, (b) cyano, and (c) silica columns. CO_2 100 bar, 4.0 mL/min, 30°C, modifier: methanol 5% to 15% in 15 min. All columns were 250 × 4.6 mm, 5 μm particles.

more pronounced. Compound 3 and mometasone furoate co-eluted at elevated temperature as did compound 6 and compound 2. Therefore, the best separation temperature is at 30°C.

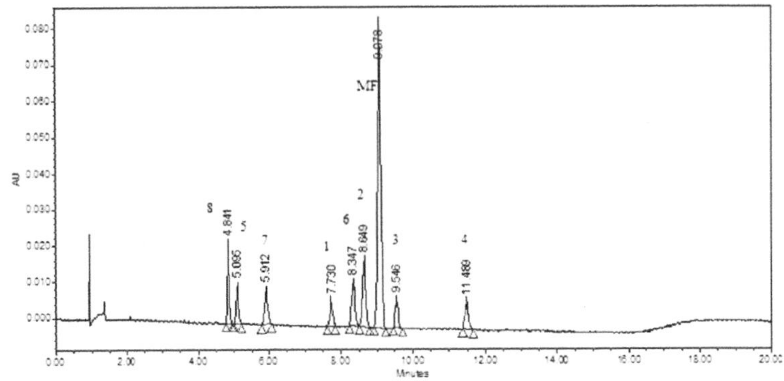

(a)

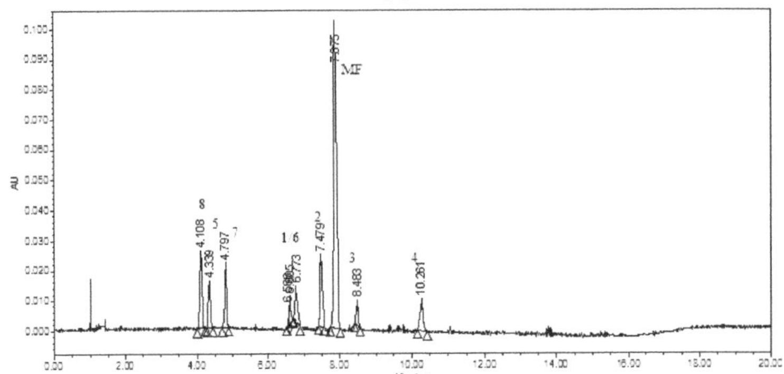

(b)

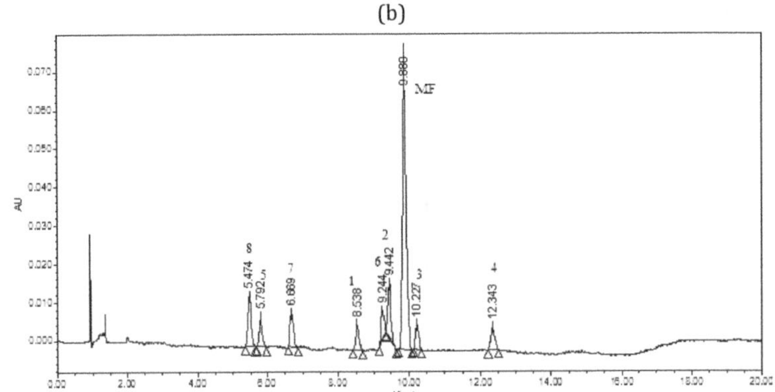

(c)

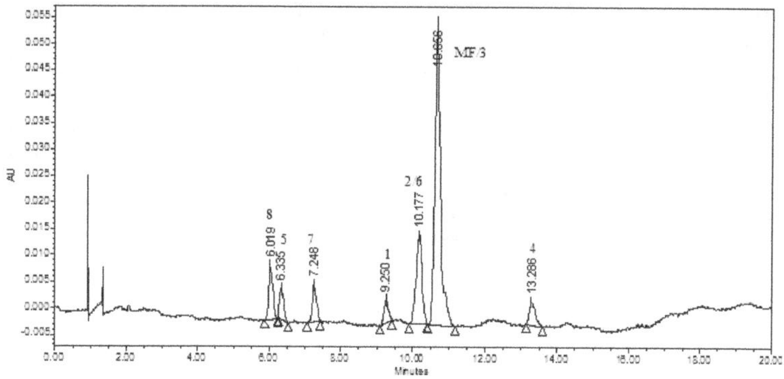

(d)

Figure 4.9 Evaluate mobile phase pressure and column temperature impact on selectivity. (a) 100 bar, 30°C; (b) 150 bar, 30°C; (c) 100 bar, 35°C; (d) 100 bar, 40°C. 4.0 mL/min, modifier: methanol 5% to 15% in 15 min, silica column, 250 × 4.6 mm, 5 μm particles.

The final optimized separation conditions are as follows: silica column, CO_2 100 bar, 30°C, 5% to 15% methanol in 15 min, total flow rate of 4 mL/min. At this optimized condition, mometasone furoate and its impurities are baseline separated and the final peak is eluted within 12 min.

The elution order of RPLC and SFC runs was compared in Fig. 4.10a. Owing to a polar stationary phase, e.g., silica, is used for this application, the SFC demonstrated a normal phase-like mechanism. SFC provides a dramatically different retention order compared to that of RPLC. The orthogonality of two chromatographic separations is also quantitatively described by calculating the correlation coefficients of retention factors from two methods. The coefficient of determination (R^2) is only about 0.2, which indicates a significant selectivity difference between two separation methods (Fig. 4.10b). The SFC method provides an orthogonal selectivity complementary to the RP-HPLC method that is very attractive during pharmaceutical impurity analysis. On the SFC method, all of the impurities and the active drug were baseline separated within 12 min, which is less than one third of the RP-HPLC method run time. This study demonstrated that SFC can be used as a complementary tool to RP-HPLC for pharmaceutical impurity profiling, separation, and quantitation.

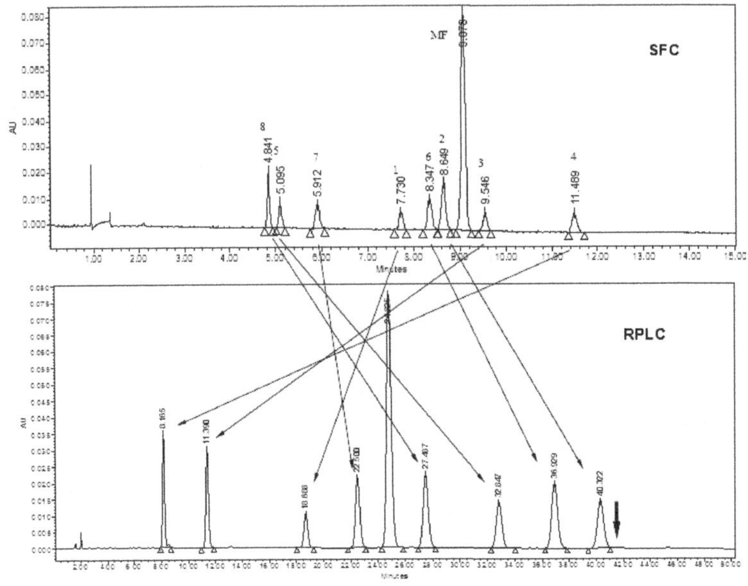

(a)

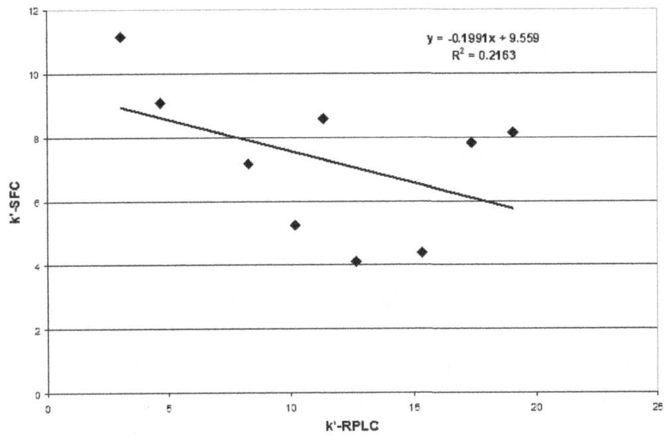

(b)

Figure 4.10 (a) The orthogonal selectivity of SFC method vs. RP-HPLC method. SFC condition: 100 bar, 30°C, 4.0 mL/min, modifier: methanol 5% to 15% in 15 min, silica column, 250 × 4.6 mm, 5 μm particles. RP-HPLC condition: 25°C, 1.5 mL/min, water/acetonitrile (58:42, v/v) to water/acetonitrile (48:52, v/v) in 60 min, Ultrasphere ODS column, 250 × 4.6 mm, 5 μm particles. (b) Comparison of retention factor (k′) in SFC method and in RP-HPLC method.

4.5 Summary

Supercritical fluid chromatography has, to date, been very successful in pharmaceutical analysis. So much so that it dominates chiral separation and purification. It provides several advantages over liquid chromatography in terms of resolution, throughput, and consumption of hazardous/expensive solvents. However, these advantages have not been fully utilized in achiral analysis. Knowledge of SFC achiral separations has been lagged behind gas and liquid chromatography, with still limited understanding on column chemistry and separation mechanisms. With further fundamental studies evolving in this area to help direct column design and method development, there is little doubt SFC will become more and more popular in achiral analysis in the near future.

References

1. Lee, M.L., Markides, K.E. (1990) *Analytical Supercritical Fluid Chromatography and Extraction* (Chromatography Conferences, Inc., Provo, UT).
2. Berger, T.A. (1995). *Packed Column SFC* (The Royal Society of Chemistry, Cambridge, UK).
3. Zelesky, T. (2008) Supercritical fluid chromatography (SFC) as an isolation tool for the identification of drug related impurities. *Am Pharm Rev*, 11, 56–60.
4. DaSilva, J.O., Yip, H.S., Hegde, V. (2009) Supercritical fluid chromatography (SFC) as a green chromatographic technique to support rapid development of pharmaceutical candidates. *Am Pharm Rev*, 12, 98–104.
5. Alexander, A.J., Staab, A. (2006) Use of achiral/chiral SFC/MS for the profiling of isomeric cinnamonitrile/hydrocinnamonitrile products in chiral drug synthesis. *Anal Chem*, 78, 3835–3838.
6. Miller, L., Potter, M. (2008) Preparative supercritical fluid chromatography (SFC) in drug discovery. *Am Pharm Rev*, 11, 112–117.
7. Zeng, L., Xu, R., Zhang, Y., Kassel, D.B. (2011) Two-dimensional supercritical fluid chromatography/mass spectrometry for the enantiomeric analysis and purification of pharmaceutical samples. *J Chrom A*, 1218, 3080–3088.

8. West, C., Lesellier, E. (2012) Chemometric methods to classify stationary phases for achiral packed column supercritical fluid chromatography *J Chemometrics*, 26, 52–65.

9. de la Puente, M.L., Soto-Yarritu, P. Lopez, Anta, C. (2012) Placing supercritical fluid chromatography one step ahead of reversed-phase high performance liquid chromatography in the achiral purification arena: A hydrophilic interaction chromatography cross-linked diol chemistry as a new generic stationary phase. *J Chrom A*, 1250, 172–181.

10. Lesellier, E. (2008) Overview of the retention in subcritical fluid chromatography with varied polarity stationary phases. *J Sep Sci*, 31,1238–1251.

11. McClain, R., Przybyciel, M., Majors, R.E. (2011) A systematic study of achiral stationary phases using analytes selected with a molecular diversity model. *LCGC North Am*, 29, 894–906.

12. Farrell, W., Aurigemma, C., Tran, P., Greig, M. (2012) Supercritical fluid chromatography-mass spectrometry for the rapid analysis of glycolic pathway components (from cell lysates). *The Column*, 8, 17–21.

13. Patel, M.A., Riley, F., Ashraf-Khorassani, M., Taylor, L.T. (2012) Supercritical fluid chromatographic resolution of water soluble isomeric carboxyl/amine terminated peptides facilitated via mobile phase water and ion pair formation. *J Chrom A*, 1233, 85–90.

14. Lazarescu, V., Mulvihill, M.J., Ma, L. (2011) A generic work flow for achiral SFC purification of complex pharmaceutical mixtures. *LCGC North Am*, 29, 438–444.

15. Zhang, Y., Hicks, M.B. (2011) Advancing SFC method development with a multi-column supercritical fluid chromatography with gradient screening. *Am Pharm Rev*, 14, 52–60.

16. de la Puente, M.L., Lopez Soto-Yarritu, P., Burnett, J. (2011) Supercritical fluid chromatography in research laboratories: Design, development and implementation of an efficient generic screening for exploiting this technique in the achiral environment. *J Chrom A*, 1218, I8551–8560.

17. Speybrouck, D., Corens, D., Argoullon, J.M. (2012) Screening strategy for chiral and achiral separations in supercritical fluid chromatography mode. *Curr Top Med Chem*, 12, 1250–1263.

18. Taylor, L.T. (2012) Packed column supercritical fluid chromatography of hydrophilic analytes via water-rich modifiers. *J Chrom A*, 1250, 196–204.

19. Brunelli, C., Zhao, Y., Hanna-Brown, M., Sandra, P. (2008) Development of a supercritical fluid chromatography high-resolution separation

method suitable for pharmaceuticals using cyanopropyl silica. *J Chrom A*, 1185, 263–272.

20. Poole, C.F. (2012) Stationary phases for packed-column supercritical fluid chromatography. *J Chrom A.*, 1250, 157–171.
21. Janssen, H.G., Lou, X. in: Caude, M., Thiébaut, D. (Eds.) (1999) *Practical Supercritical Fluid Chromatography and Extraction* (Harwood Academic Publishers, Amsterdam, NL).
22. Berger, T.A. (2011) Characterization of a 2.6 μm Kinetex porous shell HILIC column in SFC with a comparison to 3 μm totally porous silica. *J Chromatogr A*, 1218 , 4559–4568.
23. Lesellier, E. (2012) Efficiency in SFC with different superficially porous and fully porous particles ODS bonded phases. *J Chromatogr A*, 1228, 89–98.
24. Chester, T. (1999) The road to unified chromatography: the importance of phase behavior knowledge in supercritical fluid chromatography and related techniques, and a look at unification. *J Microchem*, 61, 12–24.
25. Berger, T.A. Deye, J.F. (1992) Use of solvatochromic dyes to correlate mobile phase solvent strength to chromatographic retention in supercritical fluid chromatography. *ACS Symp Ser*, 488, 132–142.
26. Strubinger, J.R., Song, H., Parcher, J.F. (1991) High-pressure phase distribution isotherms for supercritical fluid chromatographic systems. 2. Binary isotherms of carbon dioxide and methanol. *Anal Chem*, 63, 104–108.
27. Lesellier, E. (2009) Retention mechanisms in super/subcritical fluid chromatography on packed columns. *J Chrom A*, 1216, 1881–1890.
28. Ashraf-Khorassani, M., Taylor, L.T. (2010) Subcritical fluid chromatography of water soluble nucleobases on various polar stationary phases facilitated with alcohol-modified CO_2 and water as the polar additive. *J Sep Sci*, 33, 1682–1691.
29. Gyllenhaal, O., Edstrom, L., Persson, B.A. (2006) Ion-pair supercritical fluid chromatography of metoprolol and related amino alcohols on diol silica *J Chrom A*, 1134, 305–310.
30. Pinkston, J.D., Stanton, D.F., Wen, D. (2004) Elution and preliminary structure-retention modeling of polar and ionic substances in supercritical fluid chromatography using volatile ammonium salts as mobile phase additives. *J Sep Sci*, 27, 115–123.
31. Zheng, J., Glass, T., Taylor, L.T., Pinkston, J.D. (2005) Study of the elution mechanism of sodium aryl sulfonates on bare silica and a cyano

bonded phase with methanol-modified carbon dioxide containing an ionic additive. *J Chrom A*, 1090, 155–164.

32. Zheng, J., Taylor, L.T., Pinkston, J.D. (2006) Elution of cationic species with/without ion pair reagents from polar stationary phases via SFC. *Chromatographia*, 63, 267–276.
33. West, C., Lesellier, E. (2008) A unified classification of stationary phases for packed column supercritical fluid chromatography *J Chrom A*, 1191, 21–39.
34. West, C., Lesellier, E. (2010) Characterization of stationary phases in supercritical fluid chromatography with the solvation parameter model. *Adv Chromatogr*, 48, 195–253.
35. West, C., Lesellier, E. (2006) Characterization of stationary phases in subcritical fluid chromatography with the solvation parameter model IV. *J Chrom A*, 1115, 233–245.
36. West, C., Lesellier, E. (2006) Characterization of stationary phases in subcritical fluid chromatography by the solvation parameter model. *J Chrom A*, 1110, 181–190.
37. West, C., Lesellier, E. (2006) Characterisation of stationary phases in subcritical fluid chromatography with the solvation parameter model. III. Polar stationary phases. *J Chrom A*, 2006, 1110, 200–213.
38. West, C., Lesellier, E. Orthogonal screening system of columns for supercritical fluid chromatography *J Chrom A*, 2008, 1203, 105–113.
39. Wang, Z., Zhang, H., Liu, O., Donovan, B.A. (2011) Development of an orthogonal method for mometasone furoate impurity analysis using supercritical fluid chromatography. *J Chrom A*, 1218, 2311–2319.
40. Brunelli, C., Zhao, Y., Hanna-Brown, M., Sandra, P. (2008) Pharmaceutical analysis by supercritical fluid chromatography: optimization of the mobile phase composition on a 2-ethylpyridine column. *J Sep Sci*, 31, 1299–1306.

Chapter 5

Achiral Preparative Supercritical Fluid Chromatography

Vivi Lazarescu, Mark J. Mulvihill, and Lifu Ma
Department of Cancer Chemistry, OSI Pharmaceuticals LLC,
A wholly owned subsidiary of Astellas Pharma, Inc.
gvlazarescu@yahoo.com

5.1 Introduction

Historically, preparative supercritical fluid chromatography (SFC) was first envisaged by Klesper *et al.* in 1962 in their pioneering paper [1]. However, it did not fully materialize until 1982, when Perrut developed and patented [2] a prototypical preparative SFC chromatographic process, together with an eluent recycling capability for the purification of achiral petrochemicals and fatty acid intermediates (used for cardiovascular drugs). Perrut and coworkers later disclosed the feasibility applications of their integrated SFC process in several ensuing patent applications [3–6] and conference presentations [7–9]. Their innovative work was reviewed in 1990 [10] together with a utility comparison with other known chromatographic techniques (GC, HPLC, and TLC) and had set

Supercritical Fluid Chromatography: Advances and Applications in Pharmaceutical Analysis
Edited by Gregory K. Webster

ISBN 978-981-4463-00-3 (Hardcover), 978-981-4463-01-0 (eBook)
www.panstanford.com

the stage for subsequent applications of preparative SFC in industrial separation processes [11–13]. In the two years spanning from 1998 to 1999 when multiple commercial preparative SFC instruments [14–16] were available from various suppliers (Prochrom, Gilson, Berger, Jasco, and Thar), there were approximately 45 publications [17–21] relating to preparative SFC, most of which were primarily focused on the setup of industrial processes, trial-and-error equipment customizations and/or feasibility studies of various self-integrating SFC instrumentation units directly adapted from LC systems. Such studies, summarized in a few well-documented reviews [22–26], included separations of both achiral and chiral petrochemicals, natural products, pharmaceuticals, and agrochemicals on various scales, ranging from milligram in discovery laboratories to the kilogram scale in industrial processes.

Since the year 2000, with the surge of chiral drugs and the high demand for racemic separations in the pharmaceutical industry [27], packed column preparative SFC has gained wider utility, splitting into two distinct application frontiers: chiral SFC and achiral SFC. The former uses chiral stationary phases (CSPs) as packed column materials to resolve enantiomeric eluents that have identical chemical and physical properties in an achiral environment. The use of the SFC technique for chiral separations began several years after that of HPLC, which dates back to 1981 with the introduction of CSP chromatography [28]. However, the advantages for the use of supercritical fluids over liquid eluents for chiral drug separations were emerging. Low viscosity and high diffusivity of supercritical fluids allow for higher flow rates, lengthier columns, the inclusion of multiple columns, and higher sample loading capacities than other chromatographic techniques, including LC, GC, and CE [29]. Efficient separations and fast column re-equilibration are common in SFC where racemates that are inadequately resolved in normal-phase and reversed-phase HPLC can be separated. Moreover, most chiral SFC processes technically have only two fractions to be collected, which fits well with the initial preparative SFC hardware settings that are limited to collecting a maximum of six to eight fractions in a closed bed format [14–16]. The stacked injection technique with isocratic elution of the mobile phase, once used in SFC for the achiral separation of *cis/trans* isomers of phytol [12], an ingredient in the perfume industry (Fig. 5.1), presently is the most common practice for the purification of large amounts of racemic materials. Chiral

SFC has rapidly replaced LC in the past decade and now is the norm for chiral separation in both discovery laboratories and industrial manufacturing settings [30]. Commercial instrumentation suppliers for chiral SFC applications have helped to revitalize the technique by supplying a spectrum of instruments ranging from semi-preparative scale to the kilogram production scale.

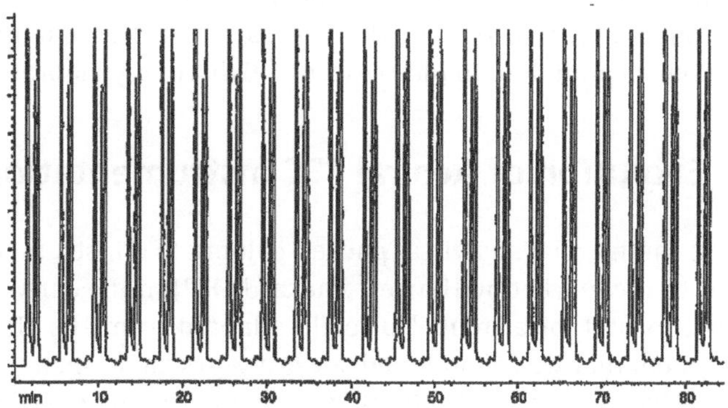

Figure 5.1 Successive laboratory-scale preparative SFC chromatography: *cis-trans* isomeric separation of phytol in 1998. Stacked injection every 4 min and 1 gram of crude material was purified within 1.5 h into four fractions. Conditions: 15 μm, 100 Å, 10 mm × 250 mm silica column; temperature: 50°C; pressure 250 bar; mobile phase: isocratic, 96.5% CO_2, 3.5% MeOH, EtOH, or IPA; flow rate: 14 mL/min [12].

However, the entry of achiral SFC into the pharmaceutical industry had been initially overshadowed by reversed-phase HPLC [30–32] and to a less extent by chiral SFC. SFC had been considered essentially analogous to normal-phase HPLC, and for a long time, the range of samples in SFC was limited to relatively nonpolar substances. The available preparative instruments on the market by the mid-2000s were inadequate to the modular HPLC settings in respect to handling more polar and basic pharmaceutical derived samples [31]. In the past 5 years, however, with advances in the instrumentation and the operational adoption of similar hardware and software as the widely used in reversed-phase HPLC, SFC has experienced a steady growth in acceptance, particularly in pharmaceutical and chemical laboratories for achiral purifications [30–32]. The incentive of employing green, sustainable SFC technologies in industrial processes as well as in analyses has also increased. To embrace this renaissance, in this chapter, we will provide an overview of the development of the achiral preparative SFC field since 2000 and its evolution from a standard technique of chiral separation to a viable tool for high-throughput purification of challenging isomeric mixtures and crude library compounds. To the best of our knowledge, this is a maiden attempt to exclusively cover achiral preparative SFC. After a historical review of commercial SFC instrumentation and stationary phases, we will focus on the integration of SFC technique as a routine purification tool to pharmaceutical discovery laboratories.

5.2 Evolution of Achiral SFC Instrumentation

The first commercial analytical packed column SFC instrument was brought to the market by Hewlett-Packard (HP) in the early 1980s, initially a prototypical model directly adapted from the HP 1084 HPLC system that was abandoned several years later when the SFC-incompatible HP model 1090 HPLC system was introduced [31]. In the following decade, several other manufacturers (e.g., Dionex, Suprex, and Gilson) followed suit providing analytical SFC systems. Many early-stage preparative SFC instruments were directly expanded and configured from the analytical models and in many cases provided dual analytical and semi-preparative configurations with a typical flow rate of 4–25 mL/min [15–16]. Prochrom, a company dedicated to HPLC and GC instrumentation, focused primarily on the preparative SFC market and offered custom-manufacturing of industrial-scale preparative SFC units subsequent to Perrut's preparative SFC system, which enabled the recycling of eluents via a cyclone separator. By 1992, instrumentation provided by Prochrom [16] ranged from laboratory scale (Super C10 and 20) to large production scale with the Super C500, which incorporated a 500 mm-diameter column capable of processing hundreds of kilograms per year.

In the 1990s, the SFC industry was experiencing a period of fluctuation, including consolidation and acquisition resulting in numerous vendors dropping out the SFC business (e.g., Dionex, Suprex) [31, 32]. This period of change was later described by experts in the field as a time when the cost to benefit was being weighed, as most of the instruments were overpriced and resulted in a lower volume of use and sales [15]. HP introduced its second-generation analytical SFC system in 1992 but sold in 1995 the rights of its SFC instrument to Terry Berger, a former HP employee, who launched Berger Instruments. Suprex was bought in 1997 by Isco, which later shifted its business activities to the fluid delivery system that upgraded an existing GC to an SFC and away from the packed column SFC area and into the "capillary SFC" field [16]. Prochrom was acquired in 1999 by Novasep, which has maintained a continuous presence in the larger preparative SFC market ever since. Most products of Novasep are for industrial processes and have been primarily provided on a custom-manufacturing basis. Jasco is another major player in the field having introduced its combined supercritical fluid extraction (SFE)-SFC system as early as 1985. This modular type of system was adapted from an LC-900 series, which enabled consumers to configure the system to meet specific application requirements such as HPLC/SFE/SFC [16].

By the end of the 1990s, semi-preparative and preparative SFC instrumentation was available from several manufacturers [16], which are summarized in Table 5.1. Thar Technology [34] entered the preparative SFC market in 1994 through a custom-manufacturing platform and in 1997 provided two systems, the SFC-50 and the SFC-200 for semi-preparative and preparative applications, respectively [16]. Berger, Jasco, and Gilson offered dual analytical and semi-preparative SFC systems that were reconfigured from their initial analytical instruments with some components removed to lower the price. Gilson exited the SFC business in 2002 [32].

There are two major differences that distinguish the SFC from the HPLC system and both revolve around the use of carbon dioxide as the mobile phase for SFC. First, since carbon dioxide is a gas at ambient temperature, SFC utilizes a chiller to maintain carbon dioxide in the liquefied state. The other difference is that the column outlet must be controlled at a pressure higher than ambient to maintain the desired fluid properties over the entire length of the column. In its early stage of development, packed column SFC suffered from pressure changes that required operators to manually adjust the pressure of the system with a relief valve. This resulted in irreproducible pressure gradients that limited the mobile phase to be isocratic or confined to a narrow range. In 1992, an automated downstream pressure control was adopted, which made the generation of pressure gradients independent from the mobile-phase flow rate and composition. A relatively simpler electronic back-pressure regulator was later introduced by some manufacturers [14–15].

Table 5.1 Select commercial preparative SFC systems in 1999

Instrument	Super C12	Series SF3	Jasco 1500	Berger SFC	Thar SFC-50
Manufacturer	Prochrom	Gilson	Jasco	Berger	Thar
Flow rate (mL/min)	45	25	20	50	50
Maximum column ID (cm)	2	2	1	1	2
Pressure Control	Analog control valve	Electronic back-pressure regulator	Automated back-pressure regulator	Electronic back-pressure regulator	PC-controlled pressure regulator
Detectors	UV	UV, ELSD	UV, CD	UV	UV, CD, ELSD
Maximum pressure (bar)	250	400	350	200	600
Maximum temperature (°C)	60	200	80	70	150
Maximum fractionation	6	8	8	5	6
Comment	Acquired by Novasep in 1999	Exited SFC market in 2002	Broader modifier composition (1–100%)	Optional column and solvent switching	cGMP validation available

A wide range of detection systems were available from the launch of SFC units, including UV, circular dichroism (CD), and evaporative light scattering detector (ELSD). High-pressure cells were required for these detectors. Fractionation, as described earlier, was typically controlled by a UV signal or a predetermined time window derived from an analytical method. Most existing preparative SFC instruments of the day employed pure carbon dioxide as the mobile phase along with cyclone separators to collect a single or a few components from an unchanging fraction feedstock, almost all in a closed beds format with a maximum of 6–8 fraction containers. The largest preparative SFC reported used pure carbon dioxide as the mobile phase, with up to 76 cm diameter columns, to separate the relatively nonpolar ethyl esters of omega-3-fatty acids from fish oils [35]. Although high trapping efficiencies (>95%) were commonly observed, cyclone separators were designed to collect large amounts of one (or two) component from multiple injections of the same sample solution, as typically is the case in chiral separation. Cyclone separators tend to be large devices, which must be disassembled and washed between samples, making them cumbersome for high-throughput purification applications.

Achiral separation, on the other hand, requires the SFC hardware to be as modular as that of a high-throughput HPLC system, being able to handle a batch of samples with an unlimited number of injections and fraction collections. The ability to collect at least hundreds of fractions is a fundamental feature necessary for the purification of compound libraries in the pharmaceutical industry. Other performance features include high recovery and negligible carryover, automation capacity, and the ability to collect fractions at atmospheric pressure into the same disposable test tubes and racks that are already employed in HPLC applications.

In 2000, Berger Instruments, which was acquired by Metter Toledo later that year, in collaboration with Merlin Instruments and Alanex reported their initial efforts to develop an SFC system suitable for high-throughput purification of combinatorial chemistry libraries [36]. This system comprised several modified LC components (a manual injector, a UV detector, a back-pressure regulator, and a Berger "separator" that prevents aerosol formation as fluid-phase methanol/carbon dioxide to expand to the gas phase and therefore separate from the liquid organic modifier). Fractions were collected at elevated pressure into a "cassette" system comprising four individual compartments, each with a glass collection tube insert, which allowed for efficient collection of up to four components per chromatogram.

Subsequently, Berger Instruments incorporated additional capabilities to their platform [37], which included an automated sample injector and fraction collector (Cartesian-style robot integrated with the core unit). On-board capacity for 128 fractions was provided and the peak detection module had been improved.

The model, designated as the Berger SFC AutoPrep (Fig. 5.2), was introduced to the market in the mid-2000s. However, due to some critical shortcomings, it was met with limited interest from the SFC field resulting in the sale of only a few units before Metter Toledo sold its Berger Instrument unit to Thar in 2007.

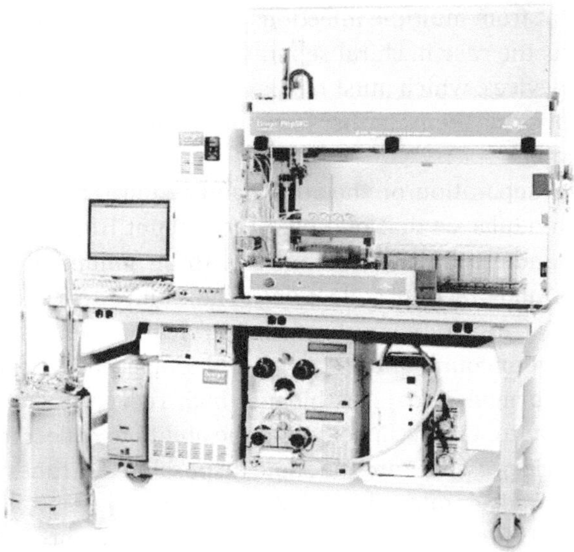

Figure 5.2 Berger SFC AutoPrep system marketed in the mid-2000s, 50 mL/min of flow rate, UV triggered fraction collection up to 4 fractions in a sealed "cassette" of 2 tube pallets (8 × 8 of 25 × 150 mm test tubes), 2 sample well plates, and as many as 96 samples.

Over the years, several pharmaceutical end users reported the customization of Berger preparative SFC units to develop an open-beds fraction collector capable of collecting at atmospheric pressure as in the HPLC systems. This presented a number of engineering and technological challenges, which were creatively tackled by analytical chemists and engineers from outside the realm of instrument manufacturers and helped propel the evolution of SFC hardware. For example, Olsen *et al.* from Abbott [38] in 2002 developed a "collection shoe," or a custom-designed fluid/gas outlet that was outfitted on a dual-arm fraction collector, modified from a Cavro miniPrep™ pipettor. This dual configuration enabled the collection of samples at atmospheric pressure. Additionally, the inclusion of a methanol wash system into the fraction collection line eliminated cross contamination between fractions and ensured high recovery. The samples purified on SFC by UV triggered collection were validated on MS from identical footprint fraction collector racks, with comparable efficiency and recovery to HPLC.

During the same period, many researchers had worked on developing an in-house mass-directed SFC system for achiral purifications. Mass-directed fraction collection offers an important advantage over UV- and time-based fraction collection methods in that it is a uniquely selective and sensitive technique for fraction collection. Fraction collection is initiated when the expected product is detected in the mass spectrometer and its ion signal exceeds an input threshold value. However, the development of a mass-directed SFC purification system was not straightforward as challenges relating to the SFC-MS interface and the fraction collection hardware had to be resolved.

Ontogen researchers in 2000 custom-designed a mass-triggered semi-preparative SFC purification system that allowed for the purification of up to four samples at one time [39]. Purified samples were deposited back into a microtiter plate by using a disposable "expansion chamber" on a customized fraction collector. The reported flow rate was 12 mL/min with a cycle time of 5.5 min. In 2001, Wang *et al.* of DuPont Pharmaceuticals [40] developed an SFC-MS purification system through the incorporation of a Gilson SF3 SFC system with a Gilson 215 liquid handler, a PE-Sciex AP150EX mass spectrometer and a modified Gilson fraction collector. A long, restrictive piece of tubing was used to connect the SFC and the mass spectrometer to counter the large pressure drop. Fraction collection was conducted at atmospheric pressure with a piece of foil cover wrapped around the top of the collection tubes. The typical sample load was 5–15 mg on a 10 mm × 150 mm column with a flow rate of 15 mL/min. Zhang *et al.* [41] from ArQule reported in 2006 their interesting customization and integration of a Berger manual preparative SFC system with a Waters ZQ mass detector and a standard 2757 fraction collector. Several technical issues including software compatibility, the SFC-MS interface had been addressed and aerosol formation during the fraction collection was greatly reduced through replacing the collection needle with a piece of Teflon tubing having a diameter of 3.2 mm. These modifications enabled the mass-directed real time fractionation SFC system as noted by the achievement of greater than 85% overall recovery for 10 drug-like compounds tested with flow rates up to 30 mL/min.

These pioneering customizations had encouraged broad interest among the end users in the use of mass-directed SFC instrumentation. To address the market need from the end users, the first commercially available gradient preparative SFC-MS system, the SFC Prep-30, was introduced by Thar in 2008. Since its introduction, the instrument has been comprehensively described in the literature [42]. Key findings, including some from the authors' hands-on experience, are that the Prep-30 provides reasonable (80–85%) compound recoveries and minimal cross contamination [43, 44]. However, its maximum flow rate of 30 mL/min has been viewed by many users as too low for broad applications. In response, Thar (acquired by Waters Corporation in 2009), introduced in 2009 its upgraded Prep-100 SFC-MS directed system (Fig. 5.3), with a maximum flow rate of 100 mL/min [45]. The commercial availability of a mass-directed SFC purification system has now made the purification of achiral pharmaceutical compounds a real possibility.

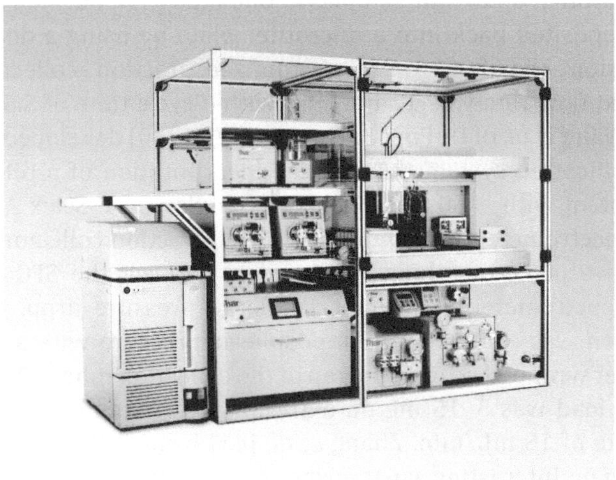

Figure 5.3 Prep-100 SFC-MS Directed System (detailed product information is listed in Table 5.2). Photo courtsey of Waters Corporation.

The use of mass (molecular weight) responses to trigger fraction collection provides isolation of only a single component from a crude

Table 5.2 Selected commercial laboratory-scale preparative SFC systems in 2011

Instrument	Prep-30	Prep-100	Prep-350	SP-2086	SP-2088
Manufacturer	Thar	Thar	Thar	Jasco	Jasco
Flow rate (mL/min)	30	100	350	20	120
Maximum column ID (cm)	2	3	5	1	3
# Column in column oven	10	6	1	6	1
Software	Masslynx & FractionaLynx	Masslynx & FractionaLynx	Masslynx & FractionaLynx	ChromNav & SF-Nav	ChromNav & SF-Nav
Detectors	MS, UV, PDA	MS, UV, PDA	UV, PDA	UV, PDA, CD	UV, PDA
Maximum pressure (bar)	350	200	320	350	500
Maximum temperature (°C)	60	70	80	60	60
Injection	CTC autosampler	Waters 2767	Stream Injector	AS-2058 autosampler	LVI system
Maximum fractionation	180 25 × 250 mm test tubes (Waters 2757)	180 25 × 250 mm test tubes (Waters 2767)	5 sealed modules	8 or an optional open bed fraction collector	8 sealed modules
Comment	First-generation preparative SFC-MS system	A tailored equipment for discovery labs	Thar acquired by Waters in 2009	MS detection unavailable	Mainly for chiral separation

reaction mixture and validation of structure during the compound purification. SFC-MS based purification process has experienced a rapid growth over the years in the pharmaceutical and fine chemical industries and has been shown to be a useful tool for the purification of chemical libraries [46].

5.3 Stationary Phases for Achiral SFC

The most common stationary phases used for packed column SFC are modified silica particles. Similar to the SFC instrumentation that was directly adapted from the early LC systems, initial packed columns used in SFC were generally normal-phase stationary phases borrowed from HPLC. The common packings of the time were 3- and 5 μm porous silica particles and functionally modified silicas of phenyl, cyano, amino, and diol derivations. For some time, no column products were solid specifically for SFC applications. All available stationary phases on the market were the same as those used more than 20 years earlier in normal-phase HPLC, with limited improvements noted [31]. They typically had low capacity, poor selectivity, and poor peak shape for SFC separations, especially for many classes of polar analytes, including phenols, acids, amines and drug-like molecules that could only be separated or partially resolved using polar additives including trifluoroacetic acid, isopropylamine, ammonium acetate, etc. This limiting situation lasted until 2001, when 2-ethylpyridine, the first stationary phase specifically designed for achiral SFC applications, was launched by Princeton Chromatography [47]. This stationary phase was embraced immediately by the end users as the 2-ethylpyridine bonded phase with its basic, aromatic surface allows for retention and separation of polar and basic analytes, such as most pharmaceutical analytes, without the use of very polar additives. A new progressive era for SFC stationary phases was launched and a variety of new phases have been developed, focusing on expanding column selectivity and providing better chromatographic performances (peak shape, loading capacity).

Today, there are approximately 100 different types of achiral analytical SFC columns (2 mm/3 mm/4.6 mm ID) available from several companies, including both new and old stationary phases.

About half of these columns are available in larger sizes (10 mm/21.2 mm/30 mm/50 mm/70 mm column ID) suitable for semi- and preparative applications. Most columns are packed with 3 or 5 μm and 5 or 10 μm particles for analytical and preparative SFC, respectively. The pores sizes range from 60 to 300 Å with the column lengths ranging from 50–250 mm. In many cases, the old practice in the field is still applied here where preparative columns especially larger than 50 mm ID have to be custom-manufactured. Princeton Chromatography currently is the largest supplier of achiral SFC stationary phases, providing more than 50 bonded phases such as 2- and 4-ethyl pyridine, pyridyl amide, diethylamino, propyl acetamide, benzamide, etc. [47]. They claim both their analytical and preparative columns are packed with the same batch of bonded phases so that the scale-up is relatively easy. Zymor, another major column supplier, under its brand ZymorSPHER, has recently developed a series of new amide, urea, and pyridine stationary phases (propyl acetamide, dicyanoimidazole, morpholine, etc.), which enhance the selectivity power of achiral separation [48]. ES Industries retails its older products under the brand name of Chromegabond and new line phases under GreenSep™. Both brands currently have about 10 bonded phases, with GreenSep™ representing specialty stationary phases, including ethyl pyridine, pyridyl amide, nitro, pentafluorophenyl and aminophenyl [49]. Phenomenex has traditional phases specifically designed for SFC and newer phases such as ether-linked phenyl and pentafluorophenyl [50]. Other column manufacturers, Kromasil, Restek, and Waters, also provide a few traditional phases in the analytical and preparative scales.

Distinct from stationary phases in HPLC, modified silica gels in SFC columns have two structural forms, end-capped and non-end-capped, representative of the saturation stage of active terminal silanol groups. The surface of silica gel contains a significant number of silanol (-SiOH) groups, which is inherently a weak acid and negatively charged in a neutral environment and can interact strongly with polar solute molecules by means of hydrogen bonding and ionic interaction. In LC stationary phases, the silanol group is conventionally modified with a functional trimethylsilyl (TMS) reagent, i.e., end-capped to avoid potential secondary reaction/interaction of the surface. Many SFC manufacturers provide non-end-capped SFC columns or mixed stationary phases that are prepared by capping reaction of the silanol group on the surface with another interacting group. Since different types of interacting groups can be used, many different mixed phases are available commercially (Fig. 5.4). Practically, vendors of stationary phases are reluctant to provide detailed column fabrication and end-capping information (only Princeton Chromatography discloses some information on its existing phases). In many cases these technical details are not publically disclosed in order to protect the company's intellectual property investment and provide a potential edge within the marketplace. Recently, McClain and Przybyciel [51] reported their systematical study of 12 SFC achiral stationary phases with 60 compounds from four chemical classes: amines, amides, alcohols, and carboxylic acids. They found no single column was identified with broad utility across the entire set of compounds profiled, but generally the non-end-capped stationary phases displayed better chromatographic performances (efficiency, peak shape, and selectivity) than those end-capped counterparts. In this regard, the silanol groups on the non-end-capped stationary phases contribute to retention and play a role in maintaining the quality of separations.

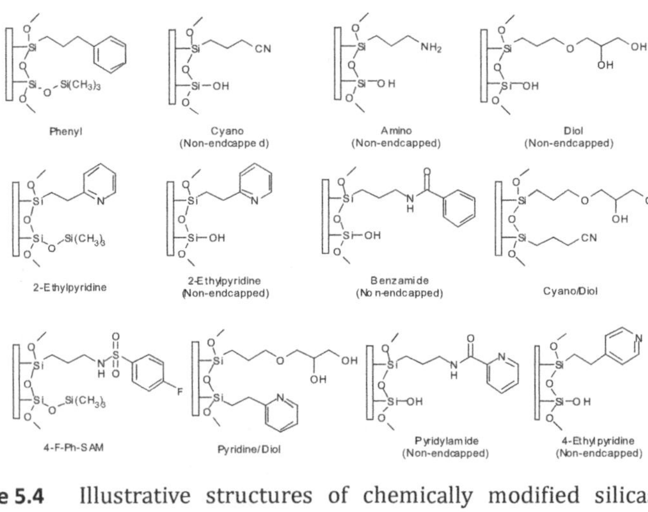

Figure 5.4 Illustrative structures of chemically modified silicas used as stationary phases in packed column SFC (available in two forms, end-capped and non-end-capped, corresponding to the terminal silanol groups end-capped or not).

Chromatographically, column chemistry is a fundamental factor of retention mechanism and directly related to the effect on separation.

Unfortunately, the lack of a fundamental understanding of the retention mechanism and separation process in SFC has limited its use. Basic studies need to be conducted in order to elucidate the key factors involved in the SFC analyte to stationary phase interaction where guidelines to choosing the optimum phase for separations are defined. Some interesting guidelines to predict retention and column selection have started to emerge. Since 2004, West and Lesellier [52–61] have reported on a series of chemometric studies that are focused on identifying the basic nature of retention for many packing materials in SFC. Their goal was to better understand the actual influence of the stationary phases when the mobile phase is very different from aqueous solutions. In recent work [59], they categorized a quantitative structure-retention relationship of their selectivity along five axes, each representing a different solvation parameter, as shown in Fig. 5.5. Almost all the traditional non-end-capped phases (amino, cyano, and diol) are clustered together, showing mostly strong hydrogen bond acidity interactions analogous to acidic bare silica. Surprisingly, 2-ethyl pyridine (non-end-capped from Princeton Chromatography) is very similar, located between cyano and amino. Diol is shown to be slightly more affected by hydrogen bond donor basicity than the others.

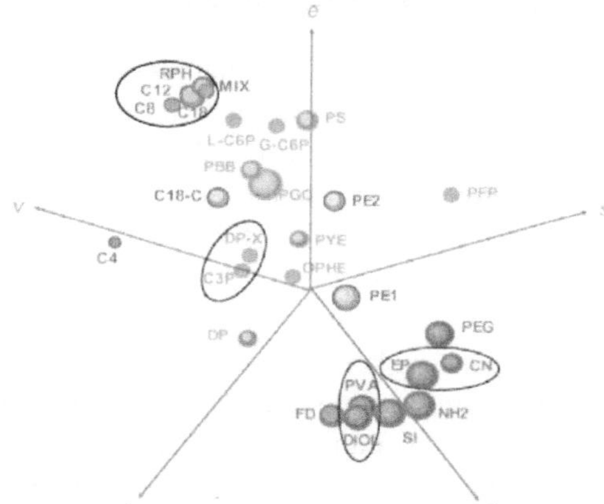

Figure 5.5 Five-dimensional representation of the solvation parameter models for 28 stationary phases. Green points, nonpolar phases; blue points, polar alkyl phases; yellow points, moderately polar aromatic phases; orange points, polar phases. Conditions: 25°C, outlet pressure, 150 bar, mobile phase, CO_2, 3 mL/min [59].

Having analyzed linear solvation energy relationship (LSERS) with 200 drug solutes, Bui *et al.* [62] from Eli Lilly reached a slightly different conclusion that hydrogen donor acidity of the solutes is dominant for positive retention, particularly for non-end-capped 2-ethylpyridine and amino columns. The factor is determined by the difference in basicity between the mobile and stationary phases. Hydrogen bond donating ability is small for non-end-capped cyano and 2-ethylpyridine stationary phases (or should be zero as one might expect). Hydrogen bond acceptor basicity of the solutes is particularly important for diol and amino columns.

Yet another team [63], including members from several pharmaceutical laboratories and column manufacturers, found the measured retention characteristics of a test library of sulfonamides on the non-end-capped 2-ethylpyridine stationary phase from Princeton Chromatography correlate with total dipole moment, molecular surface area, and the "electronic charge on the most negatively charged atom."

5.4 Method Development: Experimental

5.4.1 The Value of Pre-Purification

The key for successful achiral SFC purifications is in generating appropriate analytical data. This is why the method development for achiral preparative SFC separations should start with analytical screening of a wide range of stationary and mobile phases. The pre-purification approach is a fast and cost-saving common practice [43, 44], allowing the chromatographer to determine the stationary and mobile phases that provide the selectivity and chromatographic resolution required to isolate the desired compound from the other components of the sample and, if possible, elutes the desired product early in the chromatographic run so that appropriate sample loading can be used to increase the purification throughput and process efficiency. In the cases of purification of a large number of compounds, such as compound libraries, the implementation of this practice becomes more important and effective. Earlier work in high-throughput library purification has demonstrated the value for creating "universal" gradients based on pre-purification analysis of a subset of the compounds. With the current availability of integrated vendor solutions that create optimized purification gradients based on the pre-purification data, the value of such analyses is enhanced and strengthens the case for analysis of all samples before purification. The reader should refer to Chapter 3 for more detailed discussions on method development for achiral SFC.

5.4.2 Column Selection

Stationary phase selection is the most important element that profoundly affects the quality of the separation. However, how to choose the right support material and particle geometry from the pile of commercial bonded phases is still up to the user's knowledge and experience as no golden rules of column selection have been established for SFC at this moment. A number of column vendors have prepared indices of compounds resolved on various stationary phases and, in fact, the knowledge base such as the chemometric approaches described in Section 5.3 on the relationship between column selectivity and retention can be used as a guide for the selection. For many end users, a good collection of achiral analytical stationary phases that match with your applications is the prerequisite to start with. Usually, one or two injections indicate the likelihood of success with a particular stationary phase. Criterion for column selection should include the ability to scale-up while maintaining maximum resolution of impurities from the desired peak, and preserving peak shape with or without the use of additives in the mobile phase.

Our initial efforts to address this selection process are shown in Fig. 5.6. An initial collection of eight achiral polar stationary phases were screened and found to align with our primary applications to separate a variety of relatively polar and somewhat basic small molecules from our drug discovery laboratory. We have also used chiral stationary phases [64–66] with differing selectivity powers to separate challenging isomeric mixtures, which will be described later in this chapter. It is worth emphasizing that a selectivity pattern around a specific compound class (chemotype) will emerge, defining the most optimal column. This knowledge base would decrease the ongoing need to screen broader columns for similar compounds from the same or closely related chemotypes down the road.

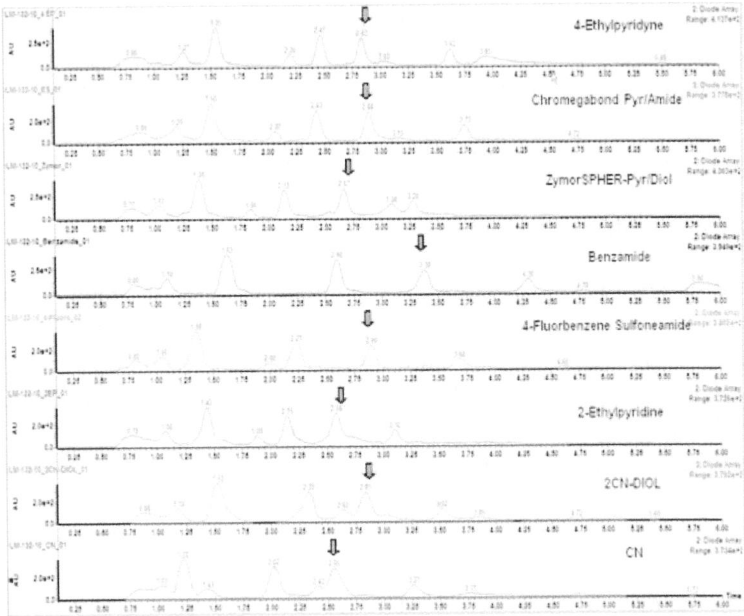

Figure 5.6 Analytical screening of eight stationary phases. Conditions: 5 μm, 4.6 mm × 150 mm columns; temperature: 40°C; mobile phase: a gradient of 10–60% MeOH with 0.2% IPA in 5 min; flow rate: 4 mL/min; injection volume: 10 μL of ~200 mg/mL in DMSO.

5.4.3 Mobile Phase Modifier and Additives

The typical mobile phase modifier used in SFC is methanol, sometimes containing an additive of either an organic acid or a base. Other common modifiers are ethanol, isopropyl alcohol (IPA), acetonitrile (ACN), and tetrahydrofuran (THF). To obtain optimal selectivity for SFC-MS analyses, the strengths and polarities of different modifiers should be considered. Changing the types of solvents can significantly influence the degree of band spacing. In addition to providing the most suitable chromatographic separations, the ideal solvent system should be compatible with positive APCI mass spectrometry. The use of solvent mixtures can also provide increased control of solvent strength, thus allowing for increased selectivity for multi-component sample analyses.

Empirical observation has shown that most small molecule drug candidates are balanced with polar groups (i.e., nitrogen atoms) and lipophilic functionality. Therefore, the addition of a small amount of basic organic compounds such as isopropylamine (IPAmine) and dimethylethylamine in a concentration of ~0.2% to the modifier was proven to be useful in improving both peak shape and loading capacity of pharmaceuticals containing basic functional groups. However, some of the basic additives such as IPAmine in higher concentrations can have a negative impact on mass spectrometric detection. Ammonium acetate is found to be an acceptable alternative if MS ionization from the MeOH/0.2% IPAmine modifier is an issue.

5.4.4 Flow Rate

Supercritical fluids have lower intermolecular forces than those in normal liquids, resulting in lower viscosities and higher diffusion rates for the mobile phase used in SFC compared to HPLC. SFC systems can therefore be run at higher flow rates than an equivalent HPLC system and with longer columns without excessive column back-pressure or loss of resolution, leading to shorter run times per sample. This throughput advantage has been reported in the use of analytical SFC and SFC/MS for the analysis of combinatorial libraries.

5.4.5 Gradient Ramp Rate

In preparatory SFC, the modifier composition is commonly programmed from as little as 5% modifier to as much as 60% modifier. The standard gradient elution methods employed are 5–50% modifier in carbon dioxide, 10–60% modifier in carbon dioxide, or 20–60% modifier in carbon dioxide, with a "hold" period of 2 min at the highest percentage of modifier at the end of each run. Varying the composition of the mobile phase alters selectivity in all chromatographic techniques. Since relative retention is dependent on the rate of change in modifier composition, larger changes in selectivity are obtained by SFC. Therefore, selectivity between different components in a mixture needs to be optimized by evaluating gradients of individual slopes. It has been found that constant-rate modifier gradients are insufficient for producing adequate resolution of fast-eluting nonpolar compounds and allowing the later polar compounds to elute in a sufficiently narrow band. Two slope gradient delivered improves separations compared to a simple linear gradient with the identical run time [67].

5.4.6 Sample Solvents

Alcohol-based solvents are often highly effective for solubilizing most pharmaceutical compounds. Several polar aprotic organic solvents, such as dimethylsulfoxide (DMSO) and dimethylformamide (DMF), are considered universal solvents for solubilizing most small drug molecules without disrupting the resulting chromatograms. At least 1–2 mL can be injected onto a 21 mm-diameter column without peak distortion. According to Berger *et al.* [31], care must be taken with DMSO and DMF. It is fairly easy to dissolve more of some solutes into the sample solvent than can be solubilized by the (less polar) mobile phase. If too high a concentration of solute is injected, the solute can precipitate resulting in over-pressure of the column and a pump shut-down. Lower polarity solvents, such as ethyl acetate, THF, chloroform, and DCM can normally be used as sample solvents without issue.

5.4.7 Triage between SFC and HPLC

In the early days of implementation for SFC purification of compound libraries, a pre-purification analysis of all samples prior to purification was necessary for determining the most appropriate technique (HPLC-UV, HPLC-MS, SFC-UV) to maximize purification success, for eliminating the added cost of purifying reactions that failed to yield enough of the desired product or for automatically creating custom purification methods and/or triggering thresholds. With the advancements in column chemistry specifically tuned for SFC and with the current availability of integrated SFC vendor solutions that create optimized purification gradients based on the pre-purification data, the triage between SFC versus HPLC will no longer be needed.

5.5 Singleton Achiral Purification of Difficult Samples for Discovery Research Support

Singleton achiral purification, formerly the subject of several studies [68, 69] that accompanied the early development of SFC instrumentation, now is in many cases a straightforward routine, run without much difficulty [70]. However, not surprisingly, separation by SFC, similar to other chromatographic techniques, is often confronted with challenging samples that contain multiple structurally similar or/and chromatographically co-eluting components and impurities. These difficult samples can be crude mixtures resulting from a complex synthetic reaction accompanied by a low yield and a high level of impurities, structural isomers such as regioisomers and diastereomers, and components whose physicochemical properties (basicity, minimal retention, aqueous stability, etc.) prevent their efficient separation on reversed-phase HPLC. Separation of these difficult samples requires more extensive analytical method development in order to effectively handle these compounds on the preparative scale. On some occasions such difficult samples may be purified by an exhaustive, time-consuming two-stage process where the primary separation is followed by a secondary purification via orthogonal chromatographic conditions or techniques such as SFC and reversed-phase HPLC. In order to enable rapid method development and fast turnaround time for pharmaceutical purification tasks, several discovery laboratories have reported interesting approaches that take advantage of the broad selectivity profile of different stationary phases.

Yan *et al.* [71] at Pfizer described in 2009 the use of SFC for the purification of pharmaceutical substances that showed little or no separation on a variety of reversed-phase HPLC columns. After analytical screening of a collection of achiral and chiral stationary phases, they identified a chiral stationary phase, CHIRALPAK AD-H for the separation of the two diastereomers and an achiral stationary phase, 2-ethylpyridine for the region-isomeric separation (Fig. 5.7).

Similarly in 2009, we established and validated an effective workflow [65, 66] using chiral as well as achiral stationary phases, in combination with compound-specific gradient methods for rapid mass-directed SFC purification of complex mixtures that displayed little to no separation on reversed-phase HPLC. The workflow included a rapid SFC analytical screening performed primarily on chiral stationary phases (CSPs) and occasionally on achiral stationary phases only when chiral columns failed to achieve resolution of the analytes.

Polysaccharide-based CSPs, 5 μm, 4.6 mm × 100 mm columns, and achiral alkyl bonded stationary phases, 5 μm, 4.6 mm × 150 mm columns, were screened in the following order: CHIRALPACK IA > IB > IC and 4-ethylpyridine > Chromegabond pyridine/amide

> CN. Only polysaccharide-bonded CSPs columns are screened because of their high compatibility with a broad range of sample solubilizing solvents (i.e., DMSO, DMF). The analytical screening is performed with a gradient of 10–60% in 5 min and modifier is 0.2% of isopropylamine (IPAamine) in methanol.

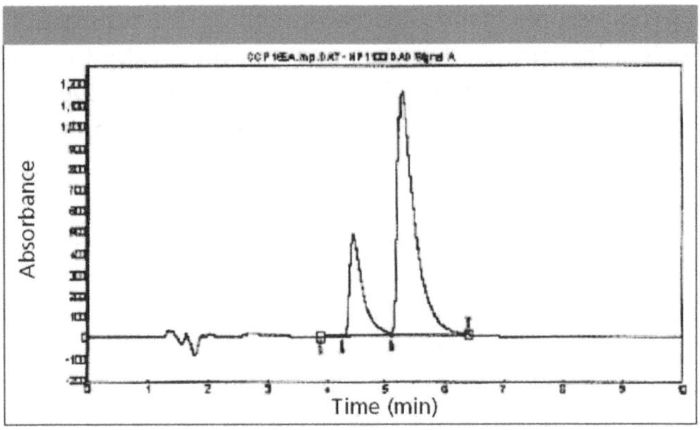

Figure 5.7 Separation of a regioisomeric pair containing an aromatic group and a secondary amine group that showed no separation on the reversed-phase HPLC column screening. The separation did occur on a 2-ethylpyridine SFC column, with 20% methanol as co-solvent at 120 bar and 35°C. Approximately 30 mg of sample was loaded on the preparative column (10 mm × 250 mm) at each injection. The two regioisomers were separated a ratio of 1:2 on a Thar Investigator system with an isolated purity of ~99% for the first peak and ~98% for the second peak [71].

Figure 5.8 highlights the differences in selectivity of the stationary phases for the separation of a diastereomeric mixture. The CHIRALPAK IA column was chosen for preparative purification based on its higher column loading and better chromatographic performance.

IB, IC and achiral 4-ethylpyridine modifier in 5 min, flow rate: 4 mL/min). Symbol highlights the peaks of interest [66].

Figure 5.9 illustrates the preparative profile of a diastereomeric mixture on a CHIRALPACK IA column. A compound-specific preparative SFC gradient of 40–60% in 4 min with a flow rate of 30 mL/min was subsequently applied and resulted in the successful isolation of the desired compounds from the impurities. The purification was successful in isolating the two isomers with purities by UV_{254} of ~99.99% and 98.66%, respectively.

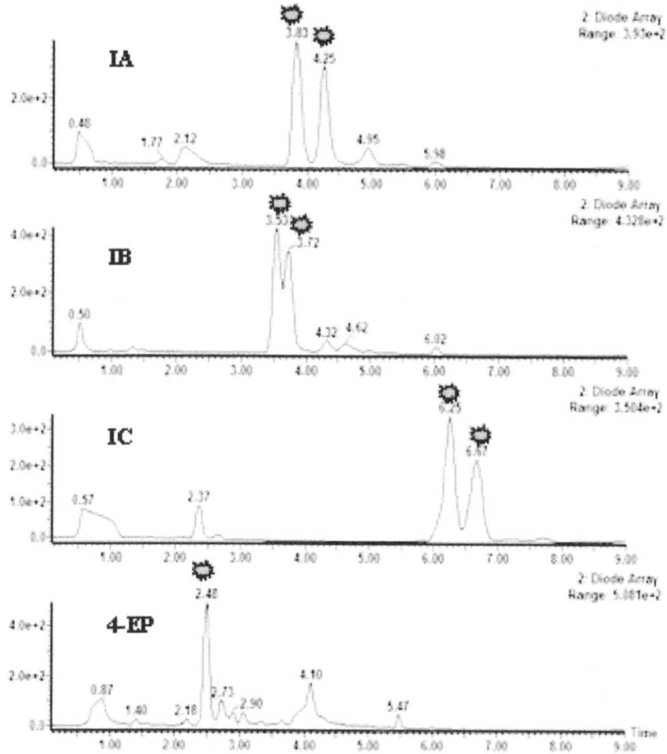

Figure 5.8 Analytical screening of stationary phases CHIRALPACK IA,

When comparing chiral to achiral columns, the chiral columns in terms of isomeric purification provide a much higher success rate, recovery, and purity of the purified products, making them more efficient and therefore offsetting associated higher costs.

5.6 Approaches for SFC Purification of Compound Libraries

Purification of crude compound libraries is a challenging and time consuming aspect of the drug discovery process that is further gaining attention as a bottleneck that would benefit from recent

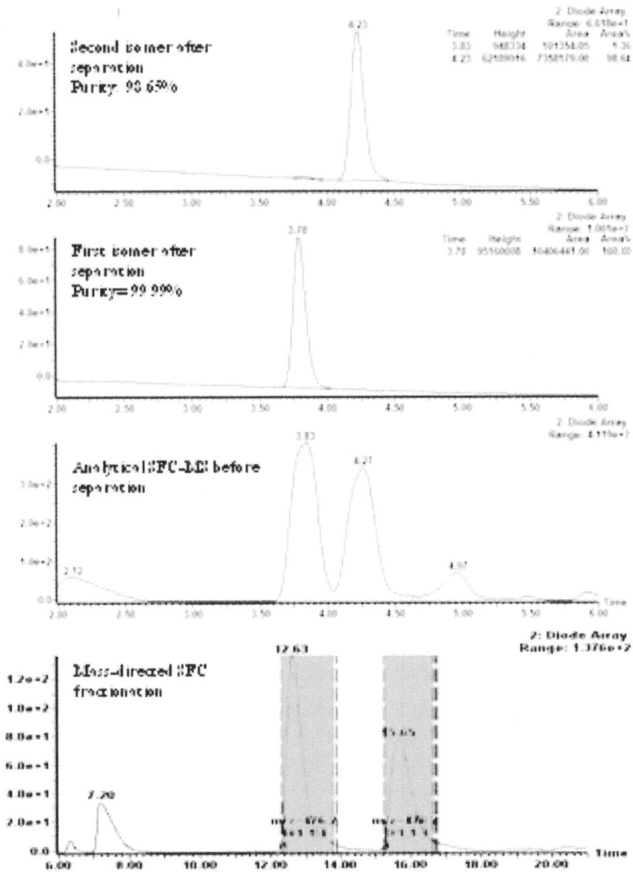

Figure 5.9 Preparative profile on CHIRALPACK IA (gradient conditions: 40–60% modifier in 18 min, flow rate: 30 mL/min) and QC of purified products [66].

advances in SFC including increases in analytical throughput. The distinct advantages for the use of preparative SFC over HPLC for high-throughput library purification are faster chromatography and greater speed of sample work-ups, with which the purified materials can be separated from the mobile phase, once the corresponding fractions have been collected. Other benefits include its "green" nature of lower solvent consumption and therefore waste disposal as well as the ability to isolate products in the free base form. Moreover, since the system does not utilize acidic or aqueous medium, the purification of compounds that would normally be susceptible to these conditions is realized. Some early studies of library purification accompanying the development of SFC instrumentation have been discussed in Section 5.2. The cases presented below represent the ongoing trend for the utility of SFC for library purification, in our view, making significant inroads to gain ground over the mainstream high-throughput reversed-phase HPLC-MS method, which is linked to higher operating costs and longer purification cycle times in laboratory preparative scale.

5.6.1 UV-Triggered or Fixed Window Fraction Collection

Farrell *et al.* [72] of Pfizer in 2001 described the purification of parallel solution-phase derived combinatorial libraries with a Berger analytical SFC/MS/CLND (nitrogen chemiluminescence) detection off-line of the preparative-scale SFC. Pre-purification analytical screening allowed for the triaging of samples for purification, and an in-house software package analyzed data for predicted quality based on an evaluation of UV and MS data for the potential of co-eluting peaks during purification. This same software package selected a collection time window for purification, which was necessary to limit the number of fractions per sample. This system accommodated the purification of samples up to 50 mg in weight. Post-purification analytical SFC/MS/CLND was used again to validate purified samples.

In 2004 Ventura *et al.* [73] at Pfizer reported a successful UV-triggered SFC purification in which analytical and preparative methods were linked to enable direct scale-up to purification. They established an automatic high-throughput purification process in which library samples in well-plates, using analytical data as a guide, were channeled via a robotic cherry-picker station to three preparative chromatographs. One HPLC unit was equipped with a UV-triggered fraction collection system, the other with a mass-triggered system and the third with a Berger SFC unit with UV-triggered fraction collection. Samples not directed toward the SFC unit included those that were insoluble in methanol, contained highly polar compounds or compounds that had no UV chromophores. Without MS-triggered fraction collection, it was not possible at that time to purify compounds that did not have good UV signals. The fractions were collected through a Bohdan auto-sampler/fraction collector and the column used for preparative SFC was a 21.2 mm × 150 mm Zyrosil Pegasus from Zymor with flow rates of 40–60 mL/min. Purity was determined from a post-preparative QC analysis and the success rates for these systems were all >90%, including recoveries that were higher than 90% as well. Figure 5.10 is a chromatographic illustration of the composition of a select sample ranging from crude analysis through SFC purification to final QC purity analysis of the product by SFC–MS. In this example, the purity of the product was improved from an initial 55% to 96% after preparative SFC.

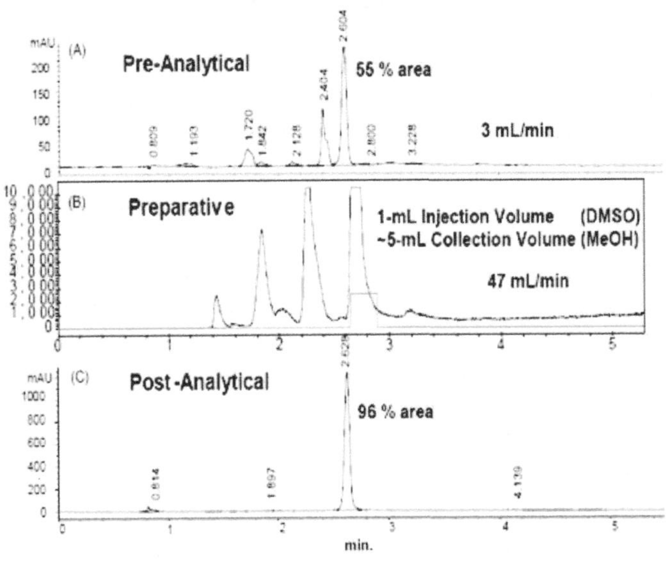

Figure 5.10 Example of purity improvement through the preparative process based on SFC. (A) SFC-MS analytical data of crude sample identifies product and specifies retention time window. (B) Preparative SFC run-rectangular trace represents time during which fraction collection occurred. (C) SFC-MS analytical data from purified well-product identified and integration gives percent purity [73].

In 2005, White and Burnett [67] at Lilly introduced an indirect SFC/MS purification approach that minimized the number of fractions per sample and simplified the post-purification process for high-throughput batch purification. In this approach, pre-purification analysis was conducted on an analytical SFC-MS system using 2-ethylpyridine as the stationary phase (4.6 mm × 150 mm column, packed with 5 μm particles), operating at 5 mL/min, with an outlet pressure of 100 bar, at 40°C. The duration of the analyses was 2 min. The separation was easily scaled up to a 8.5 min separation on a 21.5 mm × 150 mm column packed with the same stationary phase and operating at 55 mL/min, at the same temperature and outlet pressure. To ensure a linearity between the analytical and the preparative SFC systems, intersystem calibration was conducted daily by running a group of standard compounds on both systems and correlating their retention times. The correlation between the analytical and preparative retention times consistently generated a correlation coefficient of >0.99. A collection window was set along with a UV threshold, if necessary for each compound depending on their predicted preparative retention.

Figure 5.11 illustrates the UV chromatogram at 220 nm of the five-component test mix used to perform the analytical to preparative SFC retention time mapping exercise in addition to being used as the column performance check. The use of an empirical correlation involving five different compounds permitted the rapid scale-up without the need for implementing a time consuming method development process. Considerable productivity gains were achieved over HPLC, due primarily to two reasons: the faster flow velocity associated with the SFC and the faster evaporation time needed to recover the products from the collected fractions.

This approach addressed the deck capacity limitation associated with UV-based fractionation by controlling the number of fractions per sample and, hence, simplified or completely eliminated the need for post-purification fraction tracking. However, it required pre-purification SFC analysis on all samples, and the reliability of the process was largely dependent on the linear correlation between the analytical and preparative SFC systems.

The average loading capacity across the set of research compounds purified was approximately 50 mg, injecting between 0.5 and 1.0 mL of sample dissolved in a 50:50 mixture of DMSO:MeOH. Although the maximum loading attempted of 189 mg was considered overloaded, there was still good reproducibility between the analytical and preparative chromatograms.

Figure 5.12 illustrates the analytical and preparative separation of diltiazem (1), propranolol (2), and ketoprofen (3), where ketoprofen

(500 μL injection, 50 mg loading) was targeted for collection using a five-component calibration strategy.

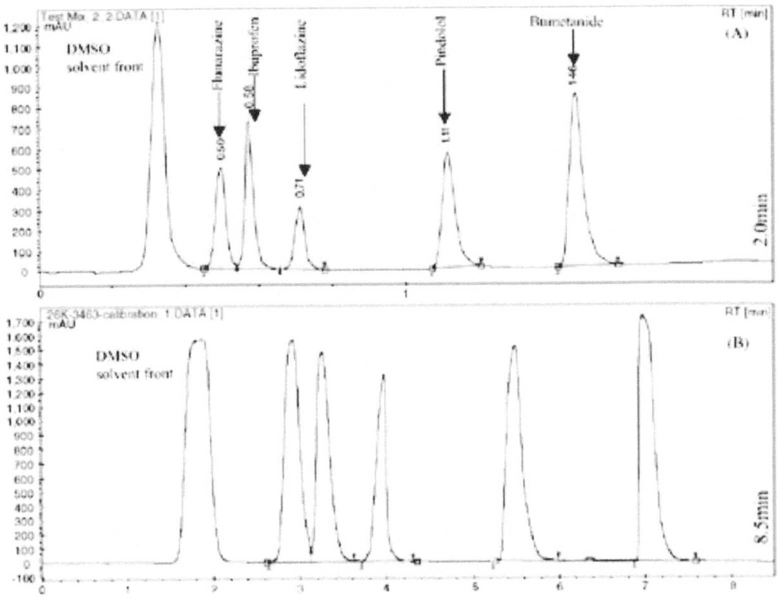

Figure 5.11 UV at 220 nm of the five-component test mix used to perform the analytical to preparative SFC retention time mapping exercise in addition to being used as the column performance check. The analytical and preparative chromatographs are displayed as A and B, respectively. Chromatogram B is a 50 mg injection [67].

A group from Abbott [74, 75] in 2004 reported the results of a comparative study where 16 diverse library sets consisting of 48-members was purified by both preparative scale SFC and reversed-phase HPLC, noting the implications of the selection of a purification technique. The preparative SFC instrument used was a custom-integrated unit as described earlier [38] in Section 5.2, consisting of a Berger manual system coupled with a Gilson auto-sampler and a UV-triggered fraction collector modified from a Cavro MiniPrep™ pipettor, which allowed collection into 18 mm × 150 mm test tubes at atmospheric pressure. This study had provided a direct comparison of HPLC to SFC in respect to the purification of compound libraries that consisted of a wide range of structural types. Figure 5.13 is an example of UV-triggered SFC purification results reported by the

Abbot group. Although they had found that no single technique offered a clear advantage chromatographically, both were found to be acceptable for the general purification of diverse structural types. On the basis of their results and other studies, Abbott Laboratories decided to increase the capacity for SFC purification due to other advantages inherent in the technology, specifically, the ease with which solvent can be evaporated from fractions and the ability to provide products in the salt-free form.

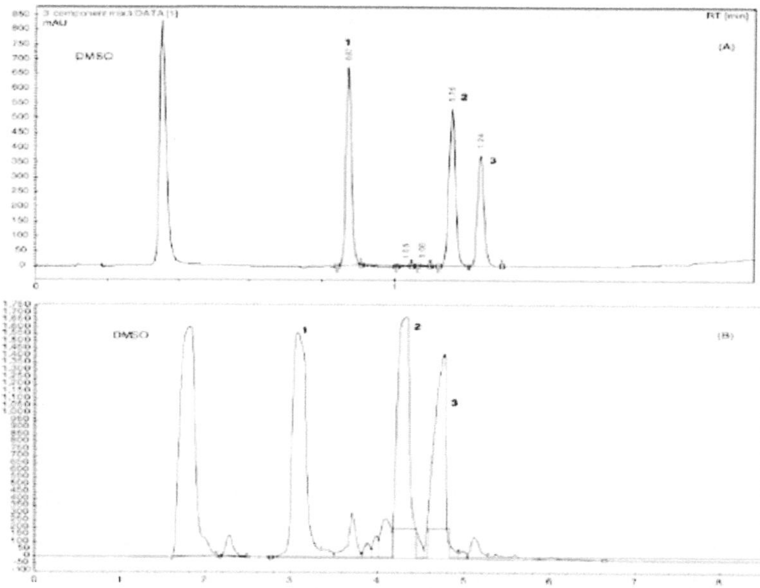

Figure 5.12 Analytical (A) and preparative (B) separation of diltiazem (1), propranolol (2) and ketoprofen (3). Preparative chromatogram is a 500 μL (50 mg loading) injection and ketoprofen was targeted for collection using the five-component calibration strategy [67].

5.6.2 Mass-Triggered Fraction Collection

Mass-directed fraction collection offers an important advantage over UV- and time-based fraction collection methods because it is uniquely selective and sensitive, allowing for only the expected products to be isolated. However, because no commercial preparative SFC-MS instruments were available prior to 2008, only a few teams, discussed below, had investigated the mass-directed SFC purification of compound libraries. Instead, some researchers focused on the customary integration of a SFC instrument with a mass spectrometer as shown in Section 5.2 and others compared the performance of HPLC and SFC on mass-based detection and analysis of compounds of pharmaceutical interest [76–78]. Very little difference has ever been found in the overall scope of reversed-phase HPLC versus SFC for a large library of compounds and several benefits have been cited using SFC (vs. HPLC) in connection with the use of APCI-MS for identification of unknowns. These findings are a good starting point for the SFC purification of compound libraries in drug discovery.

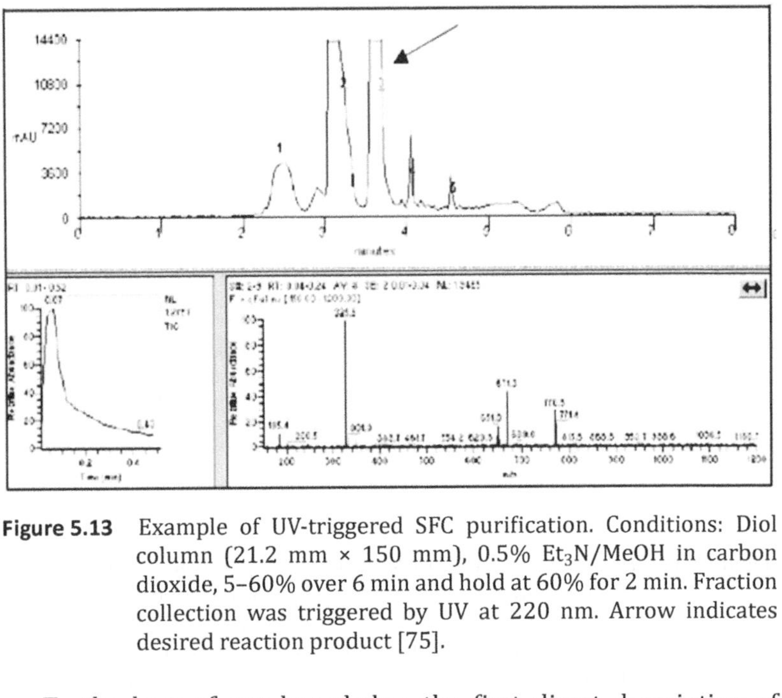

Figure 5.13 Example of UV-triggered SFC purification. Conditions: Diol column (21.2 mm × 150 mm), 0.5% Et_3N/MeOH in carbon dioxide, 5–60% over 6 min and hold at 60% for 2 min. Fraction collection was triggered by UV at 220 nm. Arrow indicates desired reaction product [75].

To the best of our knowledge, the first direct description of mass-directed SFC purification of compound libraries was reported in 2001 by Wang *et al.* [40] of DuPont Pharmaceuticals, whose semi-preparative system was a custom-modified unit as mentioned in Section 5.2. Approximately a dozen drug-like small molecules were purified and collected, through a modified Gilson 215 liquid handler, under atmospheric pressure at flow rates up to 30 mL/min and resulted in a high recovery of >60%. Standard separations were achieved using 10 mm × 150 mm columns of silica, diol or cyano based functionalization. The standard gradient used for semi-preparative separations was 10% MeOH to 60% MeOH in CO_2 in 5 min with a flow rate of 15 mL/min.

Another team at ArQule [41] reported in 2006 their use of a customized preparative SFC-MS system also described earlier in Section 5.2 for purification of library compounds that contained acid- and base-labile functional groups, which were unsuitable for reversed-phase HPLC. Preparative SFC separations were conducted using a 10 mm × 100 mm, 10 μm, 2-ethylpyridine column or 21.2 mm × 150 mm, 6 μm, cyano column under the flow rate of 5–30 mL/min with the makeup flow of methanol and 0.3% formic acid. They found the peak shape was well defined and that base-line separation between the product of interest and close-eluting impurities had been fully resolved.

One of the remaining challenges in applying SFC to complex pharmaceutical separations such as library purification is the unavailability of a versatile stationary phase, comparable to the common C18 dominantly employed in reversed-phase liquid chromatography. One would easily realize the typical column-screening strategy used for singleton achiral SFC purification cannot work for library purification, because it would become extremely impractical, if not totally impossible for a large number of compounds to process through multiple columns. With the goal of shortening purification cycle times and increasing library output, we developed in 2009 a "universal one-column-for-all" SFC approach to mimic the common C-18 practice in reversed-phase HPLC by adopting a single semi-preparative stationary phase for the purification of internal library compounds [64,79]. After screening eight stationary phases (4-ethylpyridine, Chromegabond Pyr/Amide, ZymorSPHER Pyr/Diol, benzamide, 4-fluorobenzene, sulfonamide, 2-ethylpridine, 2CN/DIOL, CN) for more than 160 library compounds, 4-ethylpyridine (4-EP) emerged as the preferred stationary phase, affording the highest percentage resolution with a ~95% overall success rate. In short, 77% of compounds were resolved and 17% were partially resolved in the analytical screen as shown in Fig. 5.14. Further resolution had been achieved in the preparative stage with the combinative use of compound-specific gradients methods and Boolean logical fractionation. Figure 5.15 illustrates examples of libraries compounds before purification and analytical QC of the isolated products on 4-Ethylpyridine column.

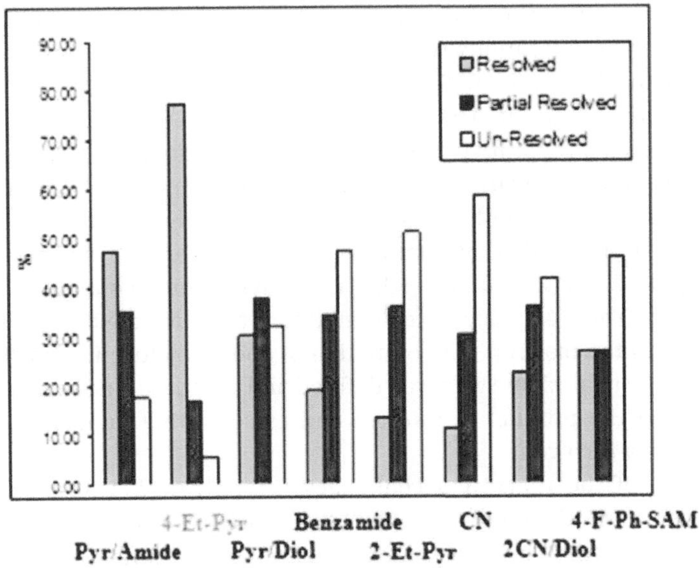

Figure 5.14 SFC column performance profile; % presents success rate for each bonded phase. Conditions: 5 µm, 4.5 mm × 150 mm columns; temperature: 40°C; mobile phase: a gradient of 10–60% MeOH with 0.2% IPA in 5 min; flow rate: 4 mL/min; injection volume: 10 µL of ~100 mg/mL in DMSO.

Separately, on the efforts to identify a "universal column" for purification of compound libraries, a team from Eli Lilly in Madrid, Spain, reported in 2010 their similar approach for overcoming the absence of a "generic phase" but concluded with slightly different results [80]. They first screened five achiral stationary phases (diol, dinitrophenyl, 2-ethylpyridine, benzene sulfonamide, and diethylaminopropyl) with the goal to identify one or two columns that could successfully resolve a majority of library compounds. Not surprisingly, no single stationary phase turned out to be ideal and 2-ethylpridine provided the highest pass rate, standing at ~42%. Subsequent studies found Luna™ HILIC, a non-pH-dependent, commonly used reversed-phase stationary phase from Phenomenex, worked for 70% of the compounds and a combination of HILIC with 2-ethylpyridine provided a pass rate of >85% across multiple libraries from ~30 different projects. As a result, achiral SFC and reversed-phase HPLC techniques have both been adopted at Eli Lilly laboratories for high-throughput library purification.

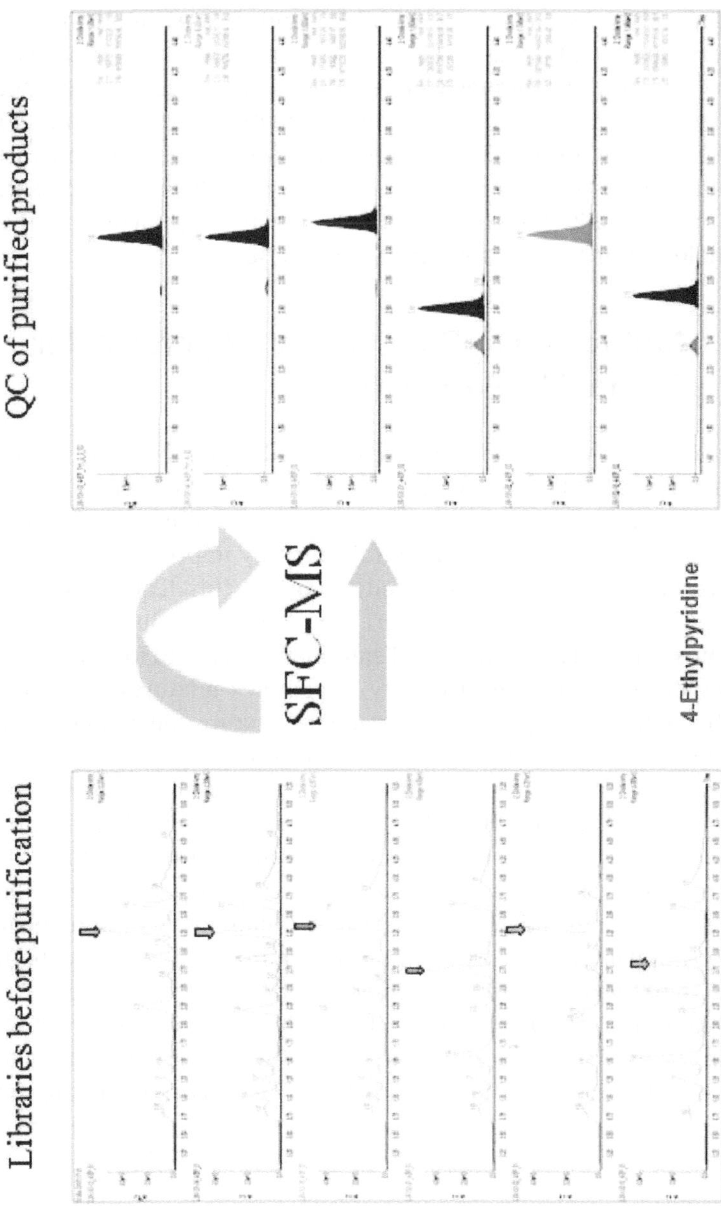

Figure 5.15 Examples of preparative chromatograms of library compounds and analytical QC of the isolated products.

Mich *et al.* [81] reported in 2010 their comparative study of 167 compounds purified by a newly installed Prep-100SFC-MS directed system and a reversed-phase HPLC-MS unit. It was found that the success rate for SFC was 27%, 25% for reversed-phase HPLC and 32% when both techniques were combined. The reason behind the observed low success rates is unclear, probably due to the complex nature of their library compounds. The author did comment that despite using a longer gradient cycle in SFC, overall batch processing time for SFC was 20% less than HPLC, demonstrating that preparative SFC holds great promise for library purification.

5.7 Multicolumn (Dual-Column) Approach for Achiral SFC Purifications

As described earlier, the low viscosity and high diffusivity of the mobile phase in SFC allows for separations to be achieved at higher flow rates and with longer columns or dual columns of mixed phase selectivity (e.g., chiral/achiral), thereby improving efficiency and resolution [82]. By taking advantage of this feature, several groups have reported their interesting results through selective combination of achiral and chiral stationary phases to form a mixed phase for separations of a variety of complex mixtures, including purification of a complex four-component mixture consisting of two enantiomeric and two diastereomeric species.

Coupling of Chiralcel OD-H and silica columns, reported by Alexander and Staab of BMS [83], significantly enhanced separation of the reaction mixture of cinnamonitrile and hydrocinnamonitrile intermediates, as shown in Fig. 5.16, which were scarcely separated by either column individually. The diastereometric/enantiometric composition of the final product was determined in one step without addition of the overall analysis time [83].

Mannerino *et al.* [84] from Regis Technologies reported in 2010 the use of a coupled achiral/chiral column combination to successfully separate a mixture that was resolved individually on either chiral or achiral column but failed on its preparative scale-up.

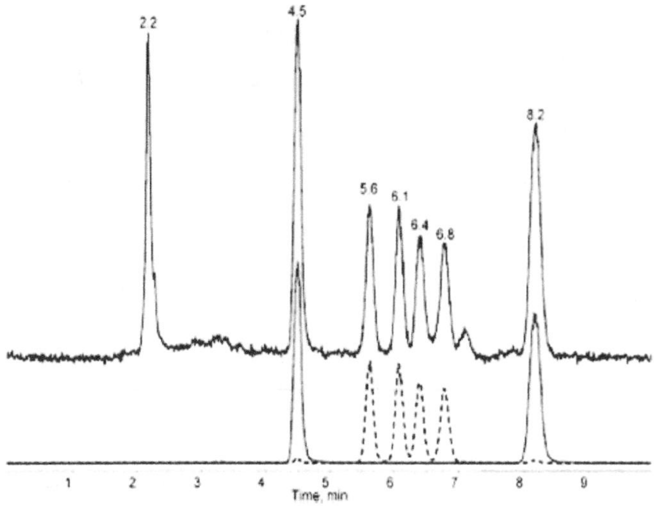

Figure 5.16 Coupled achiral and chiral SFC-MS separation of semi-purified product. Upper trace: TIC. Lower trace (dashed): product [M + H] + EIC (*m/z* 263) 4 diastereomers/enantiomers. Lower trace (solid): starting material [M + H] + EIC (*m/z* 261) another two *cis/trans* isomers [83].

Recently we also reported the feasibility evaluation of a dual tandem column approach that couples a 4-EP SFC column with a selectivity-orthogonal stationary phase to maximize both robustness and success rate applicable to all library mixtures and specifically "unresolvable mixtures" [85]. The rationale behind this approach was the result of the empirical chromatographic experience that "dissimilar stationary phases" such as benzamide, 4-fluorophenylsulfonamide (4-F-Ph-SAM), and 2CN/Diol are somewhat "orthogonal" to the basic 4-EP column in chromatographic selectivity and resolution power, presumably due to their complementary nonbasic stationary phase-retention mechanism. A chiral stationary phase, CHIRALPACK IA was also included as an appropriate stationary phase for achiral purification of challenging mixtures such as those containing structural isomers.

Figure 5.17 illustrates the typical chromatograms of a library of compounds on a 4-EP single column *versus* four achiral/achiral or achiral/chiral hyphenated dual column systems. The differences in selectivity, resolution and retention for the corresponding targets

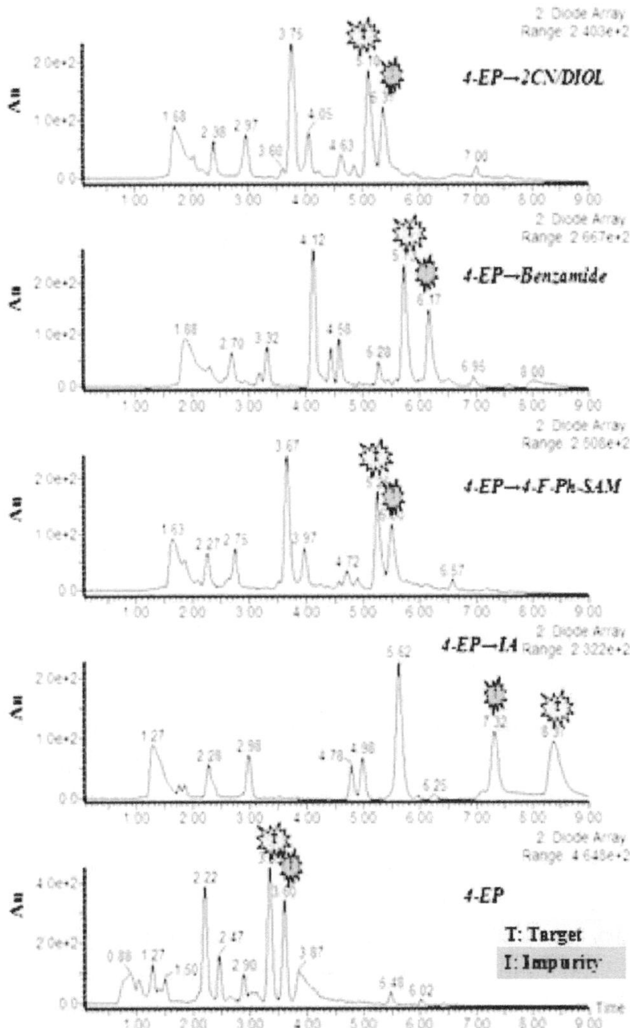

Figure 5.17 SFC-MS analytical profile of library compounds on dual columns. Conditions: 5 µm, 4.5 mm × 150 mm columns; temperature: 40°C; mobile phase: a gradient of 10–60% MeOH with 0.2% IPA in 5 min; flow rate: 4 mL/min.

on the dual column systems are summarized in Table 5.3. Of the mixtures tested, all three dual achiral/achiral column coupling systems shown revealed an increased retention, but their effect on the selectivity of the analytes was mixed. However, the dual achiral/

chiral combination of (4-EP → IA) provided the highest retention and a significant improvement in selectivity ($\alpha \geq 1.2$), which resulted in the best separation of the mixtures. Improved separation in many column combinations justifies further studies to include a more diverse set of achiral and chiral stationary phases and their combination with 4-EP for achiral SFC purification of compound libraries.

Table 5.3 Selectivity and resolution data of mixture

	4-EP	4-EP → benzamide	4 - E P → F-Ph-SAM	4-E → 2 CN/Diol	4-EP → IA
RT_{target} (T)	3.35	5.73	5.25	5.10	8.37
$RT_{impurity}$ (I)	3.60	6.17	5.50	5.37	7.32
α	1.09	1.11	1.07	1.08	1.20
Rs	1.84	2.57	1.01	1.46	3.02

Phinney *et al.* [86] developed a dual column approach to address the limited achiral selectivity of chiral stationary phases. Column coupling in SFC presents opportunities for simplifying method development by utilizing commonly available and inexpensive achiral stationary phases to modify selectivity rather than changing CSPs. Modifying the achiral selectivity of the chromatographic system is also likely to minimize the likelihood of peak co-elution and may reduce the need for overlapping achiral/chiral methods. The achiral selectivity of amino and cyano stationary phases was used to modify the resolution of compounds on a Chiralcel OD-H chiral stationary phase by combining the achiral and chiral columns in series.

Figure 5.18a illustrates the enantio-separation of four chiral β-blockers on a Chiralcel OD-H CSP using methanol as the modifier. As can be seen in the figure, peak overlap exists for enantiomers of alprenolol, andoxprenolol, as well as for atenolol and propranolol. Changes in parameters, including pressure, temperature, and modifier concentration, did not provide acceptable separation of all the components. The achiral separation of the four β-blockers on a cyano bonded phase is shown in Fig. 5.18b. Serial coupling of the cyano and Chiralcel OD-H columns produced the separation shown in Fig. 5.18c. No changes in chromatographic parameters were required to obtain this separation, and analysis time increased by only a few minutes.

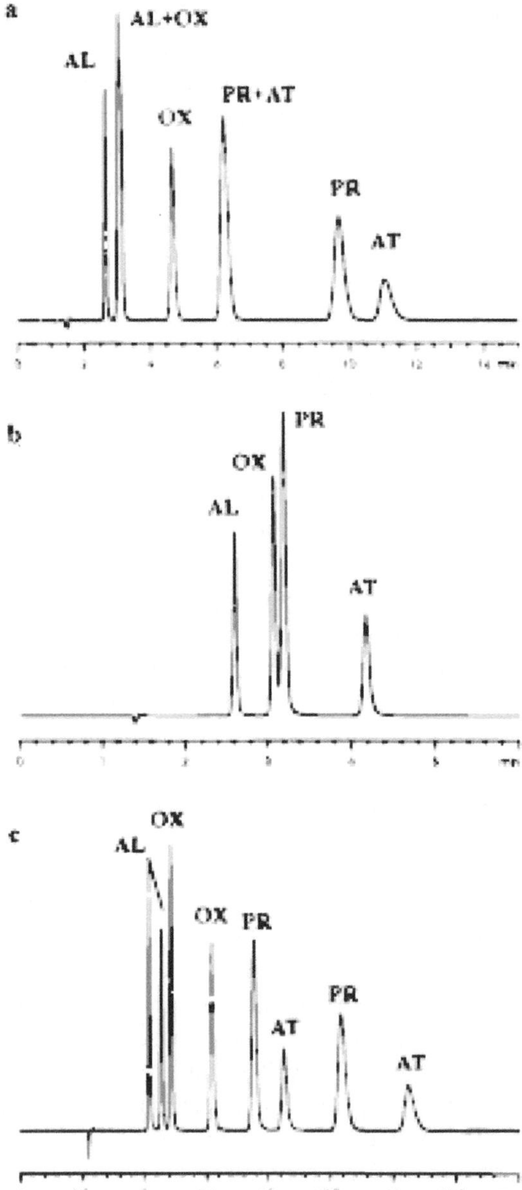

Figure 5.18 Separation of β–blocks on the Chiralcel OD-H CSP (a), the chiral cyano bonded phase (b), and the coupled cyano/Chiralcel OD-H system (c). Chromatographic conditions: 20% MeOH containing 0.5% isopropylamine in carbon dioxide; 2.0 mL/min, 150 bar; 30°C; UV at 280 nm (86)

5.8 Summary

For the first time since its invention, SFC as a chromatographic technique with its convoluted development history, has had a progressive decade with eventful strides ranging from SFC column chemistry and instrumentation to an operational movement from chiral to achiral applications. Today, the SFC is primarily used over HPLC for chiral separations. However, achiral separations by reversed–phase HPLC still dominate in the field, but SFC is rising. Applications of achiral preparative SFC are gaining ground in the purification arena as many discovery laboratories have now been equipped with preparative SFC instruments, some even having multiple newly launched units. SFC provides a selectivity that is complementary to reversed-phase HPLC, thus allowing for an orthogonal approach in method development and separation of challenging complex samples, including labor-intensive compound libraries.

The pharmaceutical industry has latched onto SFC since the beginning because it is very amenable to preparative separations including decreased turn-around time, both in terms of faster chromatographic separations and fraction dry-down; and lower solvent consumption, thus offering lower costs and very significant environmental benefits. Considering the fact that most drugs and pharmaceutical substances under research and development are achiral small molecules, one may not be surprised to find that the ever increasing utilization of the green SFC technique for pharmaceutical achiral separations can help not only alleviate the bottleneck often caused by purification issues within a drug discovery program, but also bring in operational refinements that drastically shortens timelines, reduces waste production, and cuts costs. In a world seeking renewable technologies for solving waste and disposal problems, SFC can undoubtedly contribute and make an impact on the environment we live.

Although SFC brings a set of advantages that could make it become more competitive with conventional preparative HPLC, at least for certain applications, it has encountered major roadblocks to reach wider applicability. In order for SFC to become a sustainable separations tool there needs to be a leap in the fundamental knowledge within the field [87]. This is combined with our limited understanding of the practical methods needed to apply our knowledge of the physicochemical properties of the subcritical,

critical, and/or supercritical fluids, to the development of dedicated separation processes. While the pharmaceutical industry has widely embraced SFC, the technique appears to be largely underappreciated in academia. Academic involvement in the basic research of SFC as well as financial support from governmental organizations is urgently needed to further infuse the field with new knowledge and talent. The ultimate goal of scientists and engineers in the design of chromatographic separations is to accurately predict the influence of the experimental conditions on the recovery yield, the production rate, and the solvent consumption of a chromatographic separation.

Achiral preparative SFC is still in its infancy. It may be too early and unfair to compare SFC to HPLC, a technique with over 30 years of maturation, especially in the credentials of "system robustness" and "suitability," the most crucial technical criteria necessary for a high-throughput discovery environment. Widespread adoption of preparative SFC for purification of diverse small molecules relies on the suitable commercially available instrumentation capable of providing reasonable flow rates, gradient elution, mass driven fractionation, and consistent high recoveries without cross contamination. The new laboratory-scale instrument, preparative Prep-100SFC-MS directed system, for the first time, has met most, if not all, desired technical and engineering features and can be considered, in our view, as having achieved a milestone in SFC history. Being built on a solid modular platform enables this system to be further customized and fine tuned, as necessary. Recent advances in polar achiral stationary phases suitable for pharmaceutical applications have been fruitful and broadened the realm of achiral SFC applications. With the entry into the SFC business in 2009 by two of the largest chromatography manufacturers, Waters, through its acquisition of Thar, and Agilent via its partnership with Aurora SFC systems, new waves of investments and development in SFC instrumentation and chromatographic supplies can be anticipated in the near future. With all these progresses, achiral preparative SFC is undoubtedly on a strong foundation poised for mainstream acceptance.

References

1. Klesper E., Corwin A. H., and Turner D. A. (1962) High pressure gas chromatography above critical temperatures, *J. Org. Chem.*, **27**(2), 700–706.

2. Perrut M., FR 8,209,649 (1982); EP 0,099,765 (1984); US 4,478,720 (1983).
3. Perrut M., FR 8,510,468 (1985); EP 8,640,139 (1984).
4. Perrut M., FR 2,584,618 (1985); EP 0,212,999 (1986); US 4,724,087 (1988).
5. Jusforgues P. and Perrut M. (1986). FR 8,610,809.
6. Jusforgues P. and Perrut M., FR 2,601,883 (1986); EP 0,254,610 (1988).
7. Saito M., Yamauchi Y., Honda T., and Senda M. (1988) *Proceedings of the 1st International Symposium on Supercritical Fluids, Nice (*ed. Perrut M.), Institute National Polytechnique de Lorraine, Nancy, pp. 374–380.
8. Flamant Y. and Keller U. (1988) *Proceedings of the 1st International Symposium on Supercritical Fluids, Nice (*ed. Perrut M.), Institute National Polytechnique de Lorraine, Nancy, pp. 381–388.
9. Alkio M., Harvala T., and Kompa V. (1988) *Proceedings of the 1st International Symposium on Supercritical Fluids, Nice* (ed. Perrut M.), Institute National Polytechnique de Lorraine, Nancy, pp. 389–396.
10. Berger C. and Perrut M. (1990). Review : Preparative supercritical fluid chromatography, *J. Chromatogr.*, **505**, pp. 37–43.
11. Perrut M. (1994). Advances in supercritical fluid chromatography, *J. Chromatogr. A*, **658**, pp. 293–313.
12. Jusforgues P. and Shaimi M. (1998). Preparative supercritical fluid chromatography, *Anal. Mag.*, **26**(7), pp. M 55-M60.
13. Coleman K., Boutant R., and Verillon F. (1999). Practical aspects of preparative supercritical fluid chromatography, *Isolation Purif.*, **3**, pp. 9–19.
14. Erickson B. (1994) Product review: SFE and SFC instruments, *Anal. Chem.*, 66, pp. 369A–375A.
15. Erickson B. (1997) Product review: SFC in flux, *Anal. Chem.*, **69(**21), pp. 683A–686A.
16. Harris C. M. (2002) Product review: the SFC comeback, *Anal. Chem.*, **74(3)**, pp. 87A–91A.
17. Bevan C. D. and Lellish C. J. (1995) in *Process Scale Liquid Chromatography* (ed. Subramanian G.) VCH, Weinheim, pp. 163–191.
18. Yamauchi Y. and Saito M. (1994) in *Fractionation Packed-Column SFC and SFE: Principles and Applications* (ed. Saito M., Yamauchi Y., and Okuyama T.), VCH, New York, pp. 169–178.

19. Jusforgues P. (1995) in *Process Scale Liquid Chromatography* (ed. Subramanian G.), VCH, Weinheim, pp. 153–162.
20. Chester T. L., Pinkston J. D., and Raynie D. E. (1994) Supercritical fluid chromatography and extraction, *Anal. Chem.*, **66** (12), pp. 106R –130R.
21. Taylor L. T. (1997) in *Supercritical Fluids*, ACS Symposium Series, ACS (ed. Abraham M., et al.), Washington, DC, pp. 134–153.
22. Gyllenhaal O., Karlsson A., and Vessman J. (1998) Applications of packed column supercritical fluid chromatography in the development of drugs, in *Supercritical Fluid Chromatography with Packed Columns (*ed. Anton K. and Berger C.), Marcel Dekker, New York, pp. 273–300.
23. Jusforgues P., Shaimi M., and Barth D. (1998) Preparative supercritical fluid chromatography: grams, kilograms, and tons! in *Supercritical Fluid Chromatography with Packed Columns (*ed. Anton K. and Berger C.), Marcel Dekker, New York, pp. 403–427.
24. Poole C. F. (2000) Supercritical-fluid chromatography and supercritical-fluid extraction, *J. Biochem. Biophys. Methods*, **43(**1–3), pp. 1–2.
25. Chester T. L. and Pinkston J. D. (2002) Supercritical Fluid and Unified Chromatography, *Anal. Chem.*, **74(**12), pp. 2801–2812.
26. Perrut M. (2003) Supercritical fluids applications in the pharmaceutical industry, *S.T.P. Pharma Sci.*, **13**(2), pp. 83–91.
27. Phinney K. W. (2000) Peer reviewed: SFC of drug enantiomers, *Anal. Chem.*, **72(**5), pp. 204A–211A.
28. Palmieri M. D. (1988) An introduction to supercritical fluid chromatography, *J. Chem. Ed.*, **65**(10), pp. A254-A259
29. Berger T. A. and Smith R. M. (1995) in *Packed Column SFC*, RSC Chromatography Monograph (ed. Berger T. A.), RSC, Cambridge, UK, pp. 1–21.
30. Taylor L. T. (2008) Supercritical fluid chromatography in perspective, *Chim. Oggi/Chem. Today*, **26**(5), pp. 15–18.
31. Berger T. A., Berger B., and Majors R. E. (2010) A review of column development for supercritical fluid chromatography, *LCGC North Am.*, **28**(5), pp. 344–356.
32. Mukhopadhyay R. (2008) SFC: Embraced by industry but spurned by academia, *Anal. Chem.*, **80** (9), pp. 3091–3094.
33. Taylor L. T. (2008) Supercritical fluid chromatography for the 21st century, *J. Supercrit. Fluids*, **47**(3), pp. 566–573.
34. Lefler J. L. (2006) A "peak" at chromatography in the pharmaceutical industry, *oral presentation to the 2006 CoSMoS meeting in San Diego,*

USA (slide is available for download from www.cosmoscience.org/Archive_2006.htm.

35. Lembke P. (1998) Production of high purity n-3 fatty acid-ethyl esters by process scale supercritical fluid chromatography, in *Supercritical Fluid Chromatography with Packed Columns* (ed. Anton K. and Berger C.), Marcel Dekker, New York, pp. 429–443.
36. Berger T. A., Fogleman K., Staats T., Bente P., Crocket I., Farrell W., and Osonubi M. (2000) The development of a semi-preparatory scale supercritical chromatograph for high-throughput purification of "combi-chem" libraries *J. Biochem. Biophys. Methods*, **43**, pp. 87–111
37. Personal communications in 2007 with Berger Instruments, Inc.
38. Olsen J., Pan J., Hochlowski J., Searle P., and Blanchard D. (2002) Customization of a commercially available prep scale SFC system to provide enhanced capabilities, *JALA*, **7**(4), pp. 69–74.
39. Maiefski R., Wendell D., Ripka W. C., Krakover J. D. (2001). WO 00/266622.
40. Wang T., Barber M., Hardt I., and Kassel D. B. (2001) Mass-directed fractionation and isolation of pharmaceutical compounds by packed-column supercritical fluid chromatography/mass spectrometry, *Rapid Commun. Mass Spectrom.*, **15**, pp. 2067–2075.
41. Zhang X., Towle M. H., Felice C. E., Flament J. H., and Goetzinger W. K. (2006) Development of a mass-directed preparative supercritical fluid chromatography purification system, *J. Comb. Chem.*, **8**, pp. 705–714.
42. Ridgway P. and Chen R. (2008) A case study of using Thar SFC-MS Prep-30® to purify polar, basic pharmaceutical relevant compounds, *LCGC North Am.*, September, Application Note.
43. Lazarescu V., Li Y., Mulvihill M. J., and Ma L. (2008). Mass-Directed SFC for Analysis and Preparative Purification of Chiral Small Molecules in Drug Discovery, *Poster presented to the 2nd International Conference on Packed-Column SFC in Zurich, Swiss* (slide is available for download from www.greenchemistrygroup.org/program2008.html).
44. Lazarescu V., Mulvihill M. J., and Ma L. (2009). SFC-based chiral separations: the selectivity power of dual co-solvent systems, *Poster presented to the 3rd International Conference on Packed-Column SFC in Philadelphia, USA* (slide is available for download from www.greenchemistrygroup.org/program2009.html).
45. Product information and technical parameters are available from www.tharsfc.com and www.waters.com.
46. Ebinger K., Weller H. N., Kiplinger J., and Lefebvre P. (2011) Evaluation of a new preparative supercritical fluid chromatography system for

compound library purification: the Thar SFC-MS Prep-100 system, *JALA*, **16**, pp. 241–249.

47. Information from homepage of Princeton Chromatography, Inc., www.pci-hplc.com.
48. Information from homepage of Zymor, Inc., www.liquidchromatography.com.
49. Information from homepage of ES Industries, www.esind.com.
50. Information from homepage of Phenomenex, Inc., www.phenomenex.com.
51. McClain R. and Przybyciel M. (2011) A systematic study of achiral stationary phases using analytes selected with a molecular diversity model, *LCGC North Am.*, **29**(10), pp. 894–906.
52. West C., Lesellier E., and Tchapla A. (2004) Retention characteristics of porous graphitic carbon in subcritical fluid chromatography with carbon dioxide–methanol mobile phases, *J. Chromatogr. A*, **1048**(1), pp. 99–109.
53. West C. and Lesellier E. (2006) Characterization of stationary phases in subcritical fluid chromatography by the solvation parameter model. Comparison tools, *J. Chromatogr. A*, **1110**(1–2), pp. 191–199.
54. West C. and Lesellier E. (2006) Characterization of stationary phases in subcritical fluid chromatography with the solvation parameter model. III. Polar stationary phases, *J. Chromatogr. A*, **1110**(1–2), pp. 200–207.
55. West C. and Lesellier E. (2006) Characterization of stationary phases in subcritical fluid chromatography with the solvation parameter model IV. Aromatic stationary phases, *J. Chromatogr. A*, **1115**(1–2), pp. 233–245.
56. West C. and Lesellier E. (2007) Characterization of stationary phases in supercritical fluid chromatography with the solvation parameter model. V. Elaboration of a reduced set of test solutes for rapid evaluation, *J. Chromatogr. A*, **1169(**1–2), pp. 205–219.
57. West C., Fougere L., and Lesellier E. (2008) Combined supercritical fluid chromatographic tests to improve the classification of numerous stationary phases used in reversed-phase liquid chromatography, *J. Chromatogr. A*, **1189(**1–2), pp. 227–244.
58. West C. and Lesellier E. (2008) A unified classification of stationary phases for packed column supercritical fluid chromatography, *J. Chromatogr. A*, **1191(**1–2), pp. 21–39.

59. West C. and Lesellier E. (2008) Orthogonal screening system of columns for supercritical fluid chromatography, *J. Chromatogr. A*, **1203**(1), pp. 105–113.

60. Lesellier E. (2009) Retention mechanism in super/subcritical fluid chromatography on packed columns, *J. Chromatogr. A*, **1216**(10), pp. 1881–1890.

61. West C., Ogden J., and Lesellier E. (2009) Possibility of predicting separations in supercritical fluid chromatography with the salvation parameter model, *J. Chromatogr. A*, **1216**(29), pp. 5600–5607.

62. Bui H., Masquelin T., Perun T., Castle T., Dage J., and Kuo M. (2008) Investigation of retention behavior of drug molecules in supercritical fluid chromatography using linear solvation energy relationship, *J. Chromatogr. A*, **1206**(6), pp. 186–195.

63. Cazenave-Gassiot A., Boughtflower R., Caldwell J., Coxhead R., Hitzel L., Lane S., Oakley P., Holyyoak C., Pullen F., and Langley G. J. (2008) Prediction of retention for sulfonamides in supercritical fluid chromatography, *J. Chromatogr. A*, **1189**(1–2), pp. 254–265.

64. Ma L., Lazarescu V., and Mulvihill M. J. (2010). Toward a universal approach for mass-directed SFC purification of small molecule compound libraries, *oral presentation to the 4th International Conference on Packed-Column SFC in Stockholm, Sweden,* (slide is available for download from www.greenchemistrygroup.org/program2010.html).

65. Lazarescu V., Mulvihill M. J., and Ma L. (2010). Case studies on compound-specific achiral SFC purification, *poster presented to the 4th International Conference on Packed-Column SFC in Stockholm, Sweden* (slide is available for download from www.greenchemistrygroup.org/program2010.html).

66. Lazarescu V., Mulvihill M. J., and Ma L. (2011) A generic workflow for achiral SFC purification of complex pharmaceutical mixtures, *LCGC North Am.*, **29**(5), pp. 438–444.

67. White C. and Burnett J. (2005) Integration of supercritical fluid chromatography into drug discovery as a routine support tool. II. Investigation and evaluation of supercritical fluid chromatography for achiral batch purification, *J. Chromatogr. A*, **1074**, pp. 175–185.

68. Pinkston J. D. and Chester T. L. (1995) Guidelines for successful SFC/MS, *Anal. Chem.*, **67**(21), pp. 650A–656A.

69. Bolanos B., Greig M., Ventura M., Farrel W., Aurigemma C. M., Li H., Quenzer T. L., Tivel K., Bylund J. M. R., Tran P., Pham C., and Phillipson D. (2004) SFC-MS in drug discovery at Pfizer, La Jolla, *Inter. J. Mass Spectrom*, **238**(2), pp. 85–97.

70. Miller L. and Potter M. (2008) Preparative supercritical fluid chromatography (SFC) in drug discovery, *Am. Pharm. Rev.*, **11**(4), pp. 112–117.

71. Yan T. Q., Bradow J., Chang S. P., Depianta R., and Philippe L. (2009) Approaches to singleton achiral purification of difficult samples for discovery research support, *LCGC North Am.*, **27**(4), pp. 340–349.

72. Farrell W. P., Ventura M. C., Aurigemma C. M., Tran P., Fiori K., Xiong X., Lopez R., and Osbonubi M. (2001). *Analytical and Semi-Preparative SFC for Combinatorial Chemistry* in Supercritical Fluid Chromatography, Extraction, and Processing, Myrtle Beach, South Carolina.

73. Ventura M. C., Farrell W. P., Aurigemma C. M., Tivel K., Greig M. J., Wheatley J., Vanovsky A., Milgram K. E., Dalesandro D., DeGuzman R., Tran P., Nguyen L., Chung L., Gron O., and C. Koch A. (2004) High-throughput preparative process utilizing three complementary chromatographic purification technologies, *J. Chromatogr. A*, **1036**, pp. 7–13.

74. Searle P. A., Glass K. A., and Hochlowski J. E. (2004) Comparison of preparative HPLC/MS and preparative SFC techniques for the high-throughput purification of compound libraries, *J. Comb. Chem.*, **6**, pp. 175–180.

75. Hochlowski J. (2004) High-throughput purification: triage and optimization, in *Analysis and Purification Methods in Combinatorial Chemistry* (ed. Yan B.), Wiley-Interscience, pp. 281–306.

76. Ventura M. C., Farrel W. P., Aurigemma C. M., and Greig M. J. (1999) Packed column supercritical fluid chromatography/mass spectrometry for high-throughput analysis, *Anal. Chem.*, **71**(13), pp. 2410–2416.

77. Pinkston J. D., Wen D., Morand K. L., Trey D. A., and Stanton D. T. (2006) Comparison of LC/MS and SFC/MS for screening of a large and diverse library of pharmaceutically relevant compounds, *Anal. Chem.*, **78**(21), pp. 7467–7472.

78. Rumbelow S. J. (2009) Recent developments and future challenges in supercritical fluid chromatography, *Chromatogr. Today*, Feb/Mar, pp. 15–18.

79. Ma L., Lazarescu V., and Mulvihill M. J. (2009). Mass-directed SFC purification of small molecule compound libraries in drug discovery research, *oral presentation to the 3rd International Conference on Packed-Column SFC in Philadelphia, USA.*

80. López-Soto P., Anta C., and M. L. de la Puente (2010) HILIC-diol as a generic stationary phase for achiral SFC: is SFC one step ahead of RP_

HPLC in the high-throughput purification arena, *oral presentation and poster presented to the 4th International Conference on Packed-Column SFC in Stockholm, Sweden.*

81. Mich A., Chen R., and Buehler S. (2010) A comparative study on the purification of library compounds in drug discovery using mass-directed preparative SFC and preparative RPLC, *LCGC Europe*, March, Application Note.

82. Ward T. J. and Hamburg D. M. (2004) Chiral separations, *Anal. Chem.*, **76**(16), pp. 4635–4644.

83. Alexander A. J. and Staab A. (2006) Use of achiral/chiral SFC/MS for the profiling of isomeric cinnamonitrile/hydrocinnamonitrile products in chiral drug synthesis, *Anal. Chem.*, **78**(11), pp. 3835–3838.

84. Mannerino F., Ali Z., and Wheeler J. (2010). Coupling chiral and achiral columns in series for a multi-phase approach for the separation of stereoisomers using supercritical fluid chromatography, *poster presented to the 4th International Conference on Packed-Column SFC in Stockholm, Sweden* (poster slide is available for download from www.greenchemistrygroup.org/program 2010.html).

85. Lazarescu V., Mulvihill M. J., and Ma L. (2011). Evaluation of tandem dual stationary phases for mass-directed SFC purification of small molecule compound libraries, *poster presented to the 5th International Conference on Packed-Column SFC in New York, USA* (poster slide is available for download from www.greenchemistrygroup.org/program2011.html).

86. Phinney K. W., Sander L. C., and Wise S. A. (1998) Coupled achiral/chiral column techniques in subcritical fluid chromatography for the separation of chiral and nonchiral compounds, *Anal. Chem.* **70**(11), pp. 2331–2335.

87. Guiochon G. and Tarafder A. (2011) Fundamental challenges and opportunities for preparative supercritical fluid chromatography, *J. Chromatogr. A*, **1218**(8), pp. 1037–1114.

Chapter 6

Supercritical Fluid Chromatography for Chiral Method Development Screening and Analysis

Gregory K. Webster[a] and Ted J. Szczerba[b]
[a]*AbbVie, Inc., Global Research and Development,*
1 North Waukegan Road, North Chicago, IL 60064, USA
[b]*Regis Technologies, Inc, 8210 Austin Avenue, Morton Grove, IL 60053, USA*
gregory.webster@abbvie.com, teds@registech.com

6.1 Introduction

The pharmaceutical industry addresses its need for enantiomeric selectivity through stereospecific synthesis, preparative chiral isolations or a combination of both. The path an individual company takes is more a business model decision than a scientific one. The paths a business can take are listed in Table 6.1.

Supercritical Fluid Chromatography: Advances and Applications in Pharmaceutical Analysis
Edited by Gregory K. Webster

ISBN 978-981-4463-00-3 (Hardcover), 978-981-4463-01-0 (eBook)
www.panstanford.com

Table 6.1 Paths to stereospecific drugs

Synthetic approach	Technological approach (racemic mixture resolution and isolation)
Stereoselective or asymmetric synthesis	Crystallization
Enzymatic resolution	Salt resolution
Enantioselective catalysis	Chromatography (LC, SFC)

Analytical chemistry has progressed to a point where it is used to distinguish and separate two isomers that are identical except for their ability to rotate a plane of light. Over the years, techniques such as polarimetry, circular dichroism, and spectroscopic techniques have been used to determine enantiomeric purity and establish stereoselective identity. In today's pharmaceutical industry, the vast majority of this work is now done by chiral chromatography. This need for chiral preparative separations makes chiral chromatography unique from achiral chromatography in that the preparative scale drives the innovations in the technique instead of analytical applications.

Historically, the majority of the drugs on the market have been racemic. Yet, we know the human body only uses L-amino acids and the only bioactive form of Vitamin D3 is (+)-cholecalciferol. Traditionally, it was more economical to synthesize racemic forms and formulate the drug on the basis of the enantiomeric purity. This all changed after an unfortunate series of events in the 1960s led to the discovery of the dangers of this practice. It was discovered that women who took thalidomide as a prescribed sedative during their pregnancy were prone to have significant deformities in their fetuses [1, 2]. The cause of this tragedy was found to be a specific isomer of thalidomide; the (S)- stereoisomer was found to be a teratogen. In response to this incident, regulatory agencies have rightfully required the pharmaceutical industry to characterize the stereochemistry of their drugs in development. Today, thalidomide is approved for use in the United States solely as the (R)-stereoisomer.

Today, more than 40% of the new chemical entities currently under development in the pharmaceutical industry are chiral and subject to these guidelines. The control of drug stereochemistry is significant business. As seen in Table 6.2, top selling pharmaceuticals that are chiral were worth almost $54 billion in 2004 [3].

Table 6.2 Global sales of major chiral pharmaceuticals in 2004

Drug	Global sales ($ billions)	Active ingredient	Form of API	Therapy class
Lipitor	12.0	Atorvastatin	Single Enantiomer	Cholesterol reducer
Zocor	5.9	Simvastatin	Single Enantiomer	Cholesterol reducer
Plavix	5.0	Clopidogrel	Single Enantiomer	Antithrombotic
Nexium	4.8	Esomeprazol	Single Enantiomer	Antiulcerant
Zyprexa	4.8	Olanzapine	Achiral	Antipsychotic
Norvasc	4.8	Amiodipine	Racemate	Bronchodiliator
Seretide/ Advair	4.7	Salmeterol Fluticasone	Racemate Single Enantiomer	Anti-inflammatory
Erypo	4.0	Epoetin alpha	Protein	Red blood cell stimulant
Ogastro	3.8	Lansoprazole	Racemate	Antiulcerant
Effexor	3.7	Venlafadine	Racemate	Antidepressant
Total	53.5			

6.2 Overview of Chiral SFC Separations

6.2.1 Stereoselectivity

Trained chromatographers generally approach separations by starting with a common stationary phase and adjusting the initial mobile phase parameters in a strategy based on the work of Snyder and Kirkland [4]. Chiral separations differ in this regard. Unlike achiral column chromatography, chiral chromatography begins with the need to select the stationary phase that will show stereoselective recognition for the molecule of interest. As such, the stationary phase for the needed separation must have a capability to induce partition coefficient differences between enantiomers.

The need for stereospecific chiral separations propelled the chromatography industry and academia to deliver chiral stationary phases to meet these requirements. This ultimately yielded the cyclodextrin, macrocyclic glycopeptide, polysaccharide, and Pirkle-type stationary phases available today [5]. Traditionally in chiral chromatography, chromatographers consider the stationary phase needing to meet the "three point rule" developed by Pirkle [6]. A stationary phase will show chiral selectivity if the target API has three simultaneous interactions between the chiral selector and one of the enantiomers. One of these interactions must be stereochemically dependent. Thus, for a chiral stationary phase (CSP) to be useful for the chiral separation of the intended target, a pseudo-diastereomeric complex must be formed as the target API elutes down the column. This reversible complex formation can be described in terms of the partition coefficient:

$$K_R : API_R + CSP \Leftrightarrow API_R - CSP_{Complex} \quad (6.1)$$

$$K_S : API_S + CSP \Leftrightarrow API_S - CSP_{Complex} \quad (6.2)$$

As the retention faction for the separation, k', is the distribution of the enantiomer in the mobile and stationary phase, the k_R' becomes

$$k_R' = \frac{[API_R]}{[API_R - CSP_{Complex}]} \quad (6.3)$$

and for the successful resolution of stereoisomers, k_R' and k_S' must differ significantly from each other.

6.2.2 Chiral SFC vs. Chiral HPLC

There are several studies in the literature that compare SFC and HPLC for chiral applications. In these studies, SFC results were either comparable to or had improved chromatographic performance compared with the liquid chromatographic results.

Oftentimes, however, if the HPLC separation is optimized to the extent to which the SFC separation has been, it is very likely that similar enantioselectivity can be found. The actual advantages in chromatography are sometimes overstated. Chromatographers must note that factors other than simply the nature of the CO_2 and alkane mobile phase are likely contributing to the difference in separations taking place between these modes, such as linear flow rate and choice of modifiers.

As both techniques are still normal-phase chromatography and often use the same stationary phases, this point is seemingly not emphasized enough in the commercial literature. The advantage of SFC is in its efficiency and cost over HPLC, not typically its selectivity. The perception that SFC is superior to chiral LC in all instances lacks a firm basis in empirical data [7]. Similar selectivity can be achieved in most cases by HPLC when optimized correctly.

6.2.3 Commercially Available Chiral Stationary Phases for SFC

Column technology for supercritical fluid chiral chromatography has been improving at a rapid pace. Historically, there are five broad categories of chiral columns: (1) polysaccharide-based, (2) cavity-based (i.e. using cyclodextrins, (3) antibiotics and macrocyclic glycopeptides, (4) π–π interaction–based (Pirkle-type columns), and (5) protein-based and ligand exchange–based columns. In addition to the variety of chromatographic stationary phases, column technology has moved from 250 mm and 10 μm particle sizes columns to the more efficient 5 μm particle size and shorter columns. While 5 μm particle size columns are in routine use today, 3 μm and below particle size columns are now commercially available. In achiral liquid chromatography, smaller particle sizes and shorter column lengths allow for shorter screening runs, less consumption of solvents and less generation of waste. For chiral chromatography, particularly with SFC, the need for columns with smaller than 5 μm particle sizes has not been realized to date.

The most commonly used chiral SFC columns are listed in Tables 6.3 and 6.4. Table 6.3 contains common uses for the Pirkle-type columns as they are generally lesser known. Generally, method development strategy is similar on both the polysaccharide and Pirkle-type stationary phases. There is, however, a difference in the co-solvents that may be used on each class of stationary phases. Mainly because many polysaccharide phases have traditionally been coated upon their support, there are some restrictions on the solvents that may be used on these phases. Some researchers report that SFC extends solvent compatibility limitations of coated polysaccharide phases [8]. Vendor restrictions for each column should be reviewed prior to use. There are no restrictions on solvent usage on the Pirkle-type phases. Interestingly, though bonded phases for many polysaccharide phases are commercially available today, these columns may not exhibit the same selectivity of their coated counterparts. Peak efficiency, however, is generally improved. Common co-solvents used in chiral SFC are listed in Table 6.5.

Table 6.3 Polysaccharide-based chiral columns commonly used in pharmaceutical method development screening systems

Column	Chiral phase	Chiral group bonding
tris-(3,5-dimethylphenyl) carbamoyl amylose	Amylose 3,5-dimethylphenylcarbamate 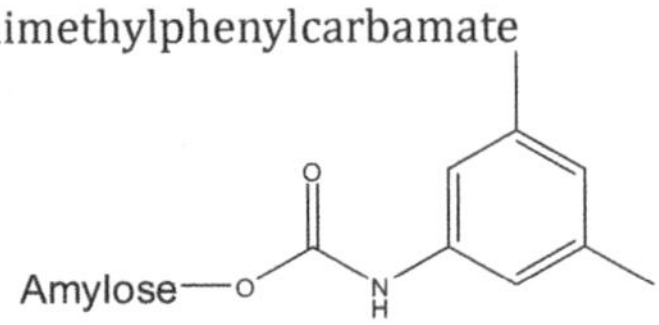	Coated and covalently bonded
tris-(3,5-dimethylphenyl) carbamoyl cellulose	Cellulose 3,5-dimethylphenylcarbamate 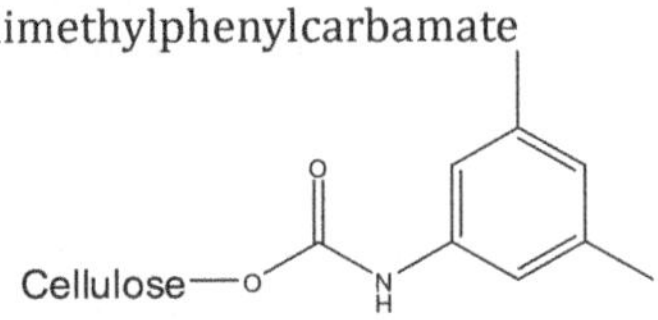	Coated and covalently bonded
tris-(5-chloro-2-methylphenyl) carbamoyl amylose	Cellulose *tris* 3,5-dichlorophenylcarbamate 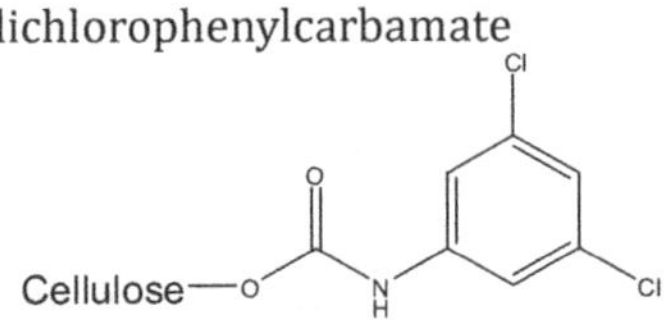	Coated and immobilized

Because of their success in liquid chromatographic chiral separations, chiral polysaccharide phases have retained their market share. Figure 6.1 illustrates the market share of chiral columns in 2005. Several studies have been conducted with individual compound and column investigations using Pirkle-type stationary phases [9–16]. Normally, the same mobile phase approaches that work with polysaccharide stationary phases also work with Pirkle-type chiral stationary phases as well. Because of the commercial success of the polysaccharide columns, many pharmaceutical laboratories simply choose not to screen Pirkle-type columns when looking for enantioseparations. This is unfortunate for unique separation

Table 6.4 π–π interaction–based (Pirkle-type columns) chiral columns commonly used in pharmaceutical method development screening systems

Column	Chiral phase	Chiral group bonding	Uses
DACH-DNB (Regis)	Contains the 3,5-dinitrobenzoyl derivative of 1,2-diaminocyclohexane 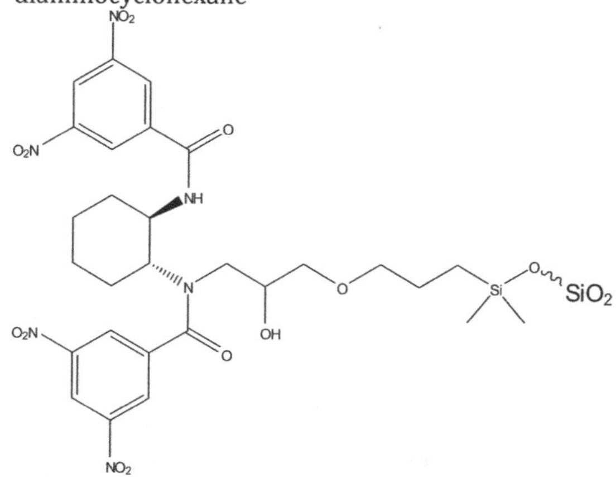	Covalently bonded	Amides, alcohols, esters, ketones, acids, sulfoxides, phosphine oxides, selenoxides, phosphonates, thiophosphineoxide, phosphineselenide, phosphine-borane, beta-lactams, organometallics, atropisomers and heterocycles

(continued)

Table 6.4 *(continued)*

Column	Chiral phase	Chiral group bonding	Uses
Leucine (Regis)	3,5-dinitrobenzoyl leucine, covalently bonded to 5 μm aminopropyl silica 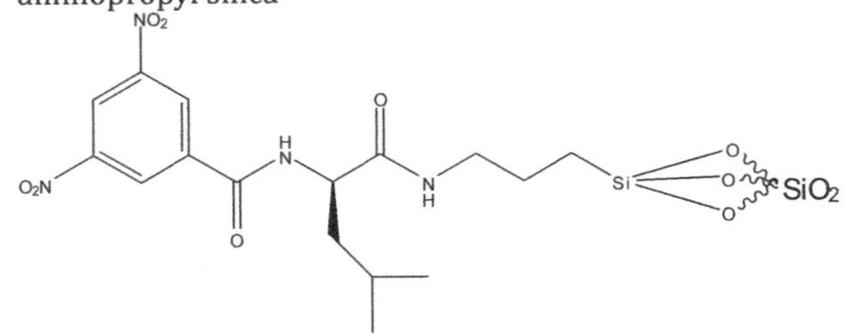	Covalently bonded	Benzodiazapines
Phenylglycine (Regis)	3,5-dinitrobenzoyl phenylglycine, covalently bonded to 5 μm aminopropyl silica 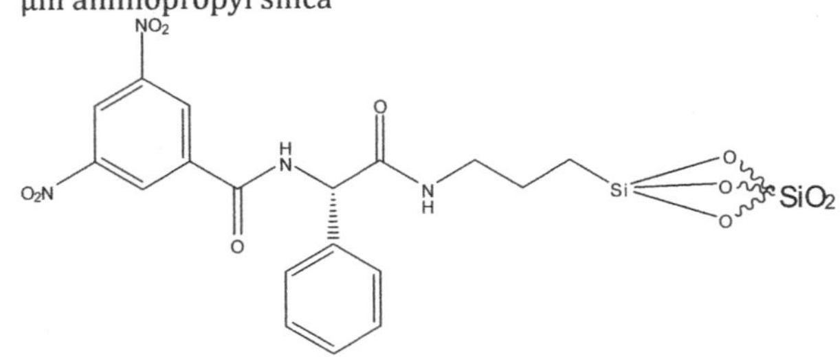	Covalently bonded	π-Basic groups, including aryl-substituted cyclic sulfoxides, bi-β-naphthol and its analogs, α-indanol and α-tetralol analogs, and aryl-substituted hydantoins

Column	Chiral phase	Chiral group bonding	Uses
Pirkle-1 J (Regis)	CSP contains an unusual β-lactam structure 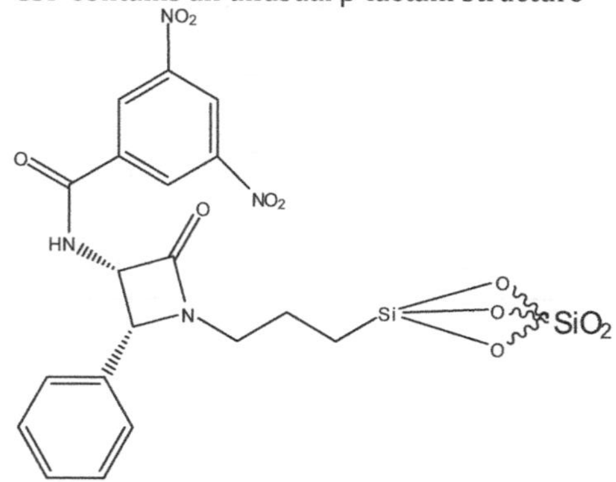	Covalently bonded	β-Blocker enantiomers; arylpropionic acid NSAIDs
ULMO (Regis)	3,5-dintrobenzoyl derivative of diphenylethylenediamine 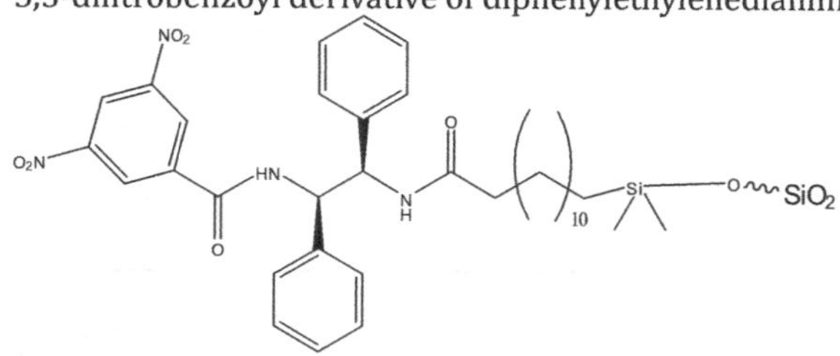	Covalently bonded	Enantiomers of aryl carbinols

(*continued*)

Table 6.4 *(continued)*

Column	Chiral phase	Chiral group bonding	Uses
Whelk-O1 (Regis)	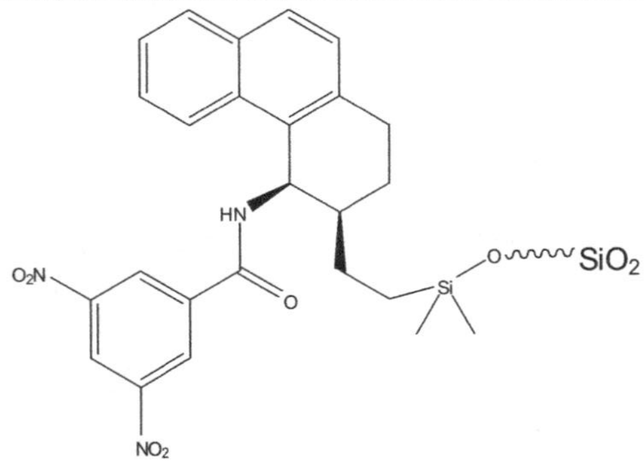	Covalently bonded	Amides, epoxides, esters, ureas, carbamates, ethers, aziridines, phosphonates, aldehydes, ketones, carboxylic acids, alcohols and non-steroidal anti-inflammatory drugs (NSAIDs).
Whelk-O2 (Regis)	Trifunctional version of the Whelk-O 1 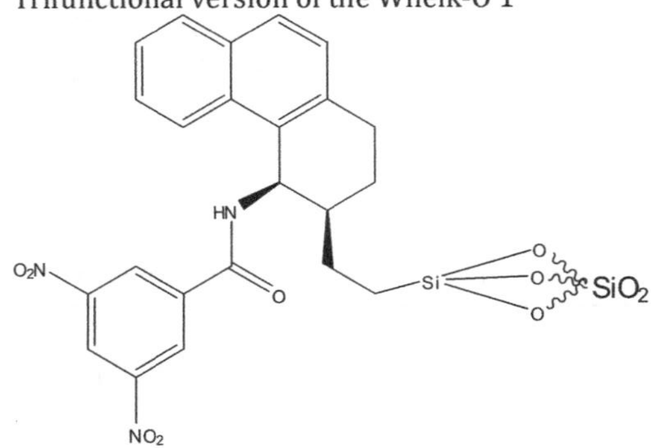	Covalently bonded	

Column	Chiral phase	Chiral group bonding	Uses
α-Burke 2 (Regis)	Dimethyl *N*-3,5-dinitro-benzoyl-amino-2,2-dimethyl-4-pentenyl phosphonate covalently bound to 5 μm silica 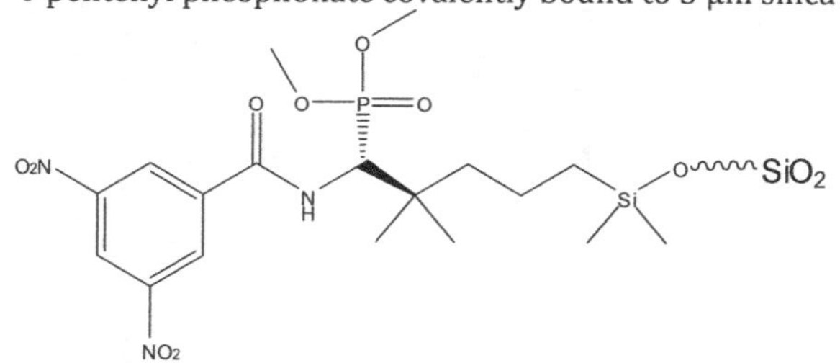	Covalently bonded	β-Blocker enantiomers
β-Gem 1 (Regis)	Covalently bonded *N*-3, 5-dinitrobenzoyl-3-amino-3-phenyl-2-(1,1-dimethylethyl)-propanoate, to 5 μm silica through an ester linkage 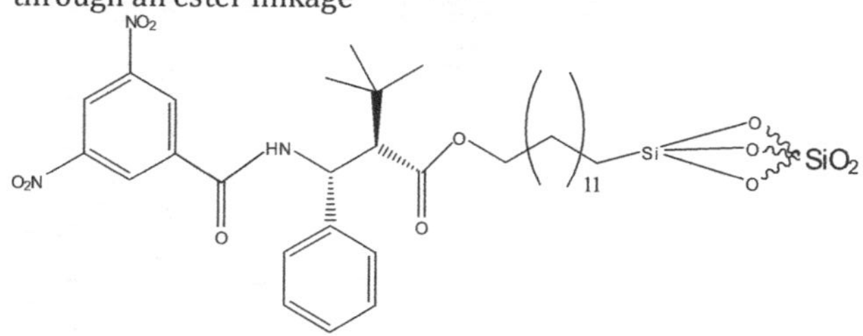	Covalently bonded	Anilide derivatives of chiral carboxylic acids, including nonsteroidal anti-inflammatory agents

opportunities may be missed. Yet, past performance of stationary phases may not be indicative of future enantioresolutions with new pharmaceutical entities. Thus, maintaining screening capabilities with alternative stationary phase choices is prudent.

Table 6.5 Typical SFC co-solvent systems for polysaccharide and Pirkle-type CSPs

Polysaccharide CSPs (coated)	Pirkle-type CSPs
Methanol	Methanol
Ethanol	Ethanol
Isopropanol	Isopropanol
Butanol	Acetonitrile
Acetonitrile	Dichloromethane
Combinations of the above solvents	Tetrahydrofuran
	Ethyl acetate
	Chloroform
	Combinations of the above solvents

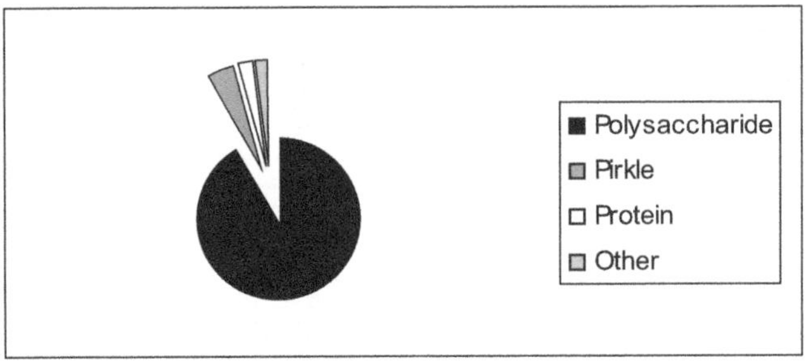

Figure 6.1 Chiral stationary phase market share in 2005.

Another interesting characteristic of Pirkle-type chiral stationary phases is that many are available in both enantiomeric and/or opposite diasteriomeric forms. This choice of enantiomeric forms allows the chromatographer to manipulate the separation for optimization. For example, in the case of an S,S-configured chiral stationary phase, the elution order of separated enantiomers can be reversed simply by switching to the R,R-configured chiral selector. This can be a significant advantage in peak collection with preparative-scale enantioseparations.

When using a new normal-phase chiral column, the chemist should rinse the column with ethanol or IPA before hook-up. Supercritical fluid CO_2 is surprisingly inefficient at removing the hexane commonly used as a storage solvent by the manufacturer. It is also recommended the chemist dedicate columns used in SFC separately from the ones used in HPLC. Many stationary phases exhibit reduced lifetimes when switching back and forth. The reason for this has not been well established.

6.2.4 Mobile Phases for Chiral SFC

When commercial SFC was still based on a capillary mode, multiple gases were used as supercritical fluid mobile phases. While CO_2 was the most common, much work was done with other gases such as nitrous oxide, ammonia and xenon [17]. Today, as SFC essentially exists solely in a packed column format, CO_2 is the only gas commonly used for SFC applications. Supercritical CO_2 (often more correctly termed "subcritical" in the way many labs use it) is compatible with a variety of solvents and can further extend the selectivity range of solvents typically found in normal-phase chromatography.

6.2.5 Co-Solvents in Chiral SFC

Generally, better chromatographic performance is found with chiral separations in the normal phase for most column manufacturers. In addition, easier solvent removal after collecting the isolated enantiomer drove the industry to normal-phase chromatography for chiral applications. It is advantageous to the chiral chromatographer that the majority of the commercially available normal-phase LC chiral stationary phases and modifiers can be used on both LC and SFC instrumentation. This flexibility allows methods developed using one mode to be transferred to the other with minor modifications.

The main limitation of supercritical CO_2 is that its solvation capabilities are limited to nonpolar analytes. As a mobile phase, supercritical CO_2 is similar to hexane. Thus, to overcome this solvation limitation, polar co-solvents are used in the SFC mobile phase. The general rule for modern SFC is that any molecule soluble in methanol can be analyzed by packed column SFC. Table 6.5 lists the co-solvents commonly used in SFC.

For most chiral separations of pharmaceutical interest, the vast majority of separations are achieved using methanol, ethanol (denatured ethanol is fine to use as long as it is denatured with methanol or isopropanol), and IPA as co-solvents. As mentioned earlier, use of high concentrations of alcohols or normal-phase solvents such as dichloromethane or ethyl acetate may ruin polysaccharide-coated stationary phases. The chemist must be careful to operate within manufacturer recommendations for each stationary phase.

Benzoin is a pharmaceutical used on minor skin sores and wounds to protect the area from irritation and infection. Its chiral SFC separation illustrates how much of an effect the choice of co-solvent can have on the resolution of enantiomers. In Figure 6.2, the benzoin selectivity and resolution is shown using methanol, ethanol, and isopropanol.

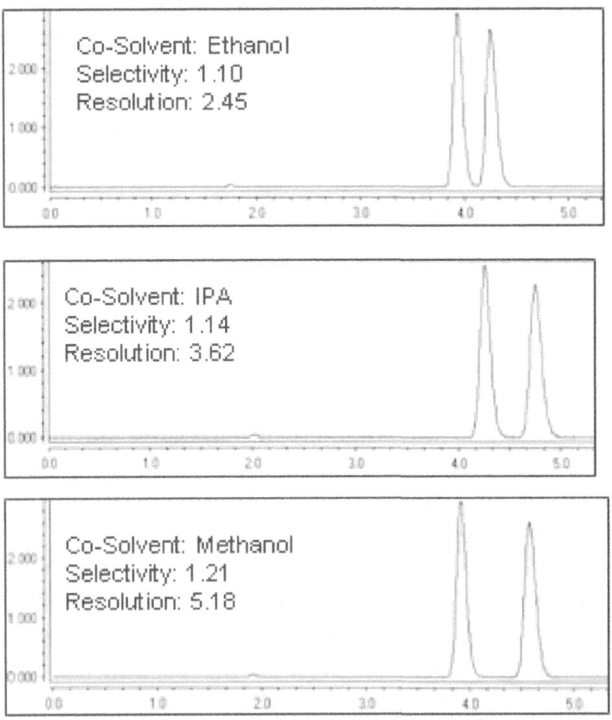

Figure 6.2 Effect of co-solvent type on Benzoin resolution (RegisPack®: 25 cm × 4.6 mm, 5 μm column using an 85:15 $\mathbf{CO_2}$/co-solvent mobile phase operating at 4.0 mL/min, 40°C and 125 bar).

As seen from the chromatographic illustrations, all three co-solvents, dual co-solvents, and various co-solvent modifiers should be tried to achieve optimal separation of a given sample. Oftentimes, the chemist must explore the use of multiple co-solvent combinations to achieve efficient chromatography. Figure 6.3 illustrates that a dual co-solvent is the only way to achieve baseline separation of all four isomers of cyclandelate, a medication that helps to widen blood vessels and improve blood flow.

6.2.5 Co-Solvent Modifiers in Chiral SFC

Ideally, the needed chiral separation is achieved using neat CO_2 and organic co-solvent. This setup yields extremely efficient recoveries in preparative SFC systems without the need for additional sample cleanup. In analytical SFC, this leads to less matrix effects in a mass spectrometer. Yet, sole use of CO_2 and organic co-solvent alone often does not lead to satisfactory analytical chromatography. A co-solvent modifier may be needed to improve peak shape and resolution.

SFC CO_2 often does not require the need of an acidic modifier because supercritical CO_2 is itself acidic. This is probably why some researchers have shown that H_2O can be a substitute for TFA as an acidic co-solvent modifier. Co-solvent modifiers are more commonly used for molecules containing amine groups.

With the exception of ESA, the co-solvent modifiers listed are volatile additives. This allows them to be compatible with the techniques used to recover analytes from preparative scale isolations and with MS work in analytical SFC. Typically, acid modifiers (acetic acid, triflouroacetic acid) work best for acidic compounds and basic modifiers (ammonium acetate, triethylamine) work best for basic compounds. The mobile phase levels of representative co-solvent modifiers are listed in Table 6.6.

Table 6.6 Typical SFC co-solvent modifiers for chiral SFC

Co-solvent modifier	Mobile phase range
Acetic acid	0.1–0.5%
Ethanesulfonic acid (ESA)	0.1– 0.2%
Trifluoroacetic acid (TFA)	0.05– 0.2%
Triethylamine (TEA)	0.1– 0.5%
Diethylamine (DEA)	0.1– 0.5%
Isopropylamine (IPAm)	0.1– 0.5%
Ammonium acetate (AA)	0.01– 0.1M

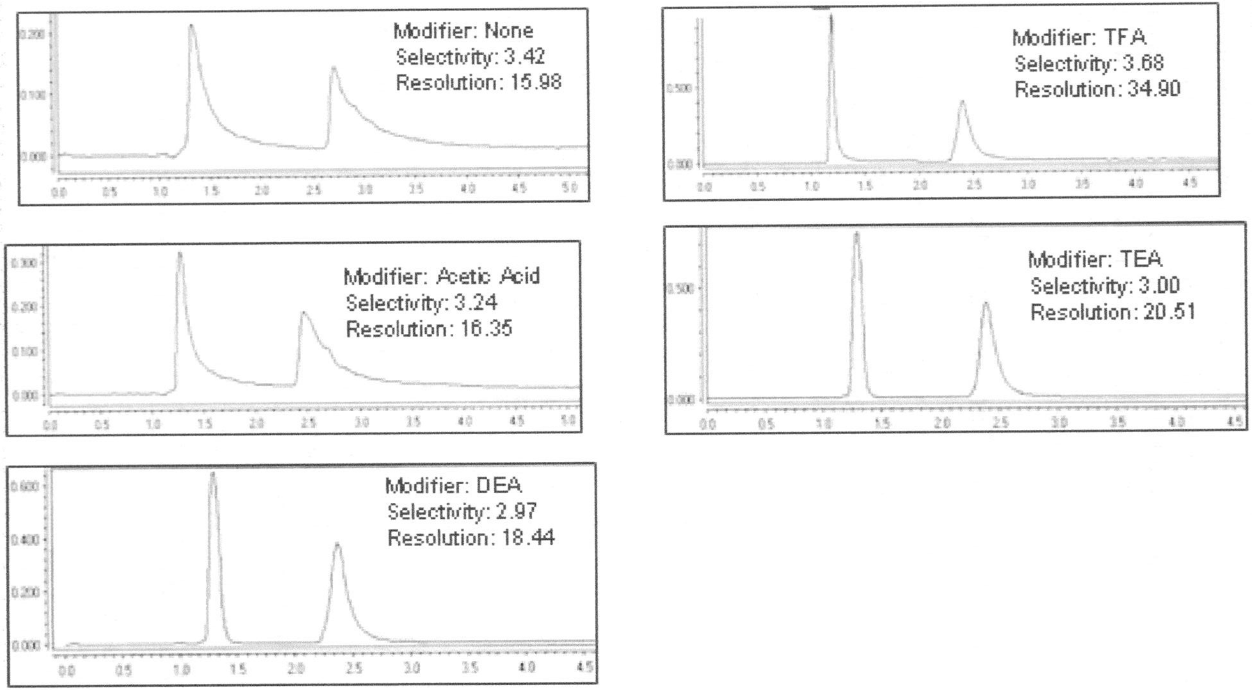

Figure 6.3 Effect of dual co-solvent modifier types on cyclandelate resolution (RegisPack® CLA-1: 25 cm × 4.6 mm, 5 μm column using a 75:25 CO_2/co-solvent + 0.5% co-solvent modifier mobile phase operating at 4.0 mL/min, 40° C and 125 bar).

Metoprolol is a pharmaceutical used as a beta-adrenergic blocking agent. The chiral SFC separation illustrates how much of an effect the choice of co-solvent modifier can have on the resolution of enantiomers. In Fig. 6.4, the benzoin selectivity and resolution is shown using acetic acid, triethylamine, diethylamine, and trifluoroacetic acid.

6.3 Chiral Method Development

As the determination of enantiomeric purity is crucial in new drug development and the number of diverse chiral compounds is increasing, the demand by the pharmaceutical industry for selective chiral methods is high. With many new CSPs on the market, it is challenging to select the most important ones for the initial screening stages and to expedite method development. With all these column and mobile phase choices, where does the chemist go?

Many address this need through the use of column screening capabilities to rapidly assess suitable stationary and mobile phase conditions that will separate the enantiomers of interest. Elaborate chiral column screening procedures for assisted method development for liquid chromatography have been developed to efficiently coordinate these activities [16–30]. While most of these screens employ liquid chromatography setups, transferring them to SFC screens is straightforward.

6.3.1 Synthetic Approach

Some laboratories are very good at matching the structure of the target of interest with the mechanism and functional groups of chiral stationary phases. Chiral method development historically began with a trial and error approach based on matching the structure of the target of interest with the mechanism of the stationary phase. This was a slow and costly process, which oftentimes did not yield a suitable chromatographic separation for regulatory agency approval. The industry as a whole has moved to replace this approach through the development of efficient chiral method development screening strategies.

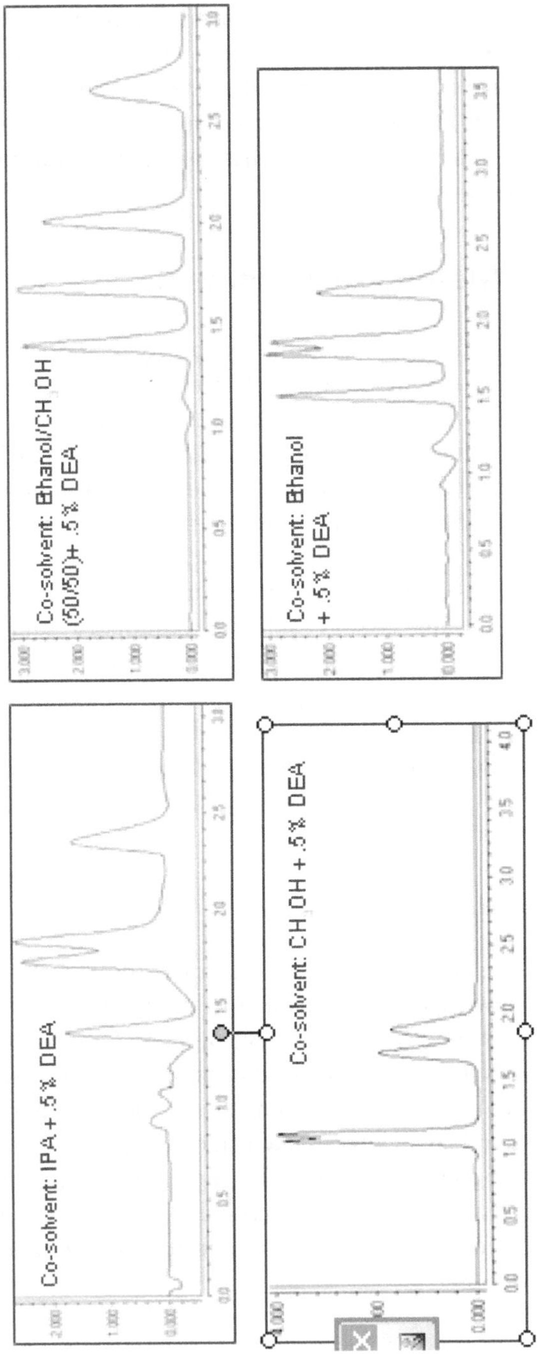

Figure 6.4 Effect of co-solvent modifier type on Metoprolol resolution (RegisCell®: 25 cm × 4.6 mm, 5 µm column using a 80:20 CO_2/ethanol + 0.5% co-solvent modifier mobile phase operating at 4.0 mL/min, 40°C and 125 bar).

Table 6.7 Typical SFC co-solvent systems for polysaccharide and Pirkle-type CSPs

Detection	UV at 254 nm
Flow rate (mL/min)	2-4
Column (°C)	40
Injection (μL)	10
Pressure (bar)	125-150
Mobile phase A	CO_2
Mobile phase B	MeOH: neat or with 0.1% modifier
Gradient[1]	2–40% linear gradient
Sample (°C)	5°C to ambient[2]

[1] Series of isocratic step can be used.

[2] Often, samples in chiral screening systems are kept at 2–8°C in autosamplers to reduce racemization during overnight screens.

6.3.2 Method Development Screens

Using dedicated chiral method development screening systems, pharmaceutical laboratories can quickly select suitable chiral stationary phases for further optimization. In our experience, many chiral compounds don't listen to this logic! Thus, it is wise for chiral separations groups to employ dedicated chiral method development screening systems and strategies to select suitable chiral chromatography conditions for initial investigations. Methods can be further optimized and tailored to the individual laboratory's needs. Many times the screen separation is suitable for much of the needs of drug development. Chiral screening approaches have been developed for normal-phase liquid [16–22, 30] and supercritical fluid [23–29] chromatography. As such, the concepts and conditions are comparable.

The basis of chiral method development screening strategies is largely dictated by whether the method is to be applied to analytical or preparative applications. Preparative chromatography tends to use volatile modifiers and mobile phases for easy removal after sample collection. Sample isolation capabilities have led to the advancement of chiral supercritical fluid chromatography since CO_2 is easily driven off. Chiral separations in preparative chromatography only need to be optimized to a point where pure samples can be collected. Conversely, while analytical chromatography can use a greater variety of modifiers and mobile phases, the separation needs to be optimized to a point where suitable resolution ($R \geq 1.5$) is achieved between the enantiomers of interest.

Whether for analytical or preparative application, finding a stationary phase that exhibits enantiomeric recognition for the molecule of interest is the point of emphasis for chiral screening. There are three modes employed in all liquid and supercritical fluid chromatographic chiral screens: (1) the presence of an unmodified, (2) acidic-modified, and (3) basic-modified mobile phase. The actual mobile phase designs differ from researcher to researcher. The stationary phases and chromatographic system tend to be those that historically have been successful at each company. One significant advantage of dedicated chiral method screens is that the systems are designed to efficiently evaluate various chiral columns and mobile phases in an unattended mode. In general, these screens will use a combination of polysaccharide and Pirkle-type stationary phase columns and only limited by the number of columns the chromatographic system can handle (Tables 6.3 and 6.4).

Whether the screen uses a gradient or runs a series of isocratic screening runs varies from laboratory to laboratory. In the beginning of column chiral chromatographic development, it was thought that chiral columns would not be rugged enough or yield enough theoretical plates for gradients runs. This notion has not proven true and many labs operate gradients with little issue.

As with most packed column SFC, the co-solvent of choice for chiral SFC screening is methanol. Methanol, ethanol, IPA, and combinations of these solvents are used as system designs allow. Some researchers advocate screening with a 1:1:1 methanol/ethanol/IPA mix initially and sort out what works best after the screening is complete. In addition, as many of today's pharmaceuticals tend to be basic compounds, the co-solvent modifiers used tend to be diethylamine, ammonium acetate and trifluoroacetic acid. For more challenging compounds, the chemist will typically move to different solvents, such as dichloromethane, as stationary phases allow. Lastly, and outside the makeup of most column screens, column coupling can be tried [31]. The diffusibility of the supercritical fluid phase is an advantage in this regard.

6.3.3 Application

It is important to remember that the first goal of the chiral method development screen is to find a stationary phase that exhibits stereoselective recognition for the analytes under study. Oftentimes, chiral screen conditions yield results that are suitable as a final chromatographic method. A broad range of results can be expected. For example, doxazosin (Figure 6.5) under the same SFC conditions yields a resolved chromatogram with a *tris*-(3,5-dimethylphenyl) carbamoyl cellulose stationary phase and does not elute on a Whelk-O1 stationary phase. Representative results for pantoprazole are shown in Figure 6.6. Both screens for this compound were successful. Each systematic screen resulted in at least a single condition of all those tried with chiral selectivity.

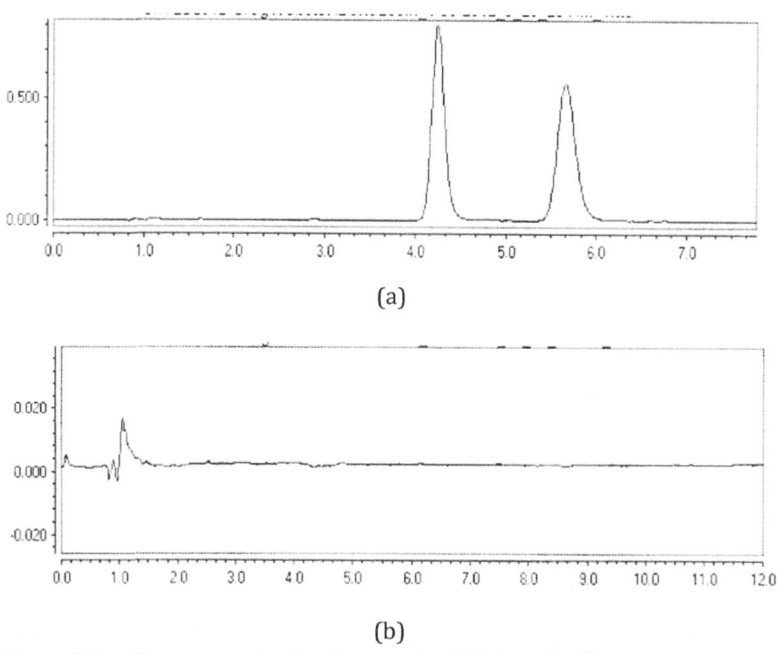

Figure 6.5 Screen results for Doxazosin. (a) RegisCell®, (b) (S,S)-Whelk-O1®: Both columns were 25 cm × 4.6 mm, 5 µm columns using a 65:/35 CO_2/IPA + 0.5% DEA mobile phase operating at 4.0 mL/min, 40°C and 125 bar.

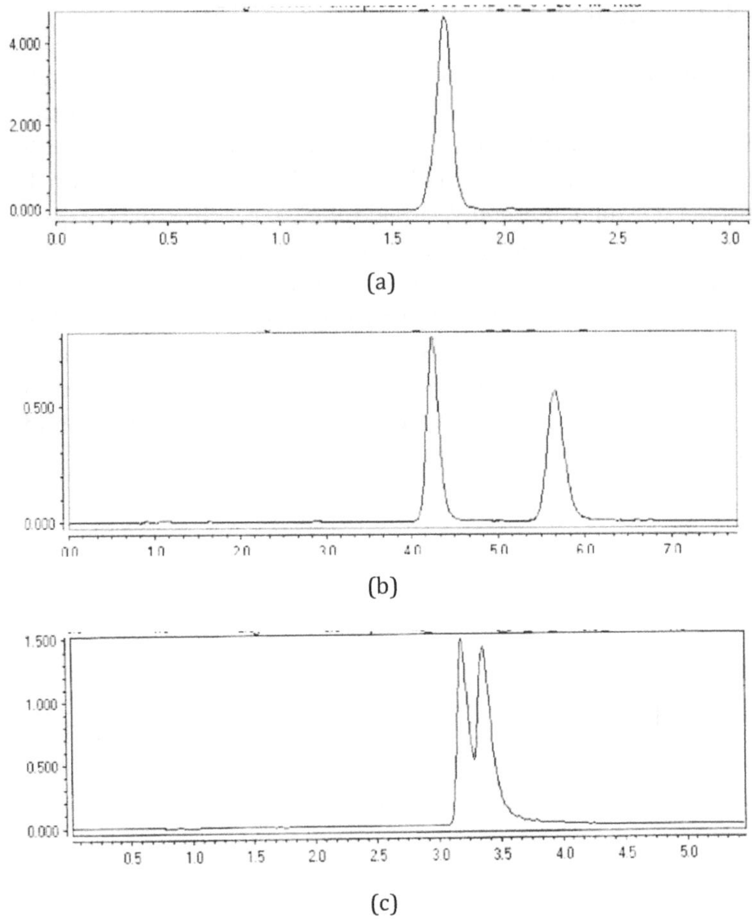

Figure 6.6 Screen results for pantoprazole. (a) RegisCell®, (b) RegisPack®, and (c) (S,S)-Whelk-O1®: All three columns were 25 cm × 4.6 mm, 5 μm columns using a 70/30 CO_2/ethanol mobile phase operating at 4.0 mL/min, 40°C and 125 bar.

6.3.3.1 Screening success rate

For both normal-phase LC and SFC chiral method development screens, most schemes seen in the literature yield a stereospecific resolution success rate typically of 80% and higher for the compounds studied to date. While chiral screening in the pharmaceutical industry primarily centers on the preparative chromatographic collection of the enantiomers, the screens routinely post methods suitable for analytical applications as well.

Literature screens often employ test compounds from single company libraries. This makes sense when considering each company is only concerned with resolving their molecules in development, not their competitors. In other words, an industry success rate is of little value to an individual company. The specific method development screen used must focus on the stereoselective resolution of the analytes of interest to that concern. Thus, it is important to evaluate and optimize each chiral method development system for the molecules of interest to each particular business concern. The variety of challenge compounds seen in this screen represents a greater diversity of molecules for polysaccharide and Pirkle-type columns to address.

As a representative example, the RegisSEP™ SFC screen will typically begin with polysaccharide stationary phases (RegisPack, RegisCell, Regis OJ, RegisPack CLA-1) and Pirkle-type stationary phases. (Whelk-O and ULMO). The supercritical fluid CO_2 mobile phase runs with methanol, ethanol and isopropanol co-solvents. The co-solvents run with diethylamine, triethylamine, acetic acid and trifluoroamine modifiers. The summary of the RegisSEP™ SFC screen for 519 chiral compounds is seen in Table 6.8.

Table 6.8 Screening summary results of 519 test compounds

Chiral stationary phase	No. of hits	%
RegisPack®	238	46
Whelk-O® 1	174	34
RegisCell®	54	10
All other CSPs	19	4
# Not separated	140	—

For the compounds presented in Table 6.8, 379 of these were separated on stationary phases applied. Of these separations, 73% resulted in a suitable separation on the first pass. One of the problems with this yield is common to all screens: the uncertainty that the sending lab actually shipped a truly racemic test compound for screening.

6.4 Summary

SFC chiral method development screening is very effective and efficient for developing chiral selective methods in the pharmaceutical industry. As an ever diverse stream of chemical entities is being developed for market, the analytical chemist must continually look for ways to optimize the screening system. This is required to ensure the best opportunity for success in yielding a robust analytical method for the new chemical entities in development. The use of SFC for chiral method development and analysis is becoming the primary technique for testing chiral pharmaceuticals in development as well as the marketplace. As new stationary phases and instrumentation advances become commercially available, SFC will remain in the forefront of chiral drug development.

References

1. Franks, M.E., Macpherson, G.R., Figg, W.D. (2004) Thalidomide: A review. *The Lancet*, **363**, 1802–1811.
2. Perri, A.J., Hsu, S. (2003) A review of thalidomide's history and current dermatological applications *Dermatol. Online J.*, **3**(5).
3. C. Aurigemma, Pfizer, "Chiral Separations: A Tutorial", Conference on Small Molecule Science 2006.
4. Snyder, L.R., Kirkland, J.J., Glajch, J.L. (1997) *Practical HPLC Method Developmen*t, 2nd Ed., John Wiley & Sons, New York, NY.
5. Webster, G.K., Kott, L. (2010) "Method Development for Pharmaceutical Chiral Chromatography," *Handbook of Modern Pharmaceutical Analysis*, 2nd Ed., Academic Press, Salt Lake City, UT.
6. L. Salam, X. Chapuisat, G. Segal, P. Hiberty, C. Minot, C. Leforrestier, P. Sautet. (1987) *J. Am. Chem. Soc.*, **109**, 2884–2887.
7. Kromidas, S. *HPLC Made to Measure*, John Wiley & Sons, New York, NY, 2006.
8. Wong, M.M., Holzheuer, W.B., Webster, G.K. (2008) A comparison of HPLC and SFC chiral method development screening approaches for compounds of pharmaceutical interest. *Crit. Pharm. Anal.*, **4**, 101–105.
9. Kennedy, J.H. (1996) Comparison of chiral separations on polysaccharide chiral stationary phases to an improved Pirkle-type phase. *J. Chromatogr.*, **725**, 219–224.

10. Siluveru, M., Stewart, J.T. (1996) Stereoselective determination of R(-)- and S(+)-prilocaine in human serum using a brush-type chiral stationary phase. *J. Pharm. Bio. Anal.*, **15**, 389–392.

11. Zhang, X., Ouyang, J., Yang, Y. (2001) A simple method for chiral separation of ephedrines using (R)-1-naphthylglycine and 3,5,-dinitrobenzoic acid as stationary phase. *Anal. Lett.*, **34**, 1851–1865.

12. Shao, B.-H., Xu, X.-Z., Wu, Q.-Z., Lu, J.-D., Fu, X.-Y. (2005) Comparative enantioseparation of 2-arylpropionic acid esters on cellulose derivative and (S,S)-Whelk-O 1 columns. *J. Liquid Chromatogr. Relat. Technol.*, **28**, 63–80.

13. Shao, B.-H., Xu, X.-Z., Wu, Q.-Z., Lu, J.-D., Fu, X.-Y. (2003) Comparison of enantioseparation and chiral recognition mechanism of racemic naproxen esters on (S,S)-Whelk-O 1 and CDMPC chiral columns. Comparison of enantioseparation and chiral recognition mechanism of racemic naproxen esters on (S,S)-Whelk-O 1 and CDMPC chiral columns. *Huaxue Xuebao*, **61**, 1635–1640.

14. Madhavan, P., Rao, B.M., Pravin, A., Kumar, P.R., Screenivasulu, M., Chandrasekhar, K.B. (2007) A validated chiral HPLC method for the determination of enantiomeric purity of R-β-amino-β-(4-methoxyphenyl) propionic acid. *Chromatographia*, **65**, 81–84.

15. Szczerba , T. (2011) Summarization of screening hits on the Whelk-O 1, RegisPack and RegisCell chiral stationary phases (CSPs). *LCGC North America*, (Suppl.), February, 24.

16. Holzheuer, W.B., Wong, M.M., Webster, G.K. (2009) Evaluation of Pirkle-type Stationary Phases in Chiral Method Development Screening for Compounds of Pharmaceutical Interest. *Crit. Pharm. Anal.*, **5**, 10–20.

17. Perrin, C., Vu, V.A., Matthijs, N., Maftouh, M., Massart, D.L., Vander Heyden, Y. (2002) Screening approach for chiral separation of pharmaceuticals. Part I. Normal phase liquid chromatography. *J. Chrom. A*, **947**, 69–83.

18. Perrin, C., Vu, V.A., Matthijs, N., Maftouh, M., Massart, D.L., Vander Heyden, Y. (2002) Screening approach for chiral separation of pharmaceuticals. Part II. Reversed phase liquid chromatography. *J. Chrom. A*, **966**, 119–134.

19. de la Puente, M.L., White, C.T., Rivera-Sagredt, A., Reilly, J., Burton, K., Harvey, G. (2003) Impact of normal-phase gradient elution in chiral chromatography: a novel, robust, efficient and rapid chiral screening procedure. *J. Chrom. A*, **983**, 101–114.

20. Andersson, M.E., Asclan, D., Clark, A., Roeraade, J., Hagman, G. (2003) 98.Evaluation of Generic Chiral Liquid Chromatography Screens for Pharmaceutical Analysis. *J. Chrom. A*, **1005**, 83–101.

21. Lipka, E., Glacon, V., Mackenzie, G., Ewing, D., Len, C., Postel, D., Vaccher, M.P, Bonte, J.P, Vaccher, C. (2003) HPLC separation and determination of enantiomeric purity of novel nucleoside analogues, on cyclodextrin chiral stationary phases using reversed and polar organic modes. *Anal. Lett.*, **37**, 385–398.

22. Esser, C.K., Black, R.M., Von Langen, D. (2004) Implementation of HPLC automation for the analysis and purification of chiral molecules. *Pharm. Discov.*, **4**, 26–32.

23. Zhang, Y., Watts, W., Nogle, L., McConnell, O. (2004) Rapid method development for chiral separation in drug discovery using multi-column parallel screening and circular dichroism signal pooling. *J. Chrom. A*, **1049**, 75–84.

24. Berger, T.A. (1995) *Packed Column SFC*, Royal Society of Chemistry, Cambridge, UK.

25. Terfloth, G. (2001) Enantioseparations in super- and subcritical fluid chromatography. *J. Chrom. A*, **906**, 301–307.

26. Anton, K.K., Berger, C., Editors (1998) *Supercritical Fluid Chromatography with Packed Columns*, Marcel Dekker, New York, NY.

27. Zhao, Y., Woo, G., Thomas, S., Semin, D., Sandra, P. (2003) Rapid method development for chiral separation in drug discovery using sample pooling and supercritical fluid chromatography–mass spectrometry. *J. Chrom. A*, **1003**, 157–166.

28. Borman, P., Boughtflower, B., Cattanach, K., Crane, K., Freebairn, K., Jonas, G., Mutton, I., Patel, A., Sanders, M., Thompson, D. (2003) Comparative performances of selected chiral HPLC, SFC, and CE systems with a chemically diverse sample set. *Chirality*, **15**, S1-S12.

29. Di, L., McConnell, O.J., Kerns, E.H., Sutherland, A.G. (2004) Rapid, automated screening method for enzymatic transformations using a robotic system and supercritical fluid chromatography. *J. Chrom. B*, **809**, 231–235.

30. Akin, A, Antosz, F.J., Ausec, J.L, Greve, K.F., Johnson, R.L, Magnusson, L.E, Ramstad, T., Secreast, S.L., Seibert, D.S., and Webster, G.K. (2007) An orthogonal approach to chiral method development screening. *Curr. Pharm. Anal.*, **3**, 53–70.

31. McDevitt, T.F., Vicente, G., Webster, G.K., Szczerba, T.J. (2009) Column Coupling to Solve a Challenging API Separation Using the Whelk-O®1 Chiral Stationary Phase. *LC/GC Application Notebook*, September, 23.

Chapter 7

Chiral Preparative Supercritical Fluid Chromatography

Manuel C. Ventura
Therapeutic Discovery, Amgen, Inc.,1120 Veterans Blvd., South San Francisco, CA 94080, USA
mventura@amgen.com

7.1 Introduction

A chiral molecule is one that does not possess an internal plane of symmetry. That is, two possible spatial forms exist in which the mirror images cannot be superimposed. This is often likened to a person's left and right hand for illustration. Chiral molecules are playing an ever-increasing role in industrial synthetic processes due to many factors. They occur in petrochemicals, flavor and are utilized in fragrances, chemical manufacturing and pharmaceuticals. By far, the pharmaceutical business produces the greatest diversity and complexity of chiral molecules. Owing to the chirality of biological building blocks, including sugars, proteins, and DNA, chiral drug molecules usually offer enhanced binding affinity to biological targets through a preferred enantiomer. Current technologies have

Supercritical Fluid Chromatography: Advances and Applications in Pharmaceutical Analysis
Edited by Gregory K. Webster

ISBN 978-981-4463-00-3 (Hardcover), 978-981-4463-01-0 (eBook)
www.panstanford.com

made possible the synthetic and separation schema to create and isolate almost any chiral molecule that can be imagined.

Attaining enantiopurity a priori without chiral chromatography can be performed in a variety of ways. For example, with classical resolution, diastereomer salts are formed through the interaction of a racemate with a chiral resolving agent. These diastereomers are intended to have different solubilities and a crystallization step can then be performed to isolate enantiomers. Another approach is kinetic resolution, in which reaction with a chiral catalyst results in relative enrichment of one enantiomer product over time. In these cases, an optimization occurs at the expense of product, time, and labor. These may be worthwhile sacrifices for advanced work at the process or production scale where long-term cost savings can be realized, but these procedures are often intolerable to the demands of a fast-paced pharmaceutical synthesis environment. The chromatographic approach to chiral separations is overwhelmingly the fastest route to achieving enantiopurity in the discovery pharmaceutical synthesis process.

7.2 Toward Useful Chiral Stationary Phases

The approach to stereoselective chromatographic isolation of chiral compound–involved technology developments on several different fronts. Efficient preparative chromatography could not really be accomplished before the development of high pressure liquid pumps, smaller stationary phase particles, and dynamic axial compression for packing columns through the 1970s and early 1980s [1]. Around the same time, the basis for chiral stationary phases still in use today was being developed by Pirkle, Okamoto, and Armstrong. The fundamental for all chiral separations is said to be a three-point interaction between the solute and the stationary phase in which at least one interaction depends on stereochemistry [2]. This interaction creates a transient diastereomer. Two enantiomers will ideally form product diastereomers with the chiral stationary phase(CSP) of differing free energy resulting in an observable retention difference. William Pirkle et al. developed a "brush-type" stationary phase for chiral recognition [3]. The discrimination of enantiomers is achieved through the bonded phase ligand's rigid structure and accessibility for π-bond interaction with solutes near its chiral centers. One can predictably reverse the elution order of enantiomers by switching from one enantiomer form of the stationary phase to the other, which may be useful for purification purposes. Derivatized polysaccharide-based phases developed by Okamoto et al. based on amylose and cellulose produce chiral selectivity due to their density of chiral centers and a highly ordered secondary structure [4]. These were coated on the silica support rather than covalently bonded, although today immobilized phases exist. Amylose and cellulose-based CSPs originally from Daicel but now available from a variety of vendors and are the most heavily utilized in the pharmaceutical industry for small-molecule chiral separations. Silica-bound cyclodextrins were developed by Armstrong et al. [5]. These CSPs utilized interactions via inclusion complexing either within the cyclodextrin ring or on the surface for chiral recognition. The low surface coverage of these phases, however, limits the loading capacity for preparative purposes. Additional classes of chiral selectors, including crown ethers, macrocyclic antibiotic, protein, and ligand exchange columns, have since been developed and are now marketed products. With the possible exception of the Vancomycin (antibiotic) phase, these are less general in application to others described here and less suitable for small-molecule purification.

7.3 SFC Application to Preparative Separation

7.3.1 Basic Background

Although the initial development and application for chiral stationary phases presumed normal-phase HPLC for the separation technology, their use for chiral separations in supercritical fluid chromatography (SFC) was soon found to be compatible and effective. SFC is similar to a normal-phase HPLC technique with the main difference that the nonpolar component consists of compressed CO_2. Desirable conditions for SFC utilize mobile phases well above the critical point, for CO_2, $T > 31.1$ K and $P > 73.8$ bar. With CO_2, critical parameters can be easily achieved with an adequate pumping system. In this state, the SFC mobile phase is a compressible fluid with a higher diffusion coefficient than HPLC. This results in faster mass transfer with the stationary phase leading to higher efficiency separations. The lower viscosity of the mobile phase also results in lower pressure drops across packed columns providing the flexibility to use longer columns and further increase separation efficiency. The same column for SFC will have a lower, flatter van Deemter curve relative to HPLC exhibiting a higher optimal linear velocity and allowing a large range of flow rates producing near optimum efficiency. As a result, separation efficiencies can often be three to five times higher than with conventional HPLC [6].

In the 1980s, users sought to take advantage of SFC's benefits for the purpose of preparative separations. The first patented preparative scale SFC was designed by Perrut [7] utilizing pure CO_2 as the mobile phase with recycling and cyclone separators for collected fractions.

7.3.2 Mobile Phases for Prep SFC

Pure supercritical CO_2 alone can rarely succeed in dissolving and eluting most compounds of interest (containing some polarity) to pharmaceuticals; so a polar mobile phase modifier at appropriate concentration is normally used. Chiral separations of polar molecules often require the introduction of sometimes a substantial fraction of polar modifier relative to CO_2. At normal near ambient operating temperatures, this results in a sub-critical state for the mobile phase. Nevertheless, this generally does not cause a loss of the enhanced properties of SFC relative to HPLC already described. There is no abrupt change in the diffusivity of the mobile phase or increase in viscosity as the trend is raised from supercritical to subcritical with increased modifier composition [8]. The most common polar modifier used in combination with CO_2 is methanol. Ethanol, isopropanol, and acetonitrile are the other most common modifiers used in chiral separation application. These modifiers adsorb to stationary phase surfaces along with CO_2 and thus change the density and polarity of the phase [9]. The fraction of organic modifier in the mobile phase is the key parameter for altering retention in SFC, whereas changing the CO_2 density was originally believed to be most impactful [10]. It follows that for the chiral SFC method development, the polar modifier percentage in CO_2 is key to altering the selectivity and retention as required for enantiomer resolution.

Mobile phase additives are also often used in chiral separations owing to the basic nature of solutes involved and the acidic nature of silica. (CO_2 also creates acidity in the mobile phase [11, 12].) Basic additives improve peak shape and can alter retention and selectivity through suppression of silica ionization, ion pair formation, and increasing base solubility in the mobile phase [13]. Common additives used in chiral applications include diethylamine, isopropylamine, and ammonia, though a wide variety of other bases are used. Additives can be delivered separately through an additional pump, but most commonly, they are simply mixed with the organic modifier. Acidic additives such as ethanesulfonic acid and trifluoroacetic acid may similarly complement separations for acidic solutes or function in the ion pair mode with base additive for certain basic solutes [14].

7.3.3 Instrumentation for Prep SFC

Many hardware components used for preparative HPLCs were found to be directly adaptable to preparative SFC. Preparative SFC requires a supply of CO_2 that can be quite significant. In an environment where systems are run for several hours a day, depending on their scale, hundreds of liters may be consumed daily; so the supply vessel must be of sufficient size. (Some preparative SFC instruments are equipped with CO_2 recycling capability to reduce consumption for extended processes.) CO_2 is supplied either by pumping out from below the liquid level in tanks, dewars or bulk vessels or by the pressurization of the vapor from the headspace of the container. Typically about 900 psi is required at the SFC pump CO_2 inlet; so unless the tank liquid is under at least this pressure, a booster pump from the CO_2 source is required. The SFC CO_2 pump head must be chilled (to around 5°C) to prevent gas formation due to heating of the pump body created by the compression of the liquid. This is achieved by using a recirculating bath for the pump head or a Peltier element to keep the liquid CO_2 cold. Flow from the modifier pump is combined with the CO_2 delivered at high pressure, and then the mixed fluid is routed to the injector.

Some systems are configured with modifier stream injection in which the mixing point with the CO_2 is set up downstream of the injector [15]. The advantage of modifier stream injection is that it prevents sample precipitation upon introduction into the mixed mobile phase, which may result in peak distortion, pressure spikes or system shutdown. The main disadvantage relative to mixed stream injection is that the sample is only delivered to the column based on the flow rate of the modifier solvent, which in some cases may be sufficiently low as to result in dispersion and reduced performance.

Usually single wavelength UV detectors are used with preparative SFC. It is the intensity, or sometimes slope, of that detected signal which will likely be used for fraction collection. UV flow cells must be capable of tolerating the high pressures inherent to SFC conditions many rated up to 400 bar. Sensitivity is lower in SFC than HPLC, though this is not usually a problem with the sample sizes relevant to preparative operation. Since the UV signals are normally so large in prep-SFC, the higher noise resulting from the refractive index change of CO_2 with an automated backpressure regulator fluctuation does not impact the ability to trigger fraction collection. Other detectors such as Evaporative light scattering detection (ELSD) and others have been used in prep-SFC but are much less prevalently utilized and will not be described in this review.

Automated backpressure regulators are the devices used to fix post-column pressure in modern prep SFC and are critical to maintaining stable system performance. There are as many varieties as the instrument vendors that utilize them, but in general they involve an electronically controlled restriction point that generates a motor movement to compensate for pressure changes in order to maintain a desired set point.

The flow from the backpressure regulator outlet is usually plumbed to waste container by default in which after decompression the mobile phase again separates into its liquid and gaseous constituents. CO_2 is allowed to exhaust through a proper channel while the liquid component is collected into a waste container. For preparative SFC, however, when fraction collection is triggered by an appropriate UV signal, this flow to waste is diverted into collection vessels, in many instruments through cyclone separators. A valve opens to one of the cyclone separators and flow is directed at a downward direction against the interior wall of the cylinder in a vortex fashion. CO_2 is vented through a pipe in the top of the cylinder while liquid containing the compound of interest is collected in the cyclone and drained through another valve at the base into the final collection container.

A typical preparative SFC instrument is depicted in the block diagram shown in Fig. 7.1.

Today commercial semi-preparative to preparative SFCs come in scales comprising flow rates between 10 g/min to 1 kg/min and are used to purify material quantities from a few milligrams up to kilograms. A variety of applications for chiral preparative SFC have been described and its prevalence continues to expand beyond the most common use for small-molecule chiral separations.

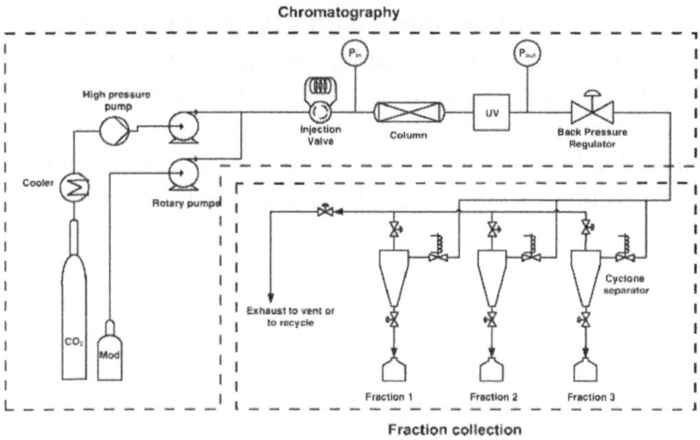

Figure 7.1 Schematic of a typical preparative SFC system showing the chromatography and the fraction collection modules [16]. Reprinted with permission.

7.4 Strategy for Chiral Preparative SFC Separation of New Drug-Like Molecules

A general procedure for designing an effective semi-preparative small-molecule chiral separation as practiced significantly in the pharmaceutical industry is described here. The retention and selectivity characteristics of racemic compounds are typically unknown in a drug discovery environment. The basic tools required include analytical and preparative scale SFC systems, solvents, and matching stationary phases on the analytical and preparative scales.

7.4.1 Analytical Method Development

A sample is dissolved in an organic solvent, preferably one of the mobile phase modifier solvents being used such as methanol, ethanol, 2-propanol, and acetonitrile. Good sensitivity on the analytical scale can usually be achieved with around 5–10 μg injected onto a typical 100 or 150 × 4.6 mm column. Smaller-dimension chiral stationary phase columns consisting of 3 μm particles, can be used and are increasingly prevalent in the industry [17]. Gradient separation methods are employed for each chiral column being screened together for a particular sample. In these methods, one of the modifiers is programmed to linearly increased in composition from usually 5%

of the combined mobile phase with CO_2, up to between 40% and 60% in variable t_G (gradient time, usually between 2 and 10 min) at flow rates between 2 and 5 mL/min. The method programmed should deliver approximately 10 column volumes during the gradient. Outlet backpressure will usually be set to a fixed value between 100 and 150 bar and the column oven set somewhere between ambient temperature and 50°C. Multiple methods, each consisting of a different column and a different mobile phase are used from which the user can extract the best combination for separation from the data. Parallel-column systems for this application are currently in the field. Finally, an isocratic method based on gradient data is run to demonstrate enantiomer resolution and these conditions can be used with appropriate scaling factors for preparative separation. Figure 7.2 shows an example of a four-column chiral gradient screen, followed by an isocratic method using the column exhibiting the best gradient separation. A scaled-up preparative method can utilize the same mobile and stationary phase with correspondingly larger dimension and higher flow rate.

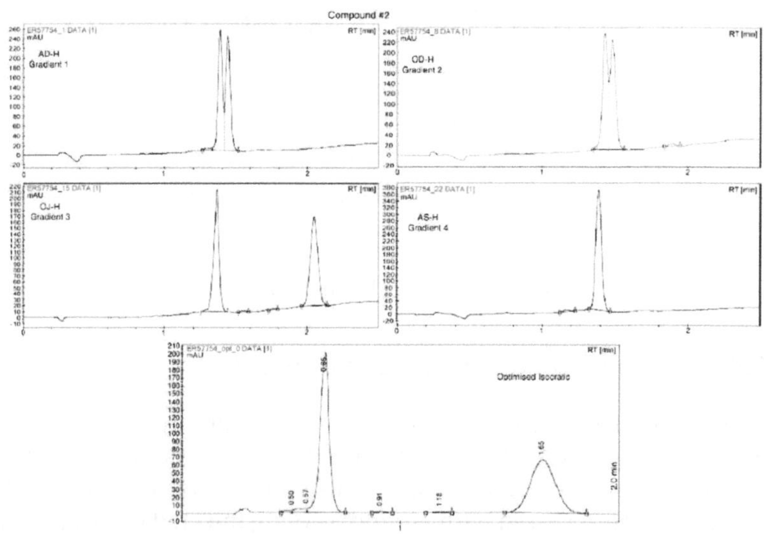

Figure 7.2 Compound 2 analyzed using the primary chiral screen identifying the OJ-H column as the column of choice for purification. The optimized isocratic conditions were 30% MeOH containing 0.2% IPAm on the OJ-H column, flow rate 5 mL/min. Outlet pressure was set to 120 bar and temperature was 40°C. Detection by UV at 220 nm [18]. Reprinted with permission.

7.4.2 Preparative Purification

As with HPLC, both mass and volume overloading effects are expected on the preparative scale with increasing sample injection size [19]. The resolution between peaks desired for separation can be increased through an increase in plate number through increasing the stationary phase length relative to that on the analytical scale. Thus, longer columns are often used in the prep run relative to the analytical run for a given sample. To reduce the volume overloading effect and to decrease the number of injections required to preparatively separate components of a given sample, that sample is usually dissolved to a maximum of concentration in a desired solvent. Methanol is most often used as a sample solvent for chiral separations since it is a strong solvent, miscible with other common SFC modifiers that may be used and it is the most common modifier solvent of all. For poorly soluble samples solvents such as dichloromethane (DCM) and THF among others can be used to increase solubility. One problem with these is that they are not recommended for use with coated polysaccharide chiral stationary phases, as they can dissolve and strip them from the silica support [20]. Immobilized chiral stationary phases are inert to these solvents and present an alternative for using stronger solvents in samples and also in the mobile phase provided chiral selectivity is adequate on these phases [21].

Initially, small injections are attempted since unknown compounds may have widely differing solubilities in CO_2 plus modifier solvent and unknown isotherms. A low-solubility compound injected can result in significant peak distortion, increased retention, and either transient or long-lasting increases in system pressure due to precipitation. Modifier stream injection has had some success in decreasing issues of insolubility by dissolving the sample in liquid prior to mixing with CO_2 as opposed to the sample solution hitting the mixed phase, perhaps high in CO_2, and then precipitating. Should modifier stream injection not be available for a user or should it be unsuccessful for a particular case, workarounds include raising the modifier composition, diluting the sample further, and considering stronger solvent alternatives to the preferred mobile phase at analytical scale. A change of one major parameter from the original desired method usually requires a compensating change of another parameter. For example, increasing the modifier composition will decrease retention and selectivity so that one may wish to compensate by increasing the column length or decreasing sample load on column. In this case, one could also possibly run into a barrier of excessive pressure on the column should the viscosity of the mobile phase be too high or the column too long. A loading study can be performed following a successful scouting run, which shows a reasonable resolution and an elution profile among the components being separated prior to a series of injections to be run in a preparative separation. The final mass injected and cycle time, representing the time between each injection on column, can be determined by a loading study to determine the injection size at which the bands for the chromatographic peaks of interest just touch [22]. This injection size and a cycle time, a window within which the peaks of interest fully elute are then programmed into a sequence that can run continuously. It can be set to finish either when the sample is expected to be consumed from sufficient numbers of injections or until such time as when a critical supply for the separation may run out. The productivity of the final preparative separation is measured by the amount of material processed per unit time per unit mass of stationary phase, kg(sample)/kg(CSP)/day or kkd.

An appropriate fraction collection triggering setup must be put in place in advance of the program to repeatedly inject the sample solution. Vendors provide a variety of mechanisms to perform fraction collection from data, but most commonly used is the threshold UV absorption. The fraction collection valves are programmed by the user to trigger open alternately and in correspondence with peak detection from the repetitive injection "stacked" separation sequence. Various fraction collection modules and types exist on commercial prep SFC systems. All involve the opening of a valve or solenoid upon reaching a preset trigger value allowing outlet flow from the backpressure regulator to direct into one of several flow channels leading to decompression in various chambers or gas-liquid separator channels, then into the desired collection flasks or bottles. This results in one final collection bottle for each chromatographic peak consisting of all dispensings associated with those peaks in a specified retention range and ultimately separate containers of the isolated compounds of interest.

7.5 Applications for Chiral Prep SFC: Successes and Challenges

7.5.1 Toward Pharmaceutical Application of Chiral Preparative SFC

Early examples for scaling up of chiral separations on preparative systems having application in pharmaceuticals were reported in the 1990s by Saito et al. and Oka et al. [23, 24]. Berger et al. described the use of the stacked (or overlapped) injection technique in SFC now essential to process scale separations of chiral pharmaceuticals [25]. With stacked injections, fractions are triggered sequentially and as specified by the operator accounting for peak retention times relative to initial injection. The next injection may be made prior to the elution of peaks from a previous injection. The frequency of injections is set to correspond to a separation time range where all the peaks of interest elute such that there is no unwanted overlapping in the process. This now commonly utilized techique as applied at Eli Lilly to pharmaceutical chiral separations was described by Craig White [18]. An example from this paper is shown in Fig. 7.3. A number of other publications, posters, and presentations have appeared describing separations performed in a similar manner [26–30].

7.5.2 Preparative Application of Chiral Stationary Phases

Just as SFC purification of enantiomers is effective with chiral stationary phases, so is the separation of diastereomers. Achiral separation of R and S diastereomers can be performed by SFC with achiral or chiral stationary phases by achiral mechanisms. One example was given by Larescu et al. using a gradient semi-preparative method with a Chiralpak IA column to separate a proprietary diastereomer using mass-directed fractionation [31]. In another example, the R/S diastereomers of spirostanol saponins was demonstrated on small scale using a long 400 × 4.6 mm total length chiral column with fractionation by ELSD [32].

Compounds with two chiral centers can be separated with a single method through chiral and achiral mechanisms with one mobile phase/stationary phase combination. Using a proprietary diastereomer White et al. demonstrated an analytical separation of four isomers with an SFC in 4 min, whereas a comparable HPLC separation took longer than 30 min [18]. Barnhart et al. utilized the coupling of two chiral columns, Chiralpak AD and Chiralcel OD to separate a mixture of four stereoisomers examining the retention properties of the tandem phases with various mobile phase modifiers including a mixed solvent (1:1 ethanol:isopropanol) [33]. Ali et al. described a large-scale SFC separation to isolate two of four diastereomers in a batch process with a 250 × 50 mm RegisPak™ column in the gradient mode [34].

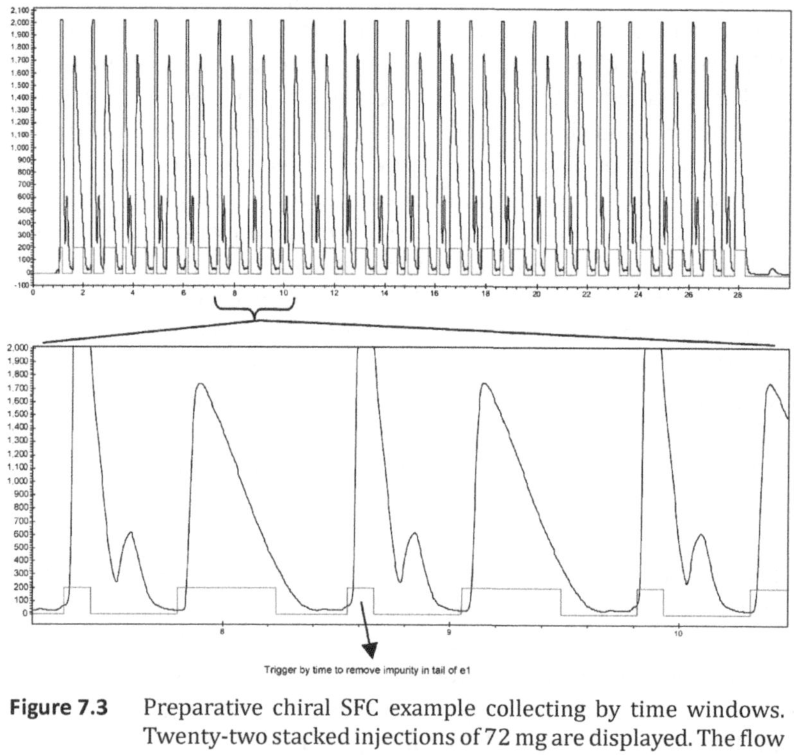

Figure 7.3 Preparative chiral SFC example collecting by time windows. Twenty-two stacked injections of 72 mg are displayed. The flow rate was 70 mL/min using the AD (10 μm) with dimensions of 25 cm × 20.0 mm diameter [18]. Reprinted with permission.

Some of the examples described highlight the advantage of coupling columns to increase chiral selectivity for preparative loading and process speed. A series of three different CSPs (Chiralpak AD, Chiralcel OD, and Chirex 2022, a brush-type phase) were used to provide a general preparative separation system for resolution of several either β-blocker or NSAID drugs. Kot et al. examined method conditions, including modifiers, additives, and temperature and their influence on chiral selectivity of the compounds being purified [35]. Preparative process examples given by Ali et al. showed productivity advantages of 2.5- and 6-fold gained by coupling of two different CSPs in cases where loading issues with single-column methods prevented a linear scale up to desirable throughput rates [36].

Achiral and chiral columns have been coupled for preparative separations of mixtures containing interfering impurities with the racemate of interest. Alexander et al. described a separation of the stereoisomers of cinnamonitrile and hydrocinnamonitrile intermediates using a silica column coupled with an OD-H chiral column [37]. A coupled column screening process to facilitate one-step chiral/achiral preparative separation was described by the author to optimize throughput for isolation of pure enantiomers with stacked injections using simple linkage of appropriate preparative columns [38]. Zeng et al. designed a sophisticated 2D SFC/SFC/MS purification system with modifications to a mass-directed prep LC and custom software for control [39]. Using mass-directed fractionation the racemate peak is isolated in the first dimension using a 2-EP achiral column, and then eluted from a trapping column with the mobile phase for the second dimension onto a chiral column for enantiomer isolation. The diagram depicting the setup of this system is shown in Fig. 7.4.

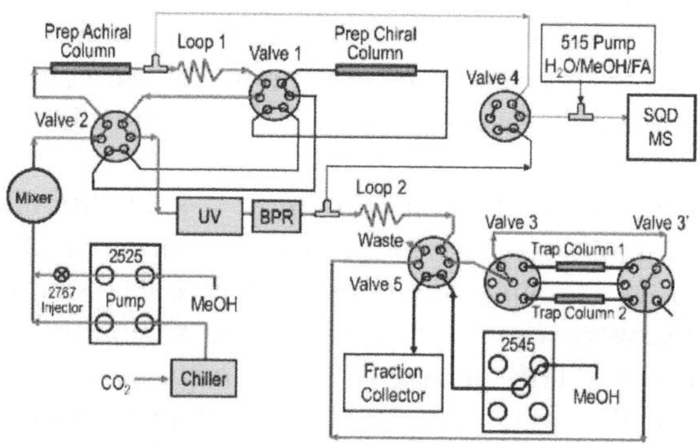

Figure 7.4 System configuration for preparative 2D SFC/SFC/MS including four-valve interface (valves 1, 2, 4, and 5) with the positions of 1A, 2A, 4A, and 5A to carry out the preparative achiral chromatography in the first dimension whilst the fraction collection valve (valve 3 and 3') is in the by-pass position [39]. Reprinted with permission.

7.5.3 Solubility and Mobile Phase Issues in Prep SFC

On the preparative scale, solubility of a sample solution in the mobile phase is a critical factor to the outcome of a chiral separation. Performance can be affected because the solubility in a CO_2-enriched mobile phase will be too low at a certain solute concentration. The solute may precipitate upon injection and entry into the column, with the mobile phase. This leads to peak distortion, or worse the precipitate may cause restricted flow and raise the system pressure sometimes leading to system shutdown. This condition may be avoided using injection techniques favorable to maintain solubility, one of which known as modifier stream injection™. In this convenient technique, the sample is injected into the modifier flow stream prior to mixing with CO_2 at the entry to the column. Miller et al. demonstrated that resolution was increased for a number of chiral standards using modifier stream injection relative to mixed stream [40]. The reduction in resolution with mixed stream injection can be explained by the temporary state of high mobile phase polarity entering the column in which some fraction of the injected sample is carried forward in this stronger solvent ahead of the fraction in equilibrium between mobile phase and stationary phase.

A method to determine solubility prior to large-scale purification at subcritical mobile phase conditions common in preparative separations was developed by Gahm et al. [41]. Using solute mixed with dry silica packed in a cartridge, solubility in CO_2 plus several modifiers could be compared analytically prior to purification. The packed cartridge is pressurized with mobile phase, equilibrated and sampled through the brief opening of a valve into an SFC system with UV detection.

One way to avoid the condition of poor solubility of racemates in the SFC mobile phase in preparative chromatography is to utilize higher solubility modifiers. Many polar samples typical in drug discovery chemistry have higher solubilities in DCM or tetrahydrofuran (THF) mixed with methanol as opposed to pure methanol. At the same time, DCM and THF are incompatible with coated polysaccharide phases commonly used for chiral separations. Miller et al. showed that utilizing combinations of DCM and/or THF mixed with methanol plus the immobilized analog of the amylose or cellulose column could, improve selectivity and performance with the increased solubility properties of these solvents [42] (see Fig. 7.5).

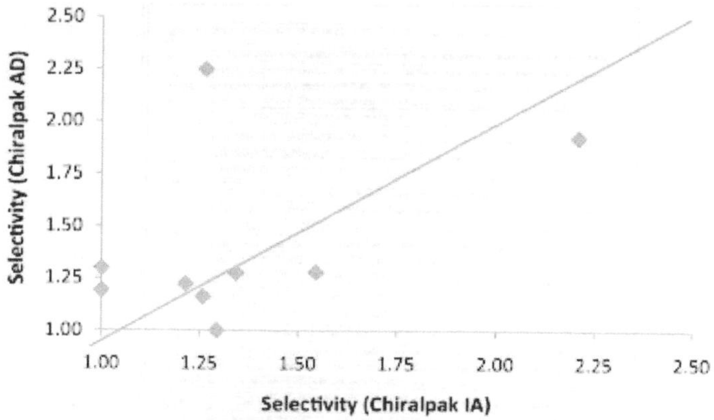

Figure 7.5 Selectivity obtained with Chiralpak IA (50/50/0.2 methanol/dichloromethane/DEA modifier) vs. selectivity obtained with Chiralpak AD (methanol w/0.2% DEA modifier) [42]. Reprinted with permission.

Another modification to the mobile phase modifier is to reduce its polarity to enhance selectivity especially for nonpolar chiral compounds. For preparative application, a reduction in polarity is not necessarily as simple as lowering the modifier composition to arbitrarily low percentages. At low modifier composition, the solute being collected can be lost in the aerosol after depressurization and is thus not recovered with the liquid phase fraction, leading to poor recovery. In such cases, the polar modifier solvent can be mixed with a substantial percentage of nonpolar solvent such as hexane or heptane. Since these are not miscible with methanol, they can be used, however, in combination with ethanol or isopropanol. Lefevbre et al. described the use of hexane for this effect showing a preparative chiral separation using 1:1 hexane:isopropanol [43]. Also described was an example of increased resolution with the hexane-diluted isopropanol relative to the same percentage of neat isopropanol. Kraml et al. described a preparative process separation utilizing 9:1 heptane:isopropanol at 20% in CO_2 to achieve the low co-solvent polarity to achieve selectivity required for the chiral compound being purified [44]. An example of the rapid repetitive injection cycle preparative method is shown in the UV chromatographic trace of Fig. 7.6.

Further mobile phase modifications in the form of volatile basic additives for preparative SFC application have been described

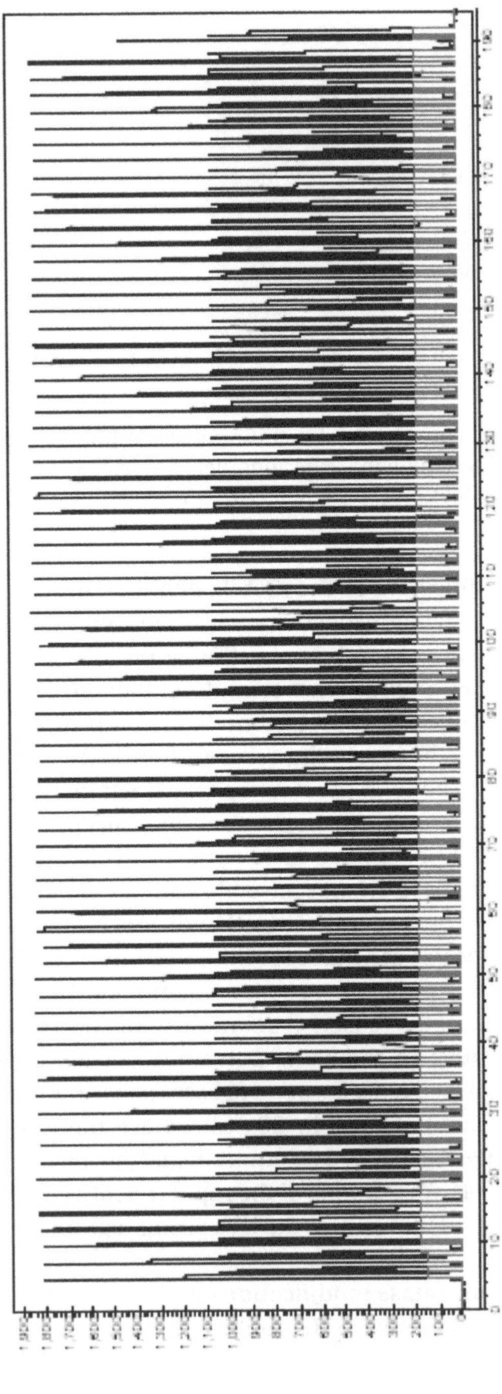

Figure 7.6 OD-H, 20% 9:1 heptane:isopropanol/CO_2, 100 bar, 60 mL/min [44]. Reprinted with permission.

recently. Ammonia or ammonium hydroxide can be used as additives for improving peak shape and retention for basic analytes to the same general effect as other less volatile additives [45, 46]. Their advantage lies in their rapid removal from the final collected purification fractions in the evaporation step as well as being less reactive with the desired solute present during concentration.

7.5.4 Recycling Chiral SFC Separation Processes

For large-scale process chiral purifications, simulated moving bed (SMB) chromatography is used and is almost always a liquid phase technique. SMB is a continuous binary process that simulates the counter current movement of the stationary phase by shifting of the inlet and outlet ports (see references for more detailed descriptions) [47–49]. Such processes increase efficiency and productivity for low selectivity and large-scale separations.

SMB with supercritical fluid (SF SMB) has been explored initially by Mazzotti et al. to utilize some unique advantages of supercritical fluids [50]. They determined theoretically that operating with a pressure gradient instead of constant or isocratic pressure throughout the system can provide a more optimal separation. Using an eight-column SF SMB unit, they demonstrated a separation of two fatty ethyl esters achieving 99.9%/100% purity in the optimized pressure gradient mode vs. 97.8%/97.7% in the isocratic mode. Today, simulations are often used to optimize the process conditions for a large-scale chiral SF SMB separation. Readers are referred to a Rajendran's description of the SF SMB process with examples utilizing triangle theory optimization [16].

More recently a related technique using supercritical fluids in a semi-continuous mode has been described and named simulated moving columns (SMC) by Zhang et al. [51]. Using switching valves, the outlet of the first column feeds into a second in series after which valves at either column outlet are switched so that the first column will again receive the flow and continue separation of the enantiomers to be resolved. A UV detector is placed at the outlet of both columns. The system, represented in Fig. 7.7, allowed a separation to be achieved in as many stages as required for sufficient resolution with only the pressure drop due to two columns at any time but with the number of theoretical plates as needed for resolution.

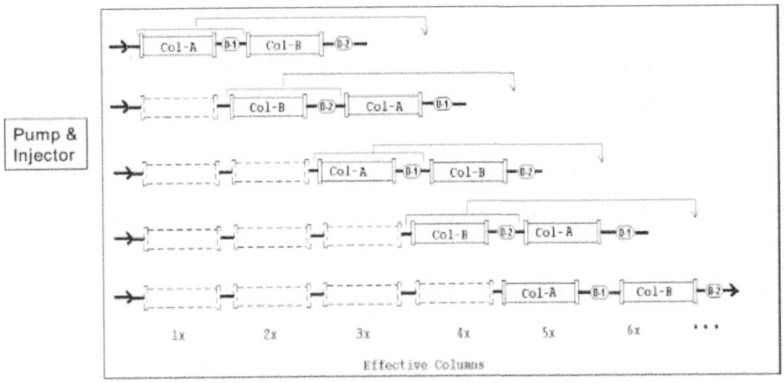

Figure 7.7 SMC simulates the movement of the columns at the direction of the mobile phase to keep the enantiomers recycling in the columns. In essence, a virtual longer column is created [51]. Reprinted with permission.

This implementation was demonstrated with 4.6 mm ID columns, but its application for preparative processes has not been published. While many of the results described above relating to continuous supercritical fluid processing seem promising, recent examples of application in industry for chiral separations were not found in the literature likely due to the cost, complexity and lack of commercial instrumentation available.

7.6 Conclusion

Chiral purification of pharmaceuticals is now the most widespread industrial application of SFC. Because of its well-known success, it has become the standard technique for semi-preparative chiral purifications in the industry and is more heavily used than chiral HPLC. Many companies in pharma or other industries have seen the long-term value of SFC despite the associated startup costs. As a green technology, less solvent is consumed and less waste is produced. If environmental concerns are not sufficient to motivate the investment in SFC technology, lower ongoing costs should ultimately persuade an operation to buy in. As vendors of SFC instrumentation and columns become more prevalent and equipment in the field more common, further improvements demanded by users will enhance the quality and capabilities of the products available.

Acknowledgment

The author would like to thank Larry Miller for helpful and timely suggestions in the preparation of this chapter.

References

1. Wellings, D. A. (2006) *A Practical Handbook of Preparative HPLC*, 1st Ed., Elsevier Science, New York.
2. Dalgliesh, C. E. (1952) The optical resolution of aromatic amino-acids on paper chromatograms. *J. Chem. Soc.*, **137**, 3940–3942.
3. Welch, C. (1994) J. Evolution of chiral stationary-phase design in the Pirkle laboratories. *J. Chromatogr. A*, **666** , 3–26.
4. Okamoto, Y., Kawashima, M., Yamamoto, K., and Hatada, K. (1984) Useful chiral packing materials for high-performance liquid chromatographic resolution cellulose triacetate and tribenzoate coated on macroporous silica gel. *Chem. Lett.*, **13**, 739–742.
5. Armstrong, D. W., and DeMond, W. (1984) Cyclodextrin bonded phases for the liquid chromatographic separation of optical, geometrical and structural isomers. *J. Chromatogr. Sci.*, **22**, 411–415.
6. Szczerba, T. J., and Berger, T.A. (2011) Going Faster, Going Greener with High Efficiency Regis Chiral Stationary Phases in Supercritical Fluid Chiral Chromatography. *LC-GC, The Application Notebook*.
7. Perrut, M. (1984) Fractionation process for mixtures by elution chromatography with liquid in supercritical state and installation for its operation. U.S. Patent No. 4,478,720.
8. Berger, T. A. (1995) *Packed Column SFC*, The Royal Society of Chemistry, Cambridge, UK.
9. Berger, T. A. (1991) Density of methanol-carbon dioxide mixtures at three temperatures: Comparison with vapor-liquid equilibria measurements and results obtained from chromatography. *J. High Resolut. Chromatogr.*, **14**, 312–316.
10. Berger, T. A., and Deye, J. F. (1990) Composition and density effects using methanol/carbon dioxide in packed column supercritical fluid chromatography. *Anal. Chem.*, **62**, 1181–1185.
11. West, K. N., Wheeler, C., McCarney, J. P., Griffith, K. N., Bush, D., Liotta, C. L., Eckert, C. A. (2001) In situ formation of alkylcarbonic acids with CO_2. *J. Phys. Chem. A*, **105**, 3947–3948.

12. Wen D., and Olesik, S. V. (2000) Characterization of pH in liquid mixtures of methanol/H_2O/CO_2. *Anal. Chem.*, **72**, 475–480.

13. Ye, Y. K., Lynam, K. G., and Stringham, R. W. (2004) Effect of amine mobile phase additives on chiral subcritical fluid chromatography using polysaccharide stationary phases. *J. Chromatogr. A*, **1041**, 211–217.

14. Gyllenhaal, O., Edstrom, L., and Persson, B-A. (2006) Ion-pair supercritical fluid chromatography of metoprolol and related amino alcohols on diol silica. *J. Chromatogr. A*, **1134**. 305–310.

15. Berger, T. A., and Fogelman, K. D. Method of sample introduction for supercritical fluid chromatography systems, US Patent 6, 576,125.

16. Rajendran, A. (2012) *Design of preparative-supercritical fluid chromatograph. J. Chromatogr. A*, **1250**, 227–249.

17. Hamman, C., Wong, M., Hayes, M., and Gibbons, P. (2011) A high throughput approach to purifying chiral molecules using 3 μm analytical chiral stationary phases via supercritical fluid chromatography. *J. Chromatogr. A*, **1218**, 3529–3536.

18. White, C., (2005) Integration of supercritical fluid chromatography into drug discovery as a routine support tool. *J. Chromatogr. A*, **1074,** 163–173.

19. Huber, U., and Majors, R. E. Principles in Preparative HPLC, Agilent publication no. 5989–6639EN.

20. Ghanem, A., Hoenen, H., and Aboul-Enein H.Y. (2006) Application and comparison of immobilized and coated amylose tris-(3,5-dimethylphenylcarbamate) chiral stationary phases for the enantioselective separation of beta-blockers enantiomers by liquid chromatography. *Talanta*, **68**, 602–609.

21. Zhang, T., Nguyen, D., and Franco, P.(2008) Enantiomer resolution screening strategy using multiple immobilised polysaccharide-based chiral stationary phases. *J. Chromatogr. A*, **1191**, 214–222.

22. Cox, G. B. (2009) Preparative chiral separations: from laboratory scale to production. *Chromatogr. Today*, **4**(2), 4–7.

23. Saito, M., Yamauchi, Y., Higashidate, S., and Okamoto, I. (1995) Preparative supercritical fluid chromatography and its applications to chiral separations. *Int. Symp. Chromatogr. 35th Anniv. Res. Group Liq. Chromatogr. Jpn.*, World Scientific, Hatano, H., Hanai, T. (Eds.), Singapore.

24. Oka, H., Shoda, S., Kato, K., Watanabe, T., Nakazawa, T. Hiroyuki, T., Kawasaki, T., and Ikeda, M. (1999) *Jpn. Chromatogr.*, **20**, 310.

25. Berger,T.A.,Smith,J.,Fogelman K.,and Kruluts,K. (2002) Semipreparative chiral separations using supercritical fluid chromatography with stacked injections, *Am. Lab.*, **34**, 14–20.

26. P. Jusforgues, P., and Shaimi, M. (1998) Preparative supercritical fluid chromatography. *Analusis*, **26**, 55–60.

27. Welch, C. J. (2005) Preparative Chiral SFC as a Green Technology for Rapid Access to Enantiopurity in Pharmaceutical Process Research, *LC/GC Europe*, **18**, 264–272.

28. Leonard, W. R. Jr, Henderson, D. W., Miller, R. A., Spencer, G. A., Sudah, O. S., Biba, M., Welch, C. J. (2007) Strategic use of preparative chiral chromatography for the synthesis of a preclinical pharmaceutical candidate. *Chirality*, **19**, 693–700.

29. Welch, C. J. (2007) Chiral chromatography in support of pharmaceutical process research, in *Preparative Enantioselective Chromatography* (G. B. Cox., ed.), Blackwell Publishing Ltd, Oxford, UK.

30. Wu, D-R., Leith, L., Li, P., and Sun, D. (2009) *Preparative Chiral SFC in Drug Discovery: Promises and challenges*, 3rd International Conference on Packed Column SFC, Philadelphia.

31. Larescu, V., Mulvihill, M.J., and Ma, L. (2011) Achiral SFC Purification of Complex Pharmaceutical Mixtures, *LC-GC North America*, **29**, 438–444.

32. McCauley, J., Kang, L., Yu, H., Ma, B., and Chen, R. (2012) Analysis and Purification of Diastereomeric Steroid Saponins from the Trigonella Foenum-Graecum Seed Extract Using SFC/ELSD. Waters App. Note #134674023.

33. Barnhart, W. W., Gahm, K. H., Thomas, S., Notari, S., Semin, D., and Cheetham, J. (2005) Supercritical fluid chromatography tandem-column method development in pharmaceutical sciences for a mixture of four stereoisomers. *J. Sep Sci.*, **28**, 619–626.

34. Ali, Z. (2009) *Challenges and Solutions to Scaling Up with SFC*, poster, 3rd International Conference on Packed Column SFC, Philadelphia.

35. Kot, A., Sandra, P., and Venema, A. (1994) Sub- and Supercritical Fluid Chromatography on Packed Columns: A Versatile Tool for the Enantioselective Separation of Basic and Acidic Drugs. *J. Chromatogr. Sci.*, **32**, 439–448.

36. Ali, Z., Kocergin, J., Edwin, V. (2009) Multicolumn Preparative SFC: An Advanced Solution to Scale-up Difficulties. *LC-GC The Peak*, 16–21.

37. Alexander, A. J., and Staab, A. (2006) Use of achiral/chiral SFC/MS for the profiling of isomeric cinnamonitrile/hydroxycinnamonitrile products in chiral drug synthesis. *Anal. Chem.*, **78**, 3835–3838.

38. Ventura, M. C. (2012) Evaluation of a Systematic Approach to Coupling of Achiral and Chiral Columns for SFC Purification. 6th International Conference on Packed Column SFC, Brussels.

39. Zeng, L., Xu, R., Zhang, Y., and Kassel, D. B. (2011) Two-dimensional supercritical fluid chromatography/mass spectrometry for the enantiomeric analysis and purification of pharmaceutical samples. *J. Chromatogr. A*, **1218**, 3080–3088.

40. Miller, L., and Sebastian, I. (2012) Evaluation of injection conditions for preparative supercritical fluid chromatography. *J. Chromatogr. A*, **1250**, 256–263.

41. Gahm, K. H., Tan, H., Liu J., Barnhart, W., Eschelbach, J., Notari, S., Thomas, S., Semin, D., Cheetham, J. (2008) Purification method development for chiral separation in supercritical fluid chromatography with the solubilities in supercritical fluid chromatographic mobile phases. *J. Pharm. Biomed. Anal.* , **46**, 831–838.

42. Miller, L. (2012) Evaluation of non-traditional modifiers for analytical and preparative enantioseparations using supercritical fluid chromatography. *J. Chromatogr. A*, **1256**, 261–266.

43. Lefebvre, P. M., Kiplinger, J. P., and Rego, M. J. (2011) Using Hexane Mixtures as Modifiers in SFC to Enhance Chiral Resolution of Racemic Pharmaceutical Compounds. poster, Pittcon, Atlanta.

44. Kraml, C. (2009) SFC Applications to Research in Chemistry and Molecular Biology at Princeton University. 3rd International Conference on Packed Column SFC, Philadelphia.

45. Ventura, M., Murphy, B., and Goetzinger, W. (2012) Ammonia as a preferred additive in chiral and achiral applications of supercritical fluid chromatography for small, drug-like molecules. *J. Chromatogr. A*, **1220**, 147–155.

46. Hamman, C., Schmidt, D. E. Jr, Wong, M., and Hayes, M. (2011) The use of ammonium hydroxide as an additive in supercritical fluid chromatography for achiral and chiral separations and purifications of small, basic medicinal molecules. *J. Chromatogr. A*, **1218**, 7886–7894.

47. Abel, S., and Juza, M. (2007) Less common applications of enantioselective HPLC using the SMB technology in the pharmaceutical industry. In *Chiral Separation Techniques, A Practical Approach* (G. Subramanian, ed.), 3rd ed., Wiley-VCH, 203–273.

48. Sa Gomes, P., Minceva, M., Pais, L. S., and Rodrigues, A. E. (2007) Advances in simulated moving bed chromatographic separation. In *Chiral Separation Techniques, A Practical Approach* (G. Subramanian, ed.), 3rd ed., Wiley-VCH, 181–202.

49. Juza, M., Mazzotti, M., and Morbidelli, M. (2000) Simulated moving-bed chromatography and its appliation to chirotechnology. *TIBTECH*, **18**, 108–118.

50. Mazzotti, M., Storti G., and Morbidelli, M.(1997) Optimal operation of simulated moving bed units under nonlinear chromatographic separation. *J. Chromatogr. A*, **769**, 3–24.

51. Zhang, Y., McConnell, O. (2004) Simulated moving columns technique for chiral liquid chromatography. *J. Chromatogr. A*, **1028,** 227–238.

Chapter 8

SFC in Process Analytical Chemistry

Yanqun Zhao
Process Analytical Chemistry, Research and Development,
Abbvie Inc., 1 North Waukegan Road, North Chicago, IL 60046, USA
Yanqun.zhao@abbvie.com

8.1 Introduction

Pharmaceutical industries have been facing an increasingly challenging environment after the booming started in the 1980s. Tougher regulatory environment, increased cost in research and development, global competition, and higher safety standards for compounds from agencies are just some of the factors contributing to the pressures felt by the pharmaceutical industry. To maintain the competitive edge, pharmaceutical companies have to adapt to the challenges and evolve themselves. One strategy is to maximizing the return on investment by reducing the cost of drug development from all aspects of the drug development process.

Process chemistry serves as the bridge between drug discovery and commercial viability. Once the compound is discovered and

Supercritical Fluid Chromatography: Advances and Applications in Pharmaceutical Analysis
Edited by Gregory K. Webster

ISBN 978-981-4463-00-3 (Hardcover), 978-981-4463-01-0 (eBook)
www.panstanford.com

identified as a new molecular entity (NME) or a new chemical entity (NCE), process chemistry is responsible to develop the scale-up and commercially viable route so that larger quantities of the developmental drug, or active pharmaceutical ingredient (API), ranging from multiple grams to hundreds of kilograms, can be made to provide supplies for toxicity studies and multiple phases of clinical trials. Process analytical chemistry supports this development process by providing process understanding and process control via method development, sample analysis, impurity identification and profiling, as well as identifying critical attributes of the process. The involvement of process analytical chemistry can start from testing and quality control of raw materials, starting materials, intermediates, in-process testing to API, and technology transfer to commercial sites. A vast variety of analytical techniques are needed to fulfill the need of the process development. Supercritical fluid chromatography (SFC) is one of these analytical techniques that process analytical chemistry utilizes to improve efficiency and reduce solvent consumptions in order to contribute to the overall goal of staying competitive and increasing the return on investment.

In general, SFC has been gaining popularity in the pharmaceutical industry for a variety of applications, from the purification of small quantities of impurities to the kilograms of intermediates, from analytical SFC method development of chiral purity determinations to chiral compound isolations. The increased awareness of the advantages and capabilities of SFC, as well as the improvement of the commercially available SFC systems, from Berger SFC™ MiniGram system to the more recent Aurora Fusion™ A5 analytical system, have helped the spread of the technology.

In this chapter, applications of SFC in process analytical chemistry are summarized. Process analytical chemistry serves as the analytical front line in the process and quality control of the API. The uniqueness of process analytical chemistry lies in the versatility of the demands to satisfy the process development needs. This uniqueness can also be reflected in the way that the SFC is used in the process analytical chemistry. Discussions of the applications will be divided into different sections.

8.2 Chiral Purity Analysis and Method Development

8.2.1 Introduction

It is well known that enantiomers can exhibit completely different physiological and biological activities, as well as pharmacodynamic and pharmacokinetic characteristics. Therefore, it is very critical to control the enantiomeric purity of drug substances. The U.S. Food and Drug Administration has been issuing guidelines for drug development requiring the analysis and control of the enantiomeric purity of drug substances [1]. During the development of chiral API, analytical methods are needed for the determination of chiral purity. Owing to the increase of the number of compounds in the earlier phase of the drug development, speed and efficiency in method development have become increasingly important.

Supercritical fluid chromatography has been used for chiral separations and has gained popularity over the years, especially for high-throughput chiral purifications and fast column screening for method development [2–8]. Advantages gained from SFC are the result of the properties of supercritical carbon dioxide such as low density and viscosity, high diffusivity, low cost, and easier solvent removal. Faster analysis, better separation efficiency, less solvent consumption, and longer column life are the typical characteristics of chiral separation with SFC. These characteristics make SFC a superior choice when it comes to chiral analysis.

8.2.2 Chiral Purity Analysis by SFC in Process Analytical Chemistry

Chiral purity analysis has been conducted using SFC in process analytical chemistry for all types of materials, from raw materials, starting materials, intermediates to in-process tests for the manufacture of intermediates and APIs.

The analysis of tartaric acid by SFC served as one of the example of raw material analysis [9]. The chiral purity analysis is needed as the incoming material test. The initial literature search returned little information on the chiral separation on tartaric acid. Method development was initiated using SFC by employing a Chiralpak AD-H column as the chiral stationary phase (CSP). As shown in Fig. 8.1, baseline resolution is achieved in less than 3 min with 20% modifier. The modifier content is ethanol containing 0.1% (v/v) trifluoroacetic acid (TFA) as additive.

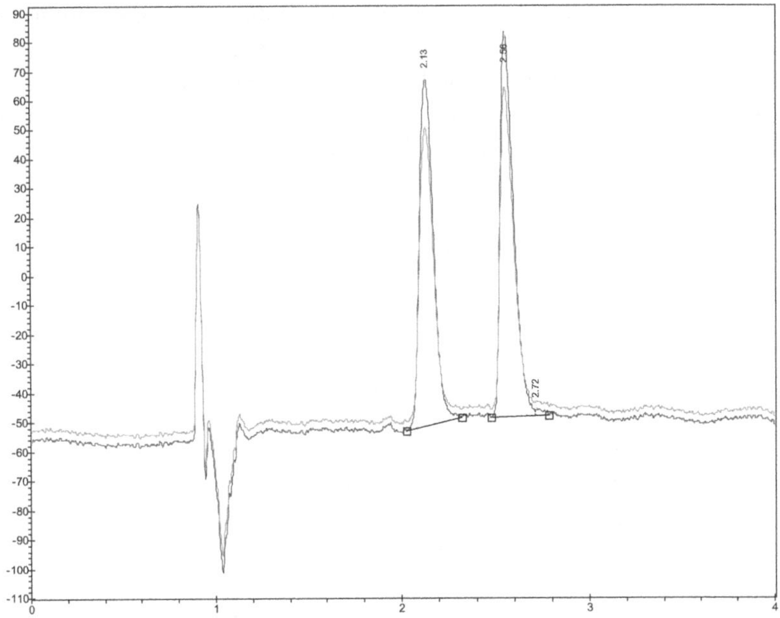

Figure 8.1 Chromatogram of tartaric acid chiral separation by SFC with Chiralpak AD column. Baseline separation is achieved in less than 3 min. The overlaid chromatograms are from two different UV detection wavelengths.

Proline and proline derivatives are widely used as the building block in pharmaceutical synthesis [10–14]. Therefore, these compounds can be considered starting materials or intermediates based on the synthetic route of the specific pharmaceutical compound. The analysis of a series of proline derivatives by SFC was reported [15]. Structures of these compounds are presented in Table 8.1. Baseline separation of these compounds was achieved using Chiralpak AD-H column with ethanol containing 0.1% TFA as modifier. The effect of the modifier content was also studied; the results are presented in Table 8.2.

Table 8.1 Structures of proline derivatives

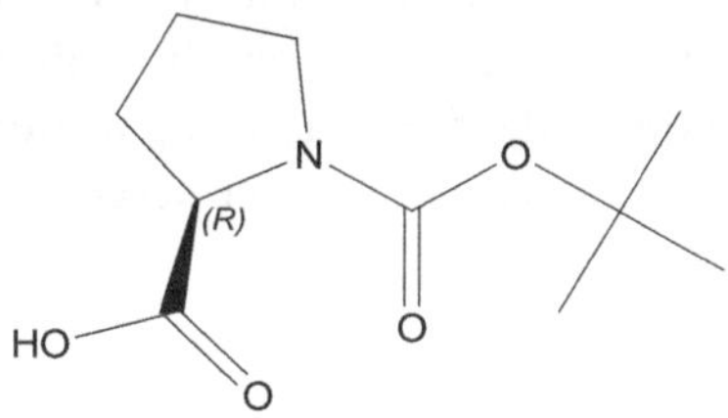 (R)-Boc-proline or D-Boc-proline	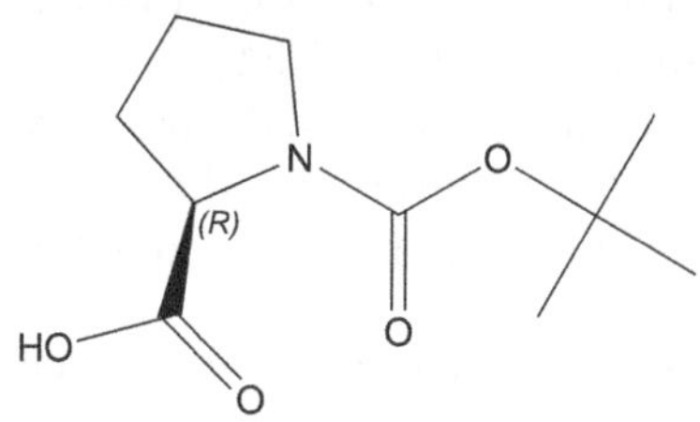 (R)-Boc-2-methyl-proline (boc acid)
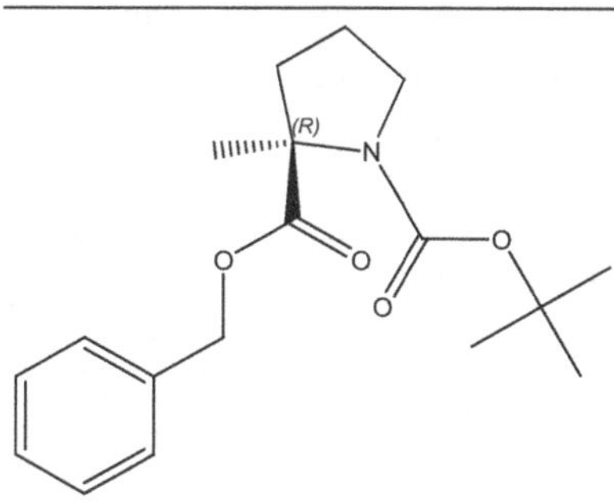 (R)-Boc-2-methylproline benzyl ester (boc ester)	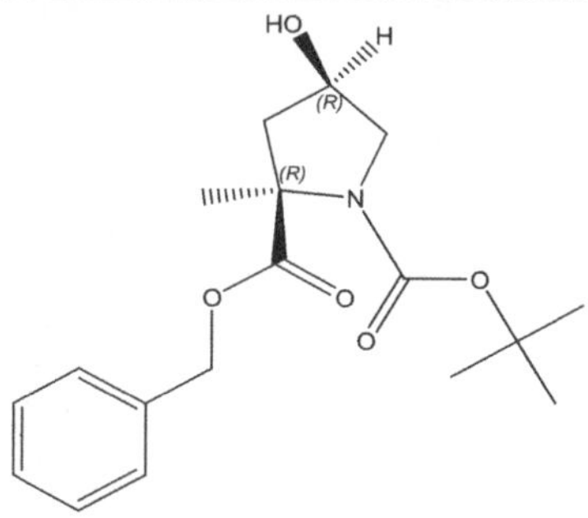 (2R, 4R)-Boc-2-methyl-4-hydroxy-proline benzyl ester (boc hydroxy ester)

8.2.3 Method Development Using SFC Screening

For quick chiral method development, SFC has been utilized for fast screening of CSPs, modifiers, and additives [5–7,16–18]. Commercially available analytical SFC systems, such as the Berger analytical system, offer the standard option of six-position column switchers and six-port valve for modifiers, which make the column and modifier screening readily available and very convenient to use.

The earlier automated SFC screening work reported by Villeneuve and Anderegg [16] in 1998 was started prior to the column selection valve becoming a commercially available standard option for SFC. They installed a six-port column selection valve and a four modifier selection valve. This system was routinely used in their laboratories for the separation request of a wide variety of chemical classes. Four columns with the highest success rate (Chiralpak AD, Chiralcel OD, Chiralpak AS, and Chiralcel OJ) were employed with four common modifiers (methanol, ethanol, isopropyl alcohol, and methanol containing 0.1% TFA) under isocratic conditions as the initial screening. Additional work was needed only if adequate separation was not achieved from the initial screening. The automated screening system was demonstrated to be very effective.

Table 8.2 Effect of modifier composition

Compound	Modifier content (%)	k'_1	k'_2	α	Rs	Elution
Boc proline	5	3.60	3.98	1.10	1.46	S < R
	7	2.06	2.16	1.05	0.58	S < R
	10	1.09	1.09	1.00	0	S = R
Boc acid	5	3.18	4.71	1.48	5.13	S < R
	7	1.82	2.55	1.40	3.70	S < R
	10	0.96	1.28	1.33	2.49	S < R
Boc ester	5	1.50	1.97	1.31	2.88	R < S
	7	1.04	1.33	1.27	2.57	R < S
	10	0.68	0.83	1.22	1.72	R < S
Boc hydroxy ester (a) 2R,4R and 2R,4S	5	6.47	8.64	1.34	5.55	2R,4R < 2R,4S
	7	3.81	4.58	1.20	3.10	2R,4R < 2R,4S
	10	2.04	2.23	1.09	1.21	2R,4R < 2R,4S
Boc hydroxy ester (b) 2R,4S and 2S,4R	5	8.64	19.0	2.20	11.5	2R,4S < 2S,4R
	7	4.58	9.67	2.11	8.52	2R,4S < 2S,4R
	10	2.23	4.44	1.99	6.82	2R,4S < 2S,4R

Screening using SFC [5,7,17] was also carried out using gradient conditions. The gradient condition helps to eliminate trial and error work to avoid unnecessary runs as a result of inappropriate modifier strength. When the modifier strength is too high, it can cause co-elution, while too low, the compound may not elute off the column.

Welch [7] and colleagues have modified the typical Berger analytical system to include an additional valve to automatically screen 10 different single chiral columns and 25 different combination of the 10 columns in tandem for more complex and difficult resolutions. An example of a mixture of four stereoisomers was given, where partial resolution (three peaks) was observed on a single Chiralpak AD-H or Chiralpak IA column. When the two columns were used in tandem, all four stereroisomers were baseline resolved using both isocratic and gradient conditions. The effectiveness of the tandem column approach is also demonstrated in the separation of the achiral starting material and racemic mixture. When a single column was used, the starting material was co-eluting with one of the enantiomers on several columns. Separation of the achiral starting material and both enantiomers was achieved when Chiralcel OJ-H and Chiralcel OF were deployed in tandem configuration. When two or more different types of columns are placed in series to improve the separation efficiency, the order of the columns can make a difference in overall separation as a result of the change of pressure at the head of the each individual column [19]. As clearly demonstrated in Fig. 8.2, the resolution and elution order were different when the order of the two tandem columns was exchanged.

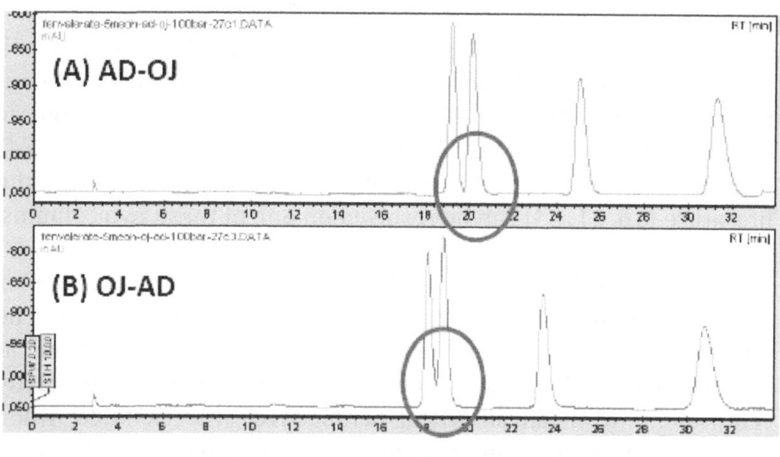

Figure 8.2 Chromatograms showing the effect on resolution and retention when the order of two tandem column was changed [19].

The parallel column screening system [18,20] increases the overall efficiency since the available CSPs are screened simultaneously. An automated parallel column screening system reported by Zeng et al. [20] utilized four columns for the parallel screening, yet each single column can be used for optimization. A software program was also developed in-house to facilitate the automated parallel screening followed with the single optimization process to streamline the chiral method development process to meet the needs of high-throughput environment.

The column screening platforms were also used for column performance evaluations, including the evaluation of the new and generic polysaccharide phases from new manufacturers that emerged onto the market in recent years after the patent expiration of the original polysaccharide phases [21] and the new 3 μm particle size chiral columns [22].

Chiral method development via a screening approach is utilized in process analytical chemistry for quick method development. Typically, a sequential column screening system, rather than a parallel screening, is often adequate for the intended use. Comparison studies of isocratic versus gradient screening conditions have been evaluated and reported [23, 24].

8.2.4 Column Selection

Among all the chiral columns that are available for SFC, the chiralpak AD column has been demonstrated to be the most effective column for separations [21, 24]. The chiralcel OD column or the generic version of this column could be classified as the second best [21, 24] column. Overall, more than 80% of the compounds reported in the literature [16, 21, 24] can be resolved on one of these two columns. Both of these columns are polysaccharide-based CSPs.

In recent years, many vendors have provided equivalent or generic columns to that of the chiralpak AD and chiralcel OD. The equivalency of some these columns have been evaluated and it can be concluded that these generic columns may not be exactly "equivalent" from an analytical perspective. For example, when the same chromatographic conditions were applied for the same analysis, same resolution might not be achieved on chiralcel OD-H column versus on the RegisCell column [21, 24]. However, this does not rule out the possibility that the same resolution can still be achieved on this equivalent column. It will require modified chromatographic conditions such as different modifier composition. Unique separations can also be found on the generic column, but not on the original column, and vice versa [21, 24].

8.2.5 Modifier and Modifier Composition

The most common modifiers are methanol, ethanol, isopropanol, and acetonitrile for the chiral analysis of typical pharmaceutical compounds. The selections of modifiers are important since different modifiers can provide different selectivity. A change from no resolution, to partial resolution and baseline resolution as a result of modifier change has been reported in many cases [15].

Modifier composition is another factor that can have a significant impact on resolution. For boc proline (see the structure in Table 8.1), near baseline resolution was achieved with 5% modifier for the two enantiomers while co-elution occurred with 10% modifier, as shown in Fig. 8.3 [15]. See Table 8.2 [15] for detailed results.

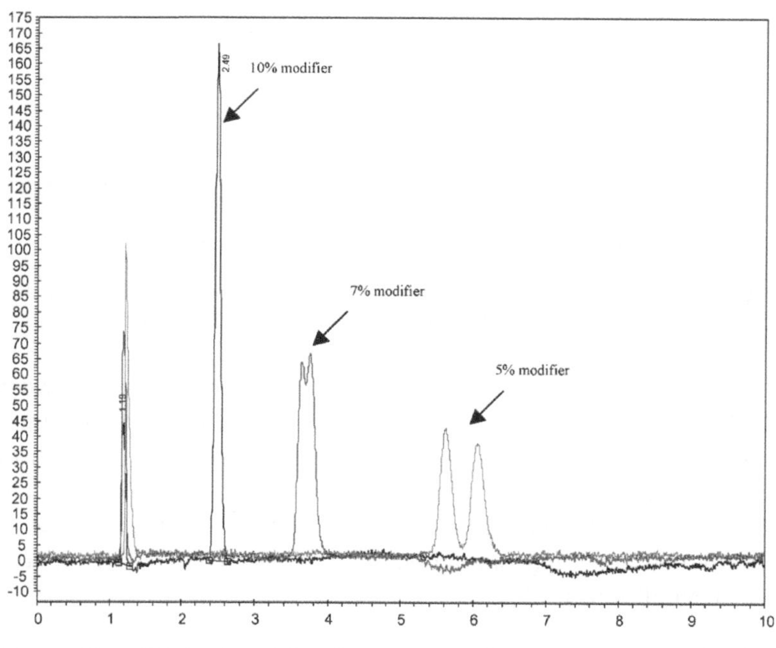

Figure 8.3 Effect of mobile phase composition on resolution for boc proline by SFC using ethanol containing 0.1% TFA as modifier.

The extent of the modifier impact is compound specific. When the modifier composition was changed from 5% to 10% for boc acid, the resolution was changed from 5.13 to 2.49. Even though the resolution was decreased with the increase of modifier content, baseline resolution is still achieved in a shorter run time with sharper peak shape. See Fig. 8.4 for the illustration of the chromatograms.

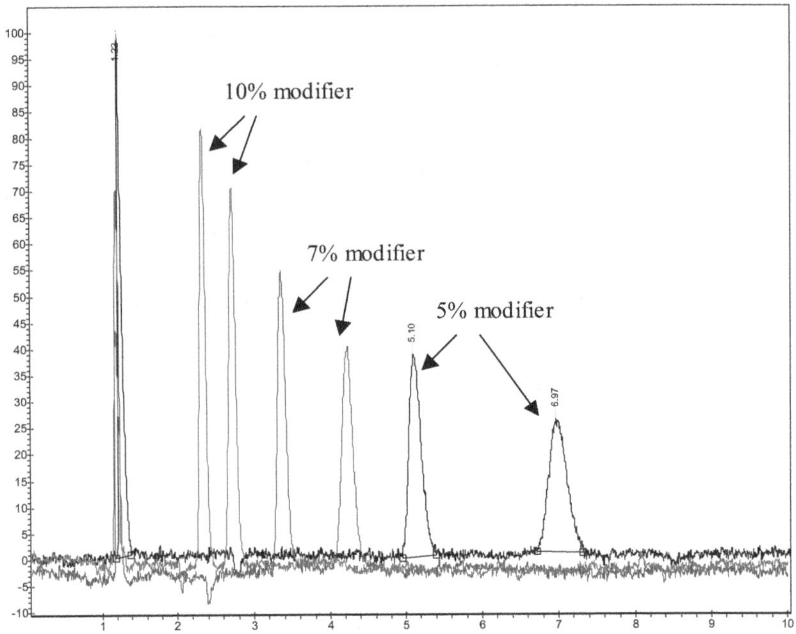

Figure 8.4 Effect of mobile phase composition on resolution for boc acid by SFC using ethanol containing 0.1% TFA as modifier.

The impact of modifier composition was compared between HPLC and SFC for the same compounds using the same columns and mobile phase system. For boc acid, the impact of mobile phase/modifier composition on resolution is less critical using SFC (see Fig. 8.4) when compared with HPLC. As shown in Fig. 8.5, a change of 1% of modifier composition resulted in the loss of resolution under HPLC conditions [25].

8.2.6 Effect of Additive

The effect of additives can be significant in chiral separation, not only for HPLC but also for SFC. This impact can be reflected on the resolution, peak shape, elution time, and even elution order. Additive effect is compound specific [24]. It has been observed that an acidic additive is often not as critical for SFC as that for HPLC since the acidity of the CO_2 may have provided the similar benefit as that of the acidic modifier.

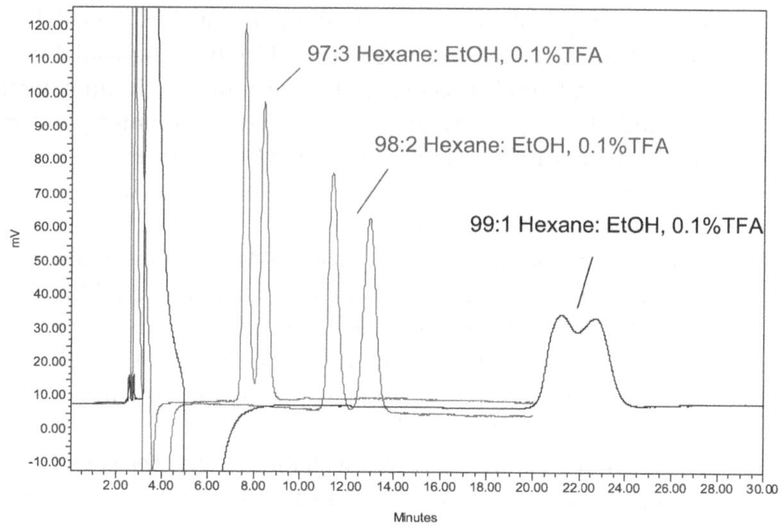

Figure 8.5 Effect of mobile phase composition on resolution for boc acid under HPLC conditions. The resolution is more susceptible to the change of the ethanol content when compared with the SFC.

Typical additives used for SFC separation include trifluroacetic acid, methanesulfonic acid, triethylamine, isopropylamine, diethylamine, and dimethyl ethylamine. Ammonium acetate, methylsulfonic acid, and ethylsulfonic acid [19] were reported to be effective additives for SFC/MS. Hamman and colleagues have reported the use of ammonium hydroxide as an additive for the achiral and chiral separation of small, basic medicinal molecules [26]. Most recently, Ventura and colleagues [27] have reported the use of ammonia as the preferred additive for chiral and achiral separations by SFC for small molecules. The advantages of using ammonium hydroxide and ammonia as additives are the reduced residue in the isolated compound when using SFC for compound isolation and purification.

8.2.7 Method Transfer between SFC and HPLC

Even though SFC is the preferred technique when it comes chiral separation, there is often a need to convert the SFC conditions to HPLC for the following considerations:

1. Instrument availability: Process analytical chemistry often provides analysis as in-process support where time is crucial. Therefore, a second unit should be available in the event of instrument problems. This second unit is not always available, creating the need to transfer SFC to HPLC where the availability of the second unit is less a challenge. In addition, the analytical method developed by process analytical chemistry often needs to be transferred internally to a QC lab, or externally to suppliers or vendors, where the availability of a SFC unit can make a HPLC method more desirable.
2. Instrument qualification: To satisfy GMP requirements, the instrument needs to be qualified. SFC instrument qualification was not done in the industry until recently with the marketing of Aurora Fusion^{TM} A5 system. A transfer of the SFC conditions to HPLC is needed if the SFC instrument is not qualified. See Section 8.3 for additional discussions.
3. Instrument sensitivity: It was well known that the sensitivity using SFC is not as good as that using HPLC due to the higher pump noise. This has changed as a result of the introduction of newer analytical SFC systems.

The transfer of SFC conditions to HPLC could be straight forward for many compounds. However, additional development time on HPLC is needed and the amount of the time needed varies. Typically, a longer run time is needed under HPLC conditions when compared with SFC. To initiate the switch from SFC to HPLC analysis, replace the CO_2 with hexanes or heptanes. When methanol is used under SFC conditions, depending on the percentage, a direct conversion in the percentage of methanol may not be possible due to the limited miscibility of methanol with hexanes or heptanes. An equal volume of ethanol may have to be added in order for the mobile phase to be a homogenous solution. However, this change of mobile phase composition can change selectivity. With the availability of qualified instruments and the improved sensitivity of new instrumentation, the need of transfer from SFC to HPLC can be expected to be reduced but not entirely eliminated.

For the chiral separation of tartaric acid by SFC, shown in Fig. 8.1, the conditions were transferred to HPLC by replacing CO_2 with hexanes [9, 28]. The HPLC mobile phase used is an 80:20 volume ratio of hexanes: ethanol, with 0.1 %(v/v) TFA. The flow rate was reduced from 3 mL/min to 1 mL/min for HPLC. A chromatogram obtained under HPLC conditions is shown in Fig. 8.6. Baseline separation is achieved, but the run time was increased from 3 min for SFC to at least 16 min by HPLC. With 20% ethanol in hexanes at 1 mL/min flow rate, the column pressure is close to 50 bar, which is the manufacturer's recommended pressure limit for this specific column. Potentially, a shorter column can be used to reduce the analysis time due to the good separation of the enantiomers.

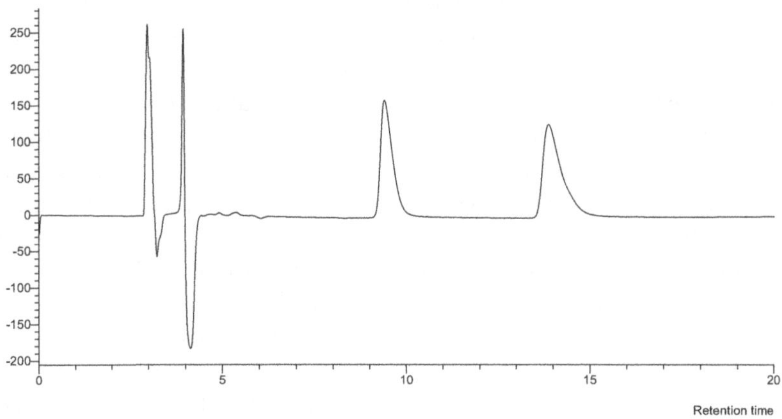

Figure 8.6 Chromatogram of tartaric acid using an AD-H column by HPLC.

8.3 SFC Instrument Qualification and Method Validation

The qualification of SFC instrument has been a challenge and the lack of the instrument qualification restricted its full utilization in the GMP environment. Recently, progress has been made in SFC instrument qualification, especially with the new SFC instrumentation. Use Aurora Fusion™ A5 as an example, since the unit is attached to the existing HPLC system, the qualification only needs to be focused on the "add-on" piece, the CO_2 preconditioning unit. Sun [29] from Amgen reported their approach in the instrument qualification process, from DQ (design qualification), IQ (installation qualification) and OQ (operational qualification) to PQ (performance qualification). The cost of the flow meter and its calibration is more than $10,000. In addition to the IQ and OQ, which can be provided by the manufacturer or the vendor, another common practice among the pharmaceutical industries is to evaluate the system using a validated method as part of the analytical instrument qualification. However, to validate a method, a qualified instrument has to be used.

Feasibility of SFC method validation for chiral separation has been reported by Kalmbach [30] from Pfizer. Flurbiprofen was used as the test compound and three commercial available SFC analytical systems, Thar Method Station II, Jasco analytical system and Aurora Fusion™ A5, were used to conduct the evaluation and comparison. The validation parameters evaluated in the study includes linearity, injection precision, LOQ (limit of quantitation) and LOD (limit of detection). Linearity is evaluated at both high and low concentration ranges. For the high concentration range, linearity is evaluated from 50% to 150% of the nominal concentration, which is 1 mg/mL, with five data points. The correlation coefficient (R^2) is greater than 0.999 from all three SFC systems. The low-level linearity ranged from 0.02% (LOD) of the nominal concentration, 0.05% (LOQ) of the nominal concentration to 5% of the nominal concentration. The correlation coefficient (R^2) is greater than 0.995 from all three SFC systems; Relative standard deviation (RSD) in retention time for both nominal concentration and LOQ level is less than 0.5% from all three systems. For injection precision, less than 1% RSD in peak area was achieved by all three systems at the nominal concentration and less than 10% at the LOQ level. For the signal to noise (S/N) evaluation of the LOD and LOQ, the Jasco system did not meet the 10:1 requirement at the LOQ even though the S/N at LOD met the 3:1 S/N requirement. Overall, the results indicate that the SFC method can be validated. Another example of method validation, which used Warfarin, also demonstrated adequate S/N of LOQ to 0.005% of the nominal concentration [19]. It should be mentioned that for both cases, the signal response at the nominal concentration is greater than 1 AU (approximately 1.3 AU for Flurbiprofen and 1.5 AU for Warfarin). For the HPLC method development, the peak height response is often kept at less than 1 AU. Although the sensitivity of the SFC is still inferior to HPLC, the response can be compensated by increasing the nominal concentration. This point is further demonstrated in an achiral impurity analysis where the sensitivity of HPLC and SFC was compared using the same compound [31]. The same sensitivity is achieved with SFC by increasing the nominal concentration 10-fold versus HPLC. Sample solubility can be a limiting factor to achieve the desired sensitivity with SFC.

In addition, Wang and coworkers [31] have also demonstrated the method validation by conducting linearity, accuracy and precision for assay and impurities for SFC method for mometasone. The limit of quantitation was also determined. These studies were conducted for both reverse-phase HPLC and SFC and the results were compared. A correlation coefficient of 0.9999 and a limit of quantitation of 0.05% (pa) were achieved on both HPLC and SFC. Accuracy and precision results were also comparable and acceptable for both HPLC and SFC. However, the nominal sample concentration for SFC is 10 times higher for SFC than HPLC to reach the same limit of quantitation .

Instrument qualification and detection sensitivity still remain the two potential hurdles for the widespread use of SFC as a routine analytical tool for chiral and achiral analysis in a GMP environment. By overcoming these two issues, it can be expected to see an increased usage of SFC.

8.4 Impurity Isolation and Material Purification

8.4.1 Impurity Isolation

Impurity isolation is sometimes conducted in process analytical chemistry for impurity identification and confirmation. When LC/NMR is available and the impurity level in the sample is at an adequate level, isolation of the desired impurity may not be necessary. However, LC/NMR may not give conclusive results when the impurity level in the sample is too low or when the impurity peak elutes too close to the main component. Impurity isolation is then needed in order to obtain the NMR spectra needed for the impurity structure confirmation or elucidation. The quantity of material needed to obtain such NMR spectra is small, such as a few hundred micrograms or a couple of milligrams, depending on the purity of the impurity isolated and the sensitivity of the available NMR instrument.

For pharmaceutical analysis, organic related impurities are typically observed by reverse-phase HPLC. The impurity can be isolated using fraction collection. However, the time needed to remove the aqueous media can be very long, sometimes, weeks, depending on the volume collected and mobile phase composition of the fraction. In addition, residual buffers or additives from the mobile phase can reduce the purity of the material or decompose the impurity. Using SFC, the isolation time can be significantly reduced since the majority of the mobile phase (CO_2) will be evaporated off at the end of the impurity isolation and the modifiers are volatile organic solvents.

Demonstrated in Fig. 8.7 is the reverse-phase HPLC trace of a mother liquor sample from a reaction mixture along with the corresponding TIC using LC/MS under scan mode with positive ionization. The mother liquor sample is used for the impurity identification since the impurity level is enriched in the mother liquors versus the isolated solid product. Impurity A (Imp A) and Impurity B (Imp B) were the peak of interest for the isolation and identification. Owing to the long 50 min run time for HPLC, isolation of the impurities using SFC was a logical alternative. Since elution order may change between reverse-phase HPLC and SFC, a correlation of peak identity needs to be established between HPLC and SFC analysis. This can be done using a few different approaches or a combination of these approaches: (1) Molecular weight of the impurity, (2) UV spectra from diode array or PDA detector, (3) relative peak size, and (4) collection using SFC and analysis using HPLC to correlate the impurity by retention time. When the mass of the impurity is already known, SFC/MS can be used for a quick impurity correlation if SFC/MS is available. When a mass detector is not available, UV spectra of the peak can be utilized as an alternative. However, when the impurities are closely related in structure to the product, which is often the case for the impurities in pharmaceutical compounds, the UV spectra approach will not work for identification purposes. Impurity fractions can always be collected first by SFC and analyzed using HPLC to confirm the impurity peak of interest by retention time.

The chromatogram of the mother liquor sample obtained using SFC is shown in Fig. 8.8. Even though the mass of the impurities was available, SFC/MS was not available for the correlation. By comparing the peak size of the impurities of interest between HPLC and SFC, four fractions were collected as labeled 1 through 4 in Fig. 8.8. The run time is only 10 min and enough samples were collected overnight for NMR analysis. Since methanol was used as the modifier, evaporation of the solvent took only a couple of hours. Peak 1 by SFC is Imp A by

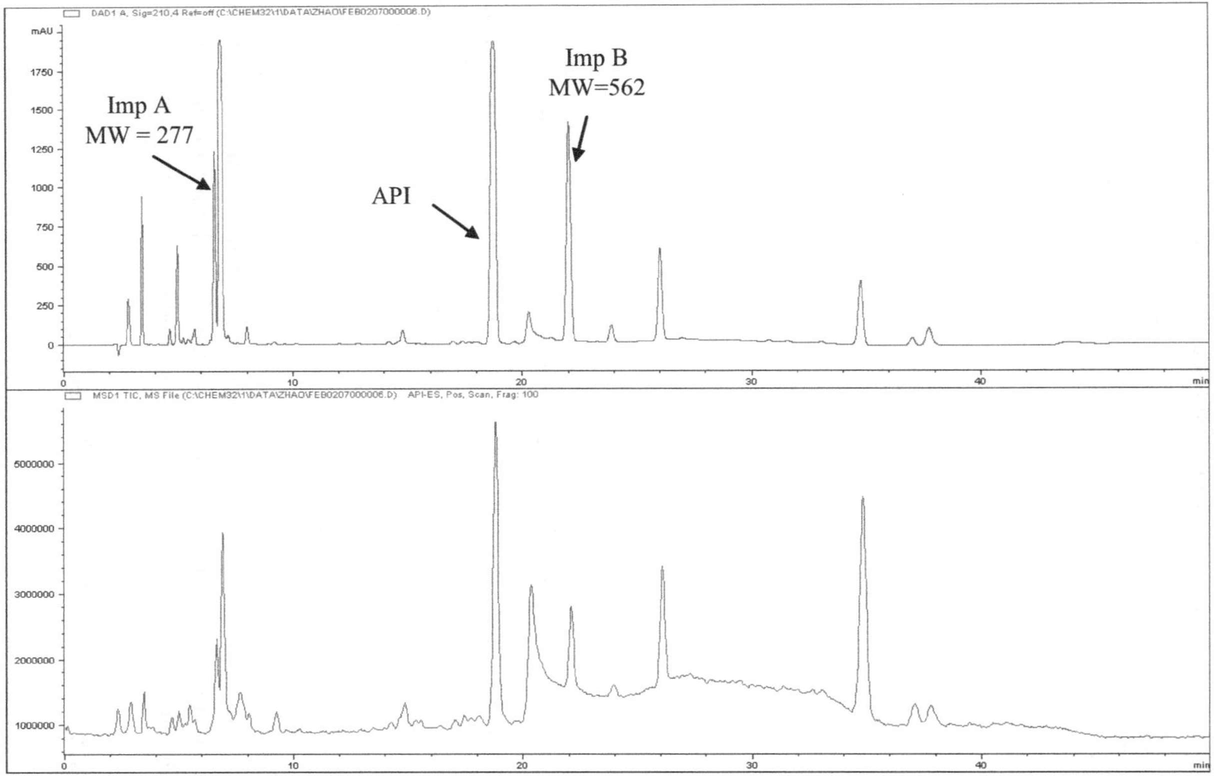

Figure 8.7 Profiles of a mother liquor sample analyzed by HPLC-UV/MS (Top = UV; Bottom = MS). Impurities A and B (labeled as Imp A and Imp B) were the peaks of interest for isolation and identification.

HPLC and Peak 2 by SFC is Imp B by HPLC and the confirmation was done by HPLC retention time and mass by LC/MS. The structures of the isolated impurities were confirmed by NMR.

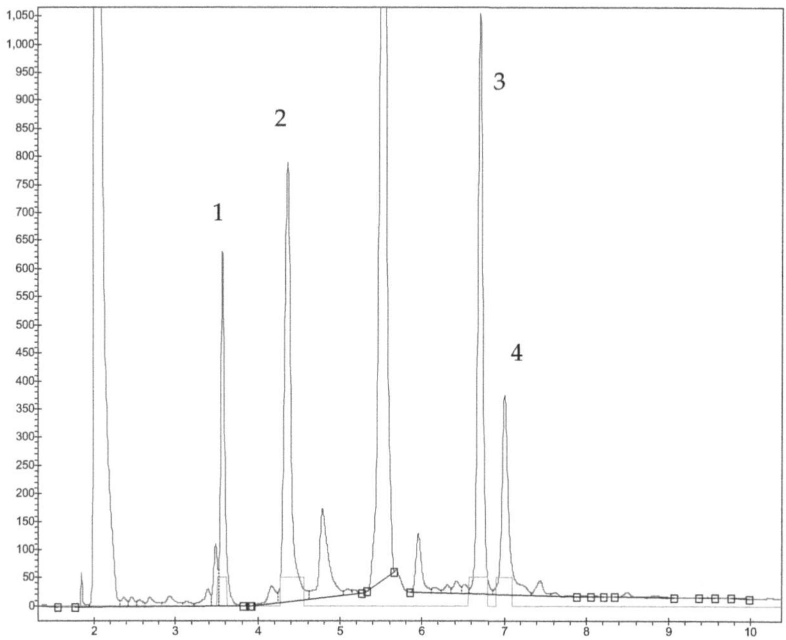

Figure 8.8 SFC chromatogram of the same mother liquor sample as that used in Fig. 8.7. Peaks 1 through 4 were the four fractions collected using a Berger SFC system. Peaks 1 and 2 corresponding to Imp A and Imp B from Fig. 8.7, respectively.

8.4.2 Material Purification

Material purification can be conducted using SFC, via different approaches, such as simulated moving bed technique or fraction collection technique to collect materials in kilogram quantities. Details of these applications can be found in the other chapter.

8.5 SFC with Mass Detection

When a mass detector is coupled with SFC for analytical purposes, specificity and improved detection sensitivity are the added benefits, especially when compared with the UV detection and for ionizable compounds with poor chromophore.

8.5.1 Using a Mass Detector

For analytical SFC-MS, samples are typically introduced to the mass detector by splitting the sample prior to the backpressure regulator [32–36], which in a way is equivalent to introducing a pressure leak into the system. See Fig. 8.9 for a simple diagram. Typical drawbacks observed with this configuration are the high noise level and as a result, a sacrifice in sensitivity. This loss of sensitivity, however, is in comparison with reverse-phase LC-MS. When compared with HPLC-UV, the sensitivity is still much improved, especially for ionizable compounds with poor chromophores [37]. The noise level is found to be related to the outlet pressure. Similar to SFC-UV, where higher outlet pressure has been reported to reduce noise levels and improve sensitivity [38], the same effect is observed for the SFC-MS. However, even though the background signal noise level is lower with a higher outlet pressure, the noise variations are higher at higher outlet pressures [39]. This is most likely due to the fact that the pump has to work harder to maintain the pressure in order to compensate for the "leak" that the system experienced as a result of sample splitting.

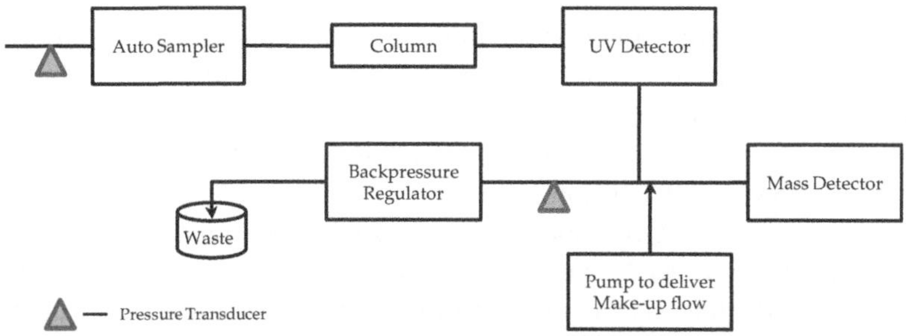

Figure 8.9 A simple schematic diagram showing the typical split connection of mass detector to the SFC/UV system along with the location of pump to deliver a backup flow if desired.

When the sample is split to the mass detector, the outlet is the spray chamber. There is a significant pressure drop between the pressure regulated region and spray chamber. This change of pressure causes CO_2 expansion. The CO_2 expansion may help with the nebulization, however, it may also cause sample solubility issues depending on the composition of the modifier and the sample solubility in the modifier. To ensure that the sample split reaches the spray chamber of the mass detector, a make-up flow is often used (see Fig. 8.9). In addition, appropriate acid or base can be added into the make-up flow to aid sample ionization in order to improve sensitivity.

Recently, Hayward and Han [40] have presented results from an alternative configuration of the mass detector. They eliminated the use of backpressure regulator by combining preheating with a fixed restrictor to hold pressure up to 150 bars. The recommendation of no backpressure regulator was first given by Pinkston [41] and the benefit of the elevated temperature was also reported by Sadoun [42]. However, Hayward and Han have demonstrated that sensitivity of SFC-MS can be improved up to 2000-fold using a heated restrictor instead of the traditional backpressure regulator and split approach. Using this new approach, the sensitivity of SFC-MS is comparable to that obtained using LC-MS. The optimal condition identified is to heat the restrictor to 120°C and maintain the pressure above 150 bars.

8.5.2 Applications

Pinkston and colleagues have reported a comprehensive comparison of HPLC-MS and SFC-MS for screening a large and diverse library of pharmaceutically relevant compounds [43], demonstrating that SFC-MS generated nearly equivalent "hits" to that of LC-MS (79.4% by LC-MS vs. 75.0% by SFC-MS).

For quantitative analysis, an analytical method for the determination of (R)- and (S)-Ketoprofen in human plasma using SFC-MS/MS has also been developed, validated and compared to the LC-MS/MS method. The method by SFC-MS/MS was found to be comparable in terms of sensitivity, linearity, accuracy, precision, specificity and ruggedness, but only 1/3 of the analysis time was needed for LC-MS/MS [44]. SFC-QqQ MS was utilized for the simultaneous determination of ten polar lipids in biological samples [45].

With the specificity of mass detection, SFC-MS has been used to distinguish the enantiomers from other achiral impurities. This is extremely useful for the monitoring the reaction mixtures. As demonstrated by Alexander and Staab [46], SFC-MS was used for the monitoring of a reaction mixture that contains a pair of E/Z isomers of cinnamonitrile derivatives and four diastereomers/enantiomers of hydrocinnamonitrile derivatives (see structures in Fig. 8.10). When a chiral column (Chiralpak AD-H) is used, one of the E/Z isomer is co-eluting with one of the diastereomers. However, detection is still possible by utilizing the extracted ion chromatogram (EIC). When an achiral column is coupled with the AD-H column, all six compounds can be resolved and the identity of the samples is confirmed by using EIC.

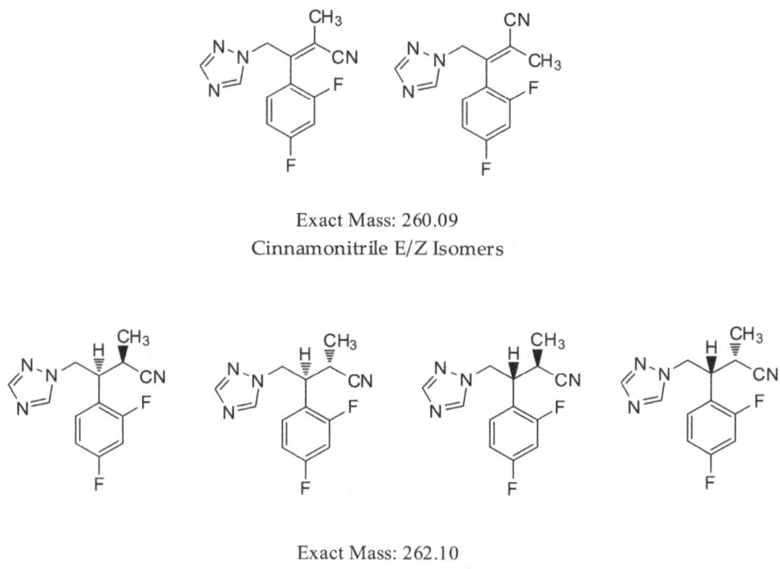

Figure 8.10 Structures of cinnamonitrile and hydrocinnamonitrile derivatives.

Chiral method development using SFC-MS as a screening tool to take advantage of the specificity from the mass detector has been reported by many authors [20, 47–48].

SFC-MS has also been utilized for achiral screening. de la Puente, Soto-Yarritu, and Burnett have reported the development and implementation of a SFC-MS screening strategy that was successfully incorporated as a routine tool for both analytical and preparative scales[49]. A greater than 85% success rate with a better than 90% recovery and greater than 95% yield was reported after the analysis of several hundred samples.

More detailed information on SFC-MS can be found in another chapter of this book.

8.6 Achiral Separations

The use of SFC in achiral separation and purification has been reported by many authors [50]. However, the use of SFC for achiral separation and sample analysis for assay and purity/impurity determinations for pharmaceutical compounds has been lagging behind for process analytical chemistry when compared with that for chiral separation. This can be attributed to the fact that HPLC is widely available to the industry and is adequate for most applications when it comes to separations as well as sample analysis. The familiarity of HPLC techniques by personnel and the wide availability of HPLC instruments make HPLC the first choice while SFC remains an orthogonal, secondary or complimentary technique for achiral analysis. Nevertheless, the capability and efficiency of SFC for achiral separations have been studied and demonstrated in the pharmaceutical industry. For majority of the cases studied, packed columns are used. For earlier achiral development work, refer to the review articles by Berger [51] and Ibanez and Senorans [52].

Brunelli and colleagues [53] at Pfizer have reported the development for the analysis of a broad spectrum of pharmaceuticals using a cyanopropyl silicagel stationary phase. They coupled five 25 cm × 4.6 mm I.D., 5.0 μm columns to generate ca. 100,000 plates. With the use of 1:3 acetonitrile: methanol mixture as the modifier, and both trifluoroacetic acid and diiospropylamine as the modifier, an optimized SFC condition was established to provide separations of these pharmaceutical compounds. The effect of additive concentrations, modifier compositions, and pressure were also studied by comparing the peak shape and elution order.

Wang and colleagues [31] from Merck & Co have reported the development of an orthogonal method by SFC for mometasone furoate (see Fig. 8.11 for structure) and related impurity analysis. After screening three normal phase columns (2-ethylpyridine, cyano, and silica) with methanol as the modifier, baseline resolution of mometasone furoate and eight related impurities were achieved on the silica column. The optimized analysis was conducted at 4.0 mL/min flow rate with a 100 bar outlet pressure. The column temperature was at 30°C with a gradient run of methanol from 5% to 15% in 15 min. The final impurity peak eluted off the column in less than 12 min, which is less than a third of the HPLC analysis time.

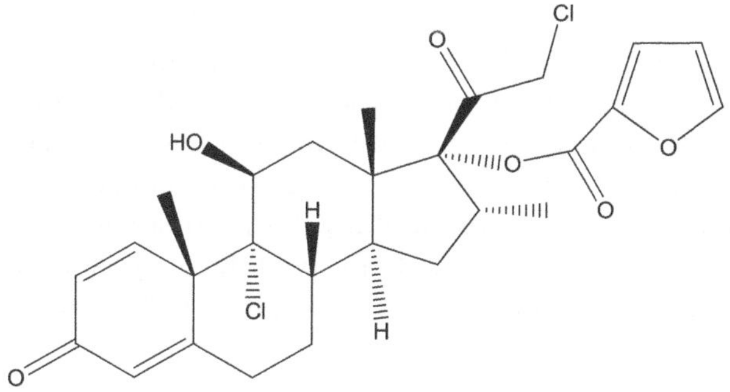

Figure 8.11 Structure of mometasone furoate.

Taking the study a step further toward the actual application of the SFC method for the pharmaceutical industry, method validation was conducted for linearity, accuracy and precision for both assay and impurity level. The limit of quantitation was also determined. These studies were conducted for both the reverse-phase HPLC method and SFC method and the results were compared. A correlation coefficient of 0.9999 and a limit of quantitation of 0.05% (pa) were achieved on both HPLC and SFC. Accuracy and precision results were also comparable and acceptable for both HPLC and SFC. However, the nominal sample concentration for SFC is 10 times higher in SFC than that from HPLC to reach the same limit of quantitation.

Strife and colleagues [54] have reported the separation of six isomers of dimethylaniline using SFC. These isomers were separated without derivatization on two 4.6 × 250 mm, 5 μm particle size, ethyl pyridine columns in tandem under isocratic conditions of 5% 10 mM ammonium acetate in methanol and 95% CO_2. A triple quadruple mass detector was used for detection with electrospray ionization. The quantitation limit was estimated to be 40 ppb based on signal to noise ratio. Even though the method was not further developed, the conditions established and the data presented demonstrated the feasibility of using SFC-MS method for trace level impurity analysis, including carcinogenic and genotoxic impurity determinations.

Patel, et al [55] even demonstrated using SFC for the separation of isomeric polypeptide pairs. In their study, two pairs of end-capped 12-mer peptides were selected. Each pair has the same molecular weight and amino acid content, but different sequence of amino acid fragment. These polypeptides were in the form of trifluoroacetate salt; therefore, the peptides are charged and water soluble. Resolution of the isomeric polypeptide pairs were achieved using HA-pyridine, 4-ethyl pyridine and amino propyl columns with methanol as the modifier and trifluoroacetic acid or isopropyl amine as additives. A mass spectrometer was used for the detection. The type of stationary phase seems the most critical parameter for the resolution. Results of this carefully designed study demonstrated the feasibilities of using SFC-MS for peptide analysis, which can be difficult even under the typical reverse-phase HPLC conditions [56].

8.7 Summary and Conclusion

SFC has been used in process analytical chemistry to support all aspects of the scale-up process for drug development. Chiral separations by SFC are the dominate applications showing the most efficiency. The advantages of SFC are also present in material purification and impurity isolations. Achiral separation by SFC is still underused in process analytical chemistry at present time, even though the potential of SFC is well known.

Acknowledgments

The author would like to thank Abbott Laboratories for promoting scientific excellence and Wayne Pritts for proofreading the chapter; Rui Chen and TharSFC for giving the permission to use Fig. 8.2. Appreciation also goes to Greg Webster and the publisher for the invitation to contribute to the chapter.

References

1. U.S. Food and Drug Administration. (1992). FDA's Policy Statement for the development of new stereoisomeric drugs, *Chirality,* **4**, pp. 338–

340. (Burke, J. E. (1964). Scattering of surface waves on an infinitely deep fluid, *J. Math. Phys.*, **6**, pp. 805–819.)

2. Berger, T. A., Fogleman, K., Staats, T., Nete, P., Crocket, I., W. Farrell, W., Osonubi, M. (2000). The development of a semi-preparatory scale supercritical-fluid chromatograph for high-throughput purification of 'combi-chem' libraries, *J. Biochem. Biophys. Methods*, **43**, pp. 87–111.
3. Terfloth, G. (2001). Enantioseparations in super- and subcritical fluid chromatography, *J. Chromatogr. A*, **906**, pp. 301–307.
4. Yan, T. Q., Orihuela, C. (2007). Rapid and high throughput separation technologies: steady state recycling and supercritical fluid chromatography for chiral resolution of pharmaceutical intermediates. *J. Chromatogr. A*, **1156**, pp. 220–227.
5. White, C. (2005). Integration of supercritical fluid chromatography into drug discovery as a routine support tool: Part I. Fast chiral screening and purification, *J. Chromatogr. A*, **1074**, pp. 163–173.
6. Maftouh, M., Granier-Loyaux, C., Chavana, E., Marini, J., Pradines, A., Heyden, Y. V., Picard, C. (2005). Screening approach for chiral separation of pharmaceuticals: Part III. Supercritical fluid chromatography for analysis and purification in drug discovery, *J. Chromatogr. A*, **1088** pp. 67 –81.
7. Welch, C. J., Biba, M., Gouker, J. R., Kath, G., Augustine, P., Hosek, P. (2007). Solving multicomponent chiral separation challenges using a new SFC tandem column screening tool, *Chirality*, **19(3)**, pp. 184–189.
8. Biba, M., Welch, C. J., DaSilva, J. O., Mathre, D. J. (2005). Impurity isolation in support of pharmaceutical process research, *Am. Pharm. Rev.*, **8**, pp. 68–72.
9. Zhao, Y., Pritts, W., and Zhang, S. (22–23 July 2009). Application of SFC in process analytical chemistry for drug development, paper presented at SFC 2009, Philadelphia, PA.
10. Lewis, A., Wilkie, J., Rutherford, T. J., Gani, D. (1998). Design, construction and properties of peptide N-terminal cap templates devised to initiate α-helixes. Part 2. Caps derived from *N*-[(2S)-2-chloropropionyl]-(2S)-Pro-(2S)-Pro-(2S,4S)-4-thioPro-OMe., *J. Chem. Soc., Perkin Trans., 1*, **22**, pp. 3777–3794.
11. Khalil, E. M., Subasinghe, N. L., Johnson, R. L. (1996). An efficient and high yield method for the *N*-tert-butoxycarbonyl protection of sterically hindered amino acids. *Tetrahedron Lett.* **37(20)**, pp. 3441–3444.
12. Thaisrivongs, S., Pals, D. T., Lawson, J. A., Turner, S. R., Harris, D. W. (1987), α-Methylproline-containing renin inhibitory peptides: in vivo

evaluation in an anesthetized, ganglion-blocked, hog renin infused rat model., *J. Med. Chem.*, **30 (3)**, p. 536–541.

13. Ma, J. S. (2003), Unnatural amino acids in drug discovery. *Chim. Oggi*, **21(6)**, pp. 65–68.
14. Petersen, M., Sauter., M. (1999), Biotechnology in the fine-chemicals industry. Cyclic amino acids by enantioselective biocatalysis. *Chimia*, **53(12)**, pp. 608–612.
15. Zhao, Y., Pritts, W. A., Zhang, S., (2008). Chiral separation of selected proline derivatives using a polysaccharide-type stationary phase by supercritical fluid chromatography and comparison with high-performance liquid chromatography, *J. Chromatogr. A*, **1189(1–2)**, pp. 245–253.
16. Villeneuve, M. S., Anderegg, R. J. (1998). Analytical supercritical fluid chromatography using fully automated column and modifier selection valves for the rapid development of chiral separations, *J. Chromatogr A*, **826**, pp. 217–225.
17. Wong, M. M., Holzheuer, W. B., Webster, G. K. (2008). A comparison of HPLC and SFC chiral method development screening approaches for compounds of pharmaceutical interest. *Curr. Pharm. Anal.*, **4 (2)**, pp. 101–105.
18. Zhang, Y., Watts, W., Nogle, L., McConnell, O. (2004). Rapid method development for chiral separation in drug discovery using multi-column parallel screening and circular dichroism signal pooling. *J. Chromatogr. A*, **1049**, pp. 75–84.
19. Chen., R. (22–23 July 2009), Advancements in SFC performance from sensitivity to throughput, paper presented at SFC 2009, Philadelphia, PA.
20. Zeng, L., Xu, R., Laskar, D. B., Kassel, D. B. (2007) Parallel supercritical fluid chromatography/mass spectrometry system for high-throughput enantioselective optimization and separation. *J. Chromatogr. A*, **1169**, pp. 193–204.
21. Pirzada, Z., Personick, M., Biba, M., Gong, X., Zhou, L., Schafer, W., Roussel, C., Welch, C. J. (2010). Systematic evaluation of new chiral stationary phases for supercritical fluid chromatography using a standard racemate library. *J. Chromatogr. A*, **1217**, pp. 1134–1138.
22. Hamman, C., Wong, M., Hayes, M., Gibbons, P. (2011). A high throughput approach to purifying chiral molecules using 3 μm analytical chiral stationary phases via supercritical fluid chromatography. *J. Chromatogr. A*, **1218**, pp. 3529–3536.

23. Zhao, Y., Warnke, M., Zhang, S., Pritts, W. A. (19–24 June 2010) Fast chiral method development using column screening by SFC/UV and SFC/MS, paper presented at HPLC 2010, Boston.

24. Zhao, Y., Warnke, M., Pritts, W. A., Zhang, S. Column screening using supercritical fluid chromatography under isocratic and gradient conditions for fast chiral analytical method development, Manuscript under preparation.

25. Zhao, Y., Pritts, W. A. (2007). Chiral separation of selected proline derivatives using a polysaccharide type stationary phase by high-performance liquid chromatography. *J. Chromatogr. A*, **1156**, pp. 228–235.

26. Hamman, C., Schmidt, D. E., Wong, M., Hayes, M. (2011). The use of ammonium hydroxide as an additive in supercritical fluid chromatography for achiral and chiral separations and purifications of small, basic medicinal molecules, *J. Chromatogr. A*, **1218(43)**, pp. 7886–7894.

27. Ventura, M., Murphy, B., Goetzinger, W. (2012). Ammonia as a preferred additive in chiral and achiral applications of supercritical fluid chromatography for small, drug-like molecules. *J. Chromatogr. A*, **1220**, pp. 147–155.

28. Zheng, J., Internal method, Process analytical chemistry, Abbvie Inc.

29. Sun., J. (2011). Critical attributes and technical challenges of successful SFC instrument qualification. *Am. Pharm. Rev.*, **May/June**, pp. 79–83.

30. Lynne Kalmbach, L. (2009). Implementation of SFC as an analytical tool in a GMP regulated environment: a three vendor evaluation, paper presented at SFC 2009, Philadelphia.

31. Wang, Z., Zhang, H., Liu, O., Donovan, B. (2011). Development of an orthogonal method for mometasone furoate impurity analysis using supercritical fluid chromatography. *J. Chromatogr. A*, **1218**, pp. 2311–2319.

32. Sjoberg, P. J. R., Markides, K. E. (1999) Capillary column supercritical fluid chromatography-atmospheric pressure ionisation mass spectrometry interface performance of atmospheric pressure chemical ionisation and electrospray ionisation. *J. Chromatogr. A*, **855(1)**, pp. 317–327.

33. Baker, T., Pinkston, J. D. (1998). Development and application of packed-column supercritical fluid chromatography/pneumatically assisted electrospray mass spectrometry. *J. Am. Soc. Mass Spectrom.*, **9(5)**, pp. 498–509.

34. Morgan, D. G., Harbol, K. L., Kitrinos, N. P. (1998) Optimization of a supercritical fluid chromatograph-atmospheric pressure chemical ionization mass spectrometer interface using an ion trap and two quadrupole mass spectrometers. *J. Chromatogr. A*, **800(1)**, pp. 39–50.

35. Garzotti, M., Hamdan, M. (2002). Supercritical fluid chromatography coupled to electrospray mass spectrometry: a powerful tool for the analysis of chiral mixtures. *J. Chromatogr. B*, **770(1–2)**, pp. 53–61.

36. Bolanos, B., Greig, M., Ventura, M., Farrell, W., Aurigemma, C. M., Li, H., Quenzer, T. L., Tivel, K., Bylund, J. M. R., Tran, P., Pham, C., Phillipson, D. (2004). SFC/MS in drug discovery at Pfizer, La Jolla. *Int. J. Mass Spectrom.* **238(2)**, pp. 85–97.

37. Zhao, Y., Yao, C., Pritts, W. A., Zhang, S. (27–31 March 2011) Supercritical fluid chromatography for chiral analysis: overcoming the sensitivity barrier by using a mass spectrometer, paper presented at the 241st ACS national meeting and exposition, 2011, Anaheim, California.

38. Berger, T. A., Berger, B. K. (2011) Minimizing UV noise in supercritical fluid chromatography. I. Improving back pressure regulator pressure noise. *J. Chromatogr. A*, **1218(16)**, pp. 2320–2326.

39. Zhao, Y., Yao, C., Pritts, W. A. (11–12 October 2010) Determination and optimization of proline derivatives by SFC/MS and comparison with SFC/UV and HPLC/UV. Paper presented at SFC west coast user meeting 2010, San Diego, California

40. Hayward, M. J., Han, Q. P. (19–21 July 2011). Interfacing SFC to various detectors with emphasis on cold electrospraymass spectrometry & achieving the sensitivity, robustness, as well as dynamic range needed for a variety of applications, paper presented at SFC 2011, New York.

41. Pinkston, J. D. (2005) Advantages and drawbacks of popular supercritical fluid chromatography/mass spectrometry interfacing approaches: a user's perspective. *Eur. J. Mass Spectrom.*, **11(2)**, pp. 189–197.

42. Sadoun, F., Virelizier, H., Arpino, P. J. (1993). Packed-column supercritical fluid chromatography coupled with electrospray ionization mass spectrometry. *J. Chromatogr. A*, **647(2)**, pp. 351–359.

43. Pinkston, J. D., Wen, D., Morand, K. L., Tirey, D. A., and Stanton, D. T. (2006) Comparison of LC/MS and SFC/MS for screening of a large and diverse library of pharmaceutically relevant compounds. *Anal. Chem.*, **78**, pp. 7467–7472.

44. Hoke, II, S. H., Pinkston, J. D., Bailey, R. E., Tanguay, S. L., and Eichhold, T, H. (2000). Comparison of packed-column supercritical

fluid chromatography–tandem mass spectrometry with liquid chromatography–tandem mass spectrometry for bioanalytical determination of (R)- and (S)-detoprofen in human plasma following automated 96-well solid-phase extraction. *Anal. Chem.*, **72**, pp. 4235–4241.

45. Lee, J. W., Takashi, Y., Uchikata, T., Matsubara, A., Fukusaki, E., Bamba, T. (2011) Development of a polar lipid profiling method by supercritical fluid chromatography /mass spectrometry. *J. Sep. Sci.*, **34**, pp. 3553–3560.

46. Alexander, A. J., Staab, A. (2006). Use of achiral/chiral SFC/MS for the profiling of isomeric cinnamonitrile/hydrocinnamonitrile products in chiral drug synthesis. *Anal. Chem.*, **78**, pp. 3835–3838.

47. Zhao, Y., Woo, G., Thomas, S., Semin, D., Sandra, P. (2003). Rapid method development for chiral separation in drug discovery using sample pooling and supercritical fluid chromatography–mass spectrometry, *J. Chromatogr. A*, **1003(12)**, pp. 157–166.

48. Zeng, L., Xu, R., Zhang, Y., Kassel, D. B. (2011). Two-dimensional supercritical fluid chromatography/mass spectrometry for the enantiomeric analysis and purification of pharmaceutical samples. *J. Chromatogr. A*, **1218**, pp. 3080–3088.

49. de la Puente, M. L., Lopez Soto-Yarritu, P., Burnett, J., (2011) Supercritical fluid chromatography in research laboratories: design, development and implementation of an efficient generic screening for exploring this technique in the achiral environment. *J. Chromatogr. A*, **1218(47)**, pp. 8551–8560.

50. White, C., Burnett, J. (2005) Integration of supercritical fluid chromatography into drug discovery as a routine support tool II. Investigation and evaluation of supercritical fluid chromatography for achiral batch purification. *J. Chromatogr. A*, **1074**, pp. 175–185.

51. Berger, T. A., (1997) Separation of polar solutes by packed column supercritical fluid chromatography. *J. Chromatogr. A*, **785**, pp. 3–33.

52. Ibañez, E., Señoráns, F. J., (2000) Tuning of mobile and stationary phase polarity for the separation of polar compounds by SFC, *J. Biochem. Biophys. Methods*, **43**, pp. 25–43

53. Brunelli, C., Zhao, Y., Brown, M., Sandra, P., (2008), Development of a supercritical fluid chromatography high-resolution separation method suitable for pharmaceuticals using cyanopropyl silica, *J. Chromatogr. A*, **1185**, pp. 263–272.

54. Strife, R. J., Mangels, M. L., Skare, J. A. (2009). Separation and analysis of dimethylaniline isomers by supercritical fluid chromatography:

electrospray ionization tandem mass spectrometry. *J. Chromatogr. A,* **1216 (41)**, pp. 6970–6973.

55. Patel, M. A., Riley, F., Wang, J., Lovdahl, M., Taylor, L. T. (2011). Packed column supercritical fluid chromatography of isomeric polypeptide pairs. *J. Chromatogr. A,* **1218**, pp. 2593–2597.

56. Mant, C. T., Cepeniene, D., Hodges, R., S. (2010). Reversed-phase HPLC of peptides: assessing column and solvent selectivity on standard, polar-embedded and polar endcapped columns. *J. Sep. Sci.,* **33(19)**, pp. 3005–3021.

Chapter 9

Analytical SFC for Impurities

Yun Huang
Analytical R&D/PharmaTherapeutics Pharmaceutical Sciences,
Pfizer Global R&D, Eastern Point Road, Groton, CT 06340, USA
yun.huang@pfizer.com

Supercritical fluid chromatography (SFC) has been historically used for non-Good Manufacturing Practice (GMP) activities in drug development because its low sensitivity and insufficient accuracy and precision could hardly meet GMP testing method requirements. In recent years, with the improvement in the sensitivity and the advancement of the instrumentation, SFC can now meet the method requirements thus to be applied for GMP release and stability testing.

This chapter presents how to implement SFC methods in GMP testing. First, the chapter provides the pioneer experience on how to qualify the SFC system for GMP use, including the system qualification protocol write up and acceptance criteria definition. Second, the chapter demonstrates the feasibility of chiral SFC methods being applied in GMP testing by passing method validation requirements with side-to-side comparison with the corresponding

Supercritical Fluid Chromatography: Advances and Applications in Pharmaceutical Analysis
Edited by Gregory K. Webster

ISBN 978-981-4463-00-3 (Hardcover), 978-981-4463-01-0 (eBook)
www.panstanford.com

HPLC methods. The successful validation of the SFC methods proves that SFC can meet the requirements for chromatographic purity methods, thus can be a potential replacement for HPLC in chiral API GMP release and stability testing. Third, the method transferability among different SFC systems has been studied and confirmed the suitability of the chiral SFC as a regulatory filing method.

Along with the continuous improvement in SFC instrumentation and further exploration on how to better apply the technology in pharmaceutical analysis, SFC has been increasingly implemented as a the fast and efficient separation tool for pharmaceutical analysis in the regulated analytical laboratories.

9.1 Introduction

The quantitation of impurities in drug substances and drug products is required in the highly regulated pharmaceutical industry to ensure the quality of drug substances and drug products. In ICH Q3A(R2) [1], impurities are classified into three categories: (1) organic impurities (process and/or drug related), (2) inorganic impurities, and (3) residual solvents. In this chapter, the impurities that arise during the manufacturing process and/or during the storage of the drug substance and the drug product are referred to as organic impurities. The impurities can be (1) starting materials, (2) intermediates, (3) by-products (specified and unspecified process-related impurities), (4) specified and unspecified degradation products, and (5) reagents, ligands and catalysts. For the chiral drug substances, the enantiomer and diastereomers are classified as specified process-related impurities.

To assure the drug product quality and consistency during clinical release and stability, the impurities are controlled through the relevant specification, which requires the identification and quantitation of the impurities. The quantitation of the impurities in the drug substance and the drug product is accomplished by using appropriate analytical methods to ensure the impurity levels meet the predefined specifications. Thus, the analytical methods are critical to assure the quality of clinical and commercial drug product. These methods must be able to accurately quantify the impurities in the drug substance and drug product. The analytical methods need to be validated to prove the suitability for the detection and quantification of impurities following ICH Q2(R1) (Validation of Analytical Procedures: Text and Methodology) [2].

A variety of analytical techniques have been applied to analyze and quantitate impurity levels in the drugs, with the high-performance liquid chromatography (HPLC) and gas chromatography (GC) being the most common. The detection sensitivity is one key parameter for analytical test method selection because the method limit of quantitation (LOQ) needs to be established at least at the reporting threshold per ICH Q3A(R2), which for drug substance is ≤0.05% when the maximum daily dose is ≤2 g/day, or ≤0.03% when the maximum daily dose >2 g/day, with the former more often applied. HPLC with UV detection has notably superior sensitivity and thus can routinely meet the requirement of LOQ at 0.05%.

In the pharmaceutical industry, HPLC has been the predominant analytical tool for the release and stability testing for drug substance and drug product including the assay and purity determination (for both drug substance and drug product), content uniformity and dissolution testing (for drug product). The development and validation of HPLC methods is time consuming and often becomes a bottleneck for lab productivity. Over the past decade, the pharmaceutical industry has been facing the major challenges on how to shorten the drug discovery and development time with R&D budgets reduced and without compromising the quality of drug products. To keep up with this industry trend and demand, pharmaceutical analytical laboratories are striving to reduce the analytical method development time, to improve the separation efficiency, to increase the sample analysis throughput, and to enhance the cost effectiveness. Therefore, high-speed chromatographic separations have been intensively studied and applied in recent years, such as two relative new techniques: ultra performance liquid chromatography (UPLC) and supercritical fluid chromatography (SFC) [3–12].

UPLC has been a new trend of liquid chromatography and gaining more and more popularity. It retains the same HPLC separation mechanism and chromatographic principles but takes advantage of the technological strides made in the preparation and packing efficiency of smaller particle (sub-2 μm) stationary phases, and chromatographic system dead-volume optimization to create a step-function improvement in chromatographic performance. UPLC provides increased resolution, faster analysis, and higher sensitivity and thus significantly reduces the method development and validation time, and increases the sample throughput. UPLC has gained widespread acceptance in the pharmaceutical industry and has started to enter the QC laboratories as standard instrumentation. UPLC is a highly likely replacement for the conventional HPLC for pharmaceutical analysis [4–7].

Supercritical fluid chromatography has been well established in the past decade owning to its distinguishing characteristics (a low viscosity more similar to that of a gas than liquid, solubility approaching the liquid phase, and high diffusivity approaching the gas phase) which result in lower back pressures even at high flow rates, faster separations and higher efficiencies. SFC has numerous practical advantages relative to reversed phase HPLC such as higher speed/throughput, more rapid equilibration and shorter cycle time [11–12]. In addition to its superiority in chiral separation, when compared to the conventional HPLC, SFC is a not only faster but also greener, which presents itself as an excellent chromatographic tool for cost saving and efficiency enhancing. As a result, SFC has rapidly become the primary choice for chiral purity assessment and purification to support drug development from early discovery to late phases [8–12].

For pharmaceutical applications, analytical SFC has been mainly used for chiral separations (1) automated chiral method screening and development; (2) high complexity separations that require column coupling techniques (e.g., chiral to chiral or chiral to achiral column couplings) where SFC is more compatible due to its lower column pressure and faster analysis speed compared to HPLC; (3) off-line chiral reaction monitoring and chiral purity checks for synthetic process development and optimization; and (4) in-process control of chiral purity in GMP drug substances, also called Active Pharmaceutical Ingredient (API). However, SFC is not limited to only chiral analysis. As a normal phase (NP) separation technique, the separation mechanism is different compared to the most popular reserved phase (RP) HPLC, which enables SFC to be developed as an orthogonal method to supplement the RP-HPLC method during method development. A secondary method from an orthogonal separation technique has been widely accepted to confirm the method specificity and ensure all impurities of interest separate from each other. In practice, the orthogonal method should be jointly developed with the primary method (mainly RP-HPLC) to provide adequate peak capacity and resolving power, and to maximize the probability of revealing all peaks. SFC has demonstrated being an orthogonal analytical tool to RP-HPLC to tackle the challenges presented in trace-level impurity analysis, which can be a drug product quality and safety concern [13–14]. SFC/MS has also gained increasing popularity especially in the quantitative analysis where RP-HPLC has found to be challenging [15–16]. In addition, SFC presents another advantage with its nonaqueous mobile phase (carbon dioxide with organic modifier, mainly methanol), leading it to being an alternative analytical tool for water labile analytes and organic solvent sample matrices [17].

Since more and more drug candidates developed in the pipeline are chiral, the demand for process development of asymmetric synthesis and chiral API manufacturing has increased dramatically over the recent years, boosting the need for chiral purity analysis. Therefore, timely development of chiral methods is one of the most challenging tasks to analytical chemists. In order to cope with the high demand, more and more analytical SFC has been implemented in the pharmaceutical analytical laboratories as one of the main chromatographic techniques for chiral method development and chiral purity analysis.

However, SFC has been historically known for its low sensitivity, insufficient accuracy, and precision and its difficulty to meet GMP release method requirement. Thus, analytical SFC has been limited its utilization to non-GMP analytical tasks and has not been widely adopted for GMP release testing [18]. As mentioned, the detector sensitivity is the key for impurity level analysis. SFC detector noise is significantly higher than that of a HPLC system when using the same detector, which limits the SFC sensitivity. The detector noise is the combined result from all the components in the SFC system. Three main sources of detector noise are examined as (1) the refractive index (RI) of pure carbon dioxide (CO_2) induced by the back pressure regulator (BPR), (2) the effect of the mobile phase fluid temperature, and (3) the percentage of modifier in the mobile phase [19]. Pressure fluctuations and refractive index changes induced by the BPR can be a significant source of detector noise in SFC. Dr. Terry Berger, in his article titled "Minimizing UV Noise in Supercritical Fluid Chromatography. I. Improving Back Pressure Regulator Pressure Noise," has detailed the sources of UV detector noise and how to minimize it in a SFC system [19]. The reduction of the UV noise has been mainly focused on the improvement in pumping as well as BRP since the noise induced by mobile phase pressure fluctuations will be propagated to the final UV detector noise. The standard OQ test specification for UV DAD detector signal noise (ASTM baseline noise) limit is ≤0.05 mAU for analytical HPLC systems. Typically, the diode array detectors (DAD) in HPLC systems can easily achieve the peak to peak noise < 0.02 mAU. Berger's study stated that a pressure noise varying 0.5 bar on a BPR contributes ~0.3–0.5 mAU UV detector noise, not meeting the requirement of <0.05 mAU for impurity quantitation. His research results demonstrated that a BPR with pressure noise of ±0.1 bar could dramatically reduce UV detector noise to <0.02 mAU with a >0.03 min (10 Hz) electronic filter.

In recent years, SFC instruments have continuously evolved along with the constant improvement in the sensitivity and the advancement in the system design, especially in the carbon dioxide delivery pumping system and the back pressure regulator. Aurora SFC Inc has developed and introduced its latest upgraded product, SFC Fusion™ A5 Evolution, to the analytical SFC market. The SFC Fusion™ A5 Evolution features a combination of improved pumping technology and a low noise, state-of-the-art BPR, resulting in reduced baseline noise in UV detectors. Agilent 1260 infinity analytical SFC system is a modular system consisting of Agilent 1260 Analytical infinity LC and the Aurora SFC Fusion™ A5 Evolution. Under optimized instrument conditions, this SFC system can meet the UV noise specification of 0.05 mAU or better. In a Pfizer Analytical Research and Development (ARD) laboratory, it has been found that the Aurora SFC Fusion™ A5 Evolution enhanced chromatographic system connected to a Agilent 1100 binary HPLC system has a notable BPR ASTM noise within ±0.1 bar and a DAD detector (Agilent Model G1315B) peak to peak noise at 0.026 mAU level. The significantly reduced UV detector noise directly boosts the UV detector sensitivity, which makes it possible to meet the level of limit of quantitation required in impurity quantitation. Thus, the analytical SFC application can be extended to the quantitative analysis of pharmaceutical impurities, especially for chiral purity determination in the QC laboratories.

Nowadays, increasing collaboration occurs between SFC manufacturers and pharmaceutical labs to qualify the SFC for the intended GMP testing use [20–22]. A great amount of feasibility studies has been performed to implement chiral SFC as the primary chiral purity tool for API release and stability testing, from which, the promising results have shown that SFC has a great potential to replace HPLC as the primary GMP release testing tool for chiral purity assessment [23–25].

This chapter focuses on how to qualify the analytical SFC system for GMP use, including the system qualification protocol write-up, the acceptance criteria definition, and a real case demonstration that chiral SFC can be successfully applied in GMP testing with comparable method performance when compared with its corresponding HPLC method. The method transferability among different SFC systems is also discussed, which is a key decision factor when choosing a SFC method as the regulatory filing method for release and stability testing.

9.2 Qualification of Analytical SFC System

9.2.1 Analytical Instrument Qualification Overview

In GMP-regulated pharmaceutical laboratories, impurity quantitation requires a validated method and needs to be performed on qualified instruments. The U.S. Food and Drug Administration has clear guidance on method validation [26] and defines validation as "establishing documented evidence that provides a high degree of assurance that a specific process will consistently produce a product meeting its predetermined specification and quality attributes." However, unlike analytical method validation and system suitability activities, currently there is no specific guidance or procedures on the analytical instrument qualification (AIQ). USP general chapter <1058> (Analytical Instrument Qualification) provides a scientific approach to AIQ and considers AIQ as one of the major requirements to generate reliable and consistent analytical data. This approach emphasized the place of AIQ in the overall process of obtaining reliable data from analytical instruments.

There are four components involved in the generation of reliable and consistent data (quality data), which is shown in Fig. 9.1 as layered activities within a quality triangle [27]. Analytical instrument qualification forms the base while the other three components are analytical method validation, system suitability tests, and quality control check samples.

Analytical instrument qualification (AIQ) is the collection of documented evidence that an instrument performs suitably for its intended purpose. The use of a qualified instrument contributes to confidence in the validity of generated data. AIQ activities are grouped into four qualification phases: design qualification (DQ), installation qualification (IQ), operational qualification (OQ), and performance qualification (PQ). Among them, OQ is pivotal to demonstrate that an instrument will function according to its operational specification in a user's environment. PQ is the documented collection of activities necessary to demonstrate that an instrument consistently performs according to predefined specifications and is appropriate for the intended use.

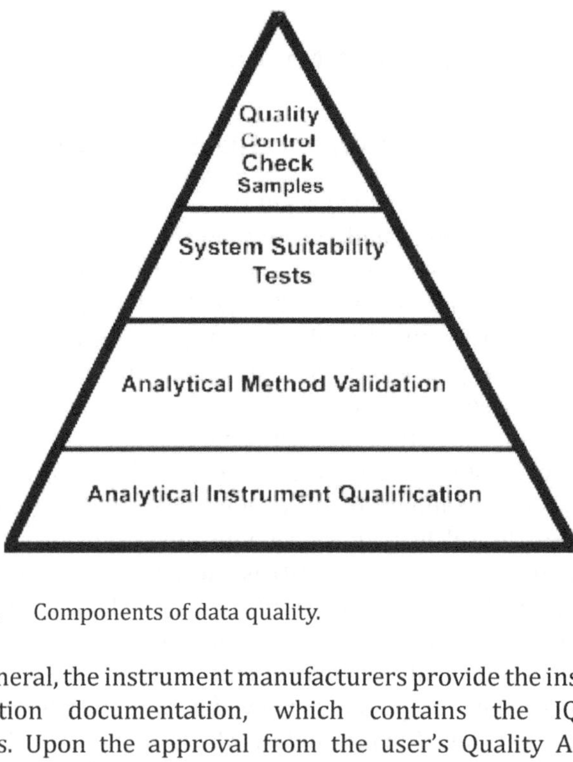

Figure 9.1 Components of data quality.

In general, the instrument manufacturers provide the instrument qualification documentation, which contains the IQ/OQ/PQ protocols. Upon the approval from the user's Quality Assurance (QA), the manufacturer installs the instrument at the user's site. After the initial installation, either the manufacturer's personnel or the user should perform the OQ/PQ tests according to the written testing protocols to verify that the instrument operates as intended.

The instrument functional and performance tests shall meet the established specifications in the user's laboratory. The documents obtained during instrument qualification should be submitted for QA approval and retained in an accessible manner prior to intended GMP use.

9.2.2 Qualification of Analytical SFC System

To implement analytical SFC in GMP-regulated analytical laboratories, it requires instrument qualification. Due to the limitation of SFC sensitivity, SFC was rarely applied in the GMP arena for release and stability testing in the past. As a result, no SFC instrument qualification protocols were developed and established from SFC manufacturers.

However, in last two years, new SFC systems with much-improved detector sensitivity have been introduced to the market, making it time to re-evaluate if the analytical SFC systems can be qualified for intended GMP use in the laboratories. Driven by both analytical and business needs, Pfizer defined the functional and operational specifications to be included in the internal design qualification (DQ) for some commercial SFC systems, and also established the internal operational qualification (OQ) and performance qualification (PQ) requirements [20–21]. As a result, several SFC systems have been qualified and implemented as GMP instruments in the company's GMP laboratories. The analytical SFC has also become the primary release and stability testing method to analyze the chiral purity during the past few years [23–25]. Amgen also completed the instrument qualification on a SFC system for GMP use in 2011 [22].

HPLC system qualification is well established and has been performed in the pharmaceutical analytical laboratories for many years. Since most of SFC components are either identical or similar to HPLC components, the HPLC qualification protocols can serve as a good reference and be applied to the similar components in SFC. For example, the HPLC pump qualification (test parameters and acceptance criteria) is a good reference to define the test protocol for SFC CO_2 delivery pump and the BRP. Table 9.1 summarizes Pfizer's internal analytical SFC instrument qualification protocols for Aurora Fusion™ A-5 SFC conversion module with Agilent 1100/1200/1260 HPLC systems.

Table 9.1 Abbreviated Pfizer internal specifications for SFC instrument qualification

Hardware	Setpoint and parameters	Acceptance criteria
CO_2 pump flow accuracy and precision	Flow rate 1 @ 1.0 mL/min Flow rate 2 @ 5.0 mL/min	Accuracy ±5%; Precision ≤1.0% RSD
Modifier pump flow accuracy and precision	Flow rate 1 @ 1.0 mL/min Flow rate 2 @ 5.0 mL/min	Accuracy ±5% from setpoint Precision ≤0.5% RSD
Gradient composition accuracy and linearity	Gradient accuracy calculated as the deviation from a linear regression line (5%, 15%, 25%, 35%, 45% and 55% modifier) at flow rate of 2 mL/min and 5 mL/min	Accuracy ±2% at each setpoint Linearity: $R^2 \geq 0.999$
Pressure regulator pressure accuracy and precision	Accuracy and precision	Accuracy ±0.3 bar Precision ≤2.0% RSD
UV detector	Baseline noise and drift, linearity, wavelength accuracy	Noise: ≤0.05 mAU Drift: ≤5.000 mAU/h Wavelength accuracy: ±2 nm Linearity: $R^2 \geq$ 0.9990
Column heater temperature accuracy and stability	Accuracy	Accuracy: ±2.0°C from setpoint (<60°C) Stability: ±1.0°C
Injection	Precision	≤1.0% RSD in area (6 replicates)
Injection carryover	Inject a blank sample immediately after a nominal concentration standard injection	Area carryover ≤0.2% of nominal standard

Pfizer conducted the test for instrument qualification on Agilent 1260 infinity analytical SFC system against the predefined specifications in early 2011 [20]. The qualification test results met the acceptance criteria listed in Table 9.1.

In OQ/PQ, the overall system performance is verified by the execution of an analytical SFC test method and comparing key performance attributes against the pre-defined specification set by the user's intended GMP use (e.g., GMP impurity level testing). This test method is an essential part for SFC system qualification. Pfizer has designed and developed a method to verify if the system performance can meet the method validation requirements stated in ICH Q2(R1) "Validation of Analytical Procedures: Text and Methodology" for impurity level testing. The test method has been included as an important part of SFC instrument qualification for Pfizer internal GMP use. The test method is listed in Table 9.2.

Table 9.2 Analytical SFC test method for instrument qualification

Method parameter	Detail
Column	Zorbax Rx-Sil, 5 μm, 4.6 × 150 mm
Column temperature	40°C
Mobile phase	A = carbon dioxide (CO_2); B = methanol
Gradient program	Initial at 5%B for 0.5 min; 5 to 45%B over 4.0 min; 45% to 5% B in 0.1 min; hold at 5%B to 6 min.
Flow rate	4 mL/min
Pressure	150 bar
UV detection (DVD)	254 nm
Injection volume	5 μL (full loop injection)
Chromatographic data system	ChemStation vB.04.03, linked to Empower 2 CDS via a SAT/IN box for data processing and report.
Standard mix solution of amcinonide, carbamazepine, flavone, ketoprofen, and sulfamethazine	5 concentration levels from 0.25–0.70 mg/mL (nominal concentration at 0.5 mg/mL)

The following factors were considered when selecting a system performance test method for the instrument evaluation and qualification purposes:

1. Column: commercially available achiral column with well-known robustness.
2. Testing articles: a standard mixture solution that covers a wide pK_a range (acid, neutral and base) and easy to obtain (i.e., commercially available). This mixture serves as the system

separation and resolution power indicator. For example, Pfizer chose a five-component solution containing amcinonide, carbamazepine, flavone, ketoprofen, and sulfamethazine.

3. Simple modifier without any additives, e.g., methanol.
4. A wide range gradient program instead of manufacturer preferred isocratic condition to fully evaluate the performance of pumping system for accurate and consistent delivery of CO_2 and the modifier.
5. The method is robust, easy to conduct, and reproducible.

The following are performed and used to assess SFC instrument performance:

1. Linearity: prepare five different concentration levels for all five components, analyze and calculate, respectively.
2. System precision: expressed as the relative standard deviation (%RSD) for the peak area, and retention time of each component to verify the consistence of pumping system.

Pfizer conducted the system performance qualification on Agilent 1260 infinity analytical SFC system by using the analytical method described above. The system performance qualification results of specificity, linearity and system precision are summarized in Fig. 9.2 and Tables 9.3 and 9.4 [20–21].

Table 9.3 Linearity at a range of 0.25 – 0.70 mg/mL (nominal conc. of 0.5 mg/mL)

Parameter (acceptance criteria)	r^2 (≥0.9990)	*y*-Int. (b) (≤2.0%)*	%Response factor to nominal (98–102%)				
			50%	75%	100%	113%	120%
Flavone	0.99999	1.2	101.06	100.49	100	99.91	99.66
Ketoprofen	1.00000	0.3	100.28	100.13	100	100.04	99.97
Amcinonide	0.99999	0.6	99.63	99.65	100	100.15	100.13
Carbamazepine	0.99999	1.2	101.00	100.65	100	99.88	99.79
Sulfamethazine	0.99982	1.4	100.73	101.04	100	100.15	99.35

**y*-Intercept (expressed as peak response) ≤ 2.0% of nominal sample concentration;.

As shown in Table 9.3, the linearity results across a minimum of five concentrations demonstrated that each standard met the acceptance criteria for the correlation coefficient, *y*-intercept and response factor (three key parameters for linearity assessment).

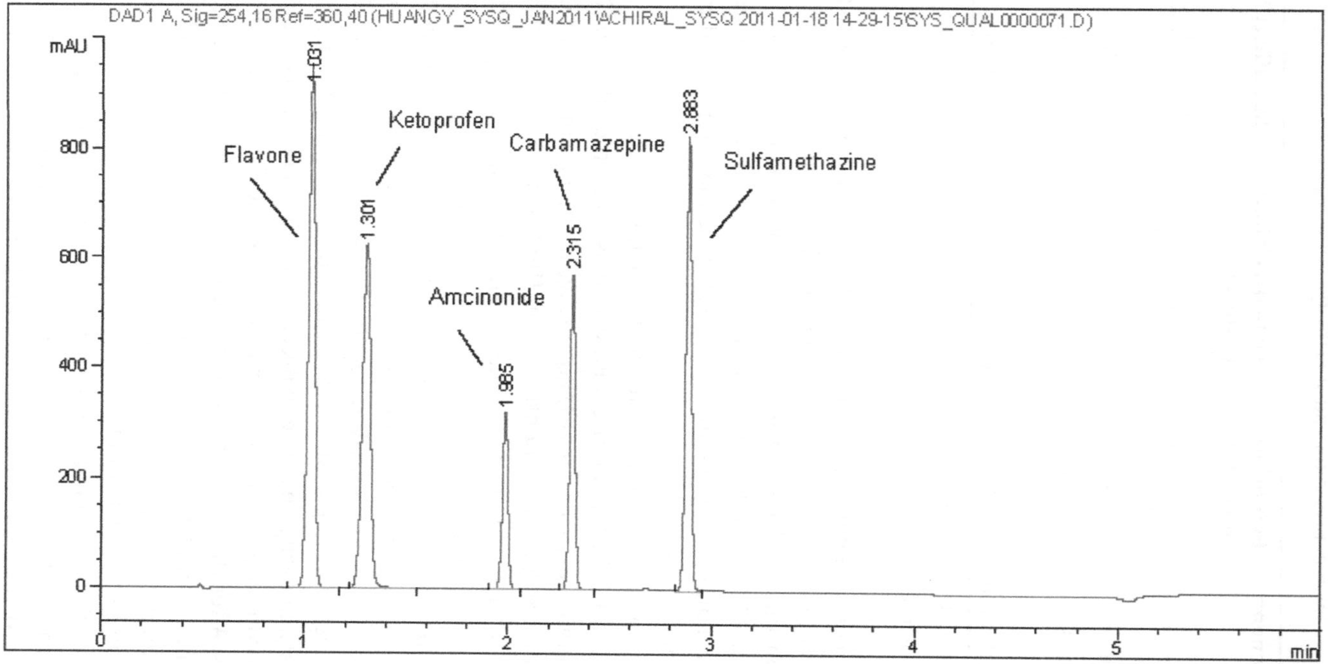

Figure 9.2 Chromatograms of standard mix at nominal concentration of 0.5 mg/mL.

Table 9.4 System precision and specificity

Components	Retention time (RT)	System precision[a] (6 injections)		Specificity[b]
		%RSD	%RSD (area)	USP resolution
Flavone	1.04	0.44	0.44	–
Ketoprofen	1.31	0.34	0.38	3.81
Amcinonide	2.03	0.11	0.41	10.94
Carbamazepine	2.37	0.11	0.41	6.25
Sulfamethazine	2.99	0.04	0.32	11.76

[a]Acceptance criteria ≤1.0%.

[b]Acceptance criteria: ≥ 1.5.

Five components were well separated from each other with USP resolution ≥1.5 for all components. The system precision of retention time and peak area was ≤1.0% for all components. The system precision results also indicated that both CO_2 delivery pump, modifier pump and auto—injector performed consistently and met the specifications well.

In summary, the Pfizer internal instrument qualification for Agilent 1260 infinity analytical SFC system was completed according to predefined instrument OQ/PQ testing protocol. For system OQ, the hardware similar or identical to an HPLC system (e.g., modifier pump, UV detector, column heater and injector) was qualified according to HPLC OQ specifications for the same components. The CO_2 pump was qualified following the similar specification applied to HPLC pump. The BPR specification was defined based on achieving satisfactory S/N on SFC UV detector. The system PQ were carried out through running a SFC test method to verify that the method specificity, linearity and system precision could meet the acceptance criteria comparable to those achieved by HPLC for impurity level analysis. After the successful completion of internal OQ/PQ, Agilent 1260 Infinity analytical SFC system was considered to be qualified for the intended impurity quantitative analysis and ready for GMP release and stability testing.

To fulfill the demand of instrument qualification requests from the pharmaceutical companies, the SFC manufacturers are moving toward providing the IQ/OQ/PQ protocols and performing the OQ/

PQ tests to qualify their analytical SFC systems (launched or to be launched) for GMP use. In the meantime, standardized IQ/OQ/PQ protocol across the industry is required to harmonize the SFC system qualification practice for GMP use. The success in analytical SFC OQ/PQ has dramatically extended its applications not only to GMP assay analysis but also to low-level impurity quantitation, which is essential to make chiral API clinical release and stability testing by SFC feasible.

9.3 Analytical SFC as Primary Chiral Purity Tool for Clinical Release and Stability Testing

The advantages of SFC on the speed and efficiency of analysis was explored extensively in the industry. Qualification of the latest marketed SFC systems to meet GMP requirements gives a clear landscape of what the current analytical SFC can offer to pharmaceutical impurity analysis. With the advanced SFC systems achieving satisfactory sensitivity, it is applicable for pharmaceutical impurity quantitation, especially for chiral analysis.

SFC has been widely used in early drug discovery for its demonstrated advantages over HPLC in chiral separation. For chiral API, the assessment and control of the enantiomeric impurity is critical, and generally an SFC method is developed during the drug discovery phase. When the chiral API progresses to later phases of development, the SFC chiral assay method usually remains in use because the changes in the synthetic route will mostly affect the achiral impurity profile but with minimal change to the enantiomeric impurity. In other words, the method for achiral impurity profile may need to be re-developed to reflect the new synthetic route; however, the method for the enantiomeric impurity usually is not typically affected and the same method can be propagated to later phases, unless new achiral impurities interfere with the chiral analysis, which is fairly rare. Thus, being able to apply SFC in the final API release and stability testing is the most cost- and time efficient for chiral impurity analysis as the same method can be used from the drug discovery to the drug development and commercialization phases.

The chiral method development and validation approaches are illustrated as follows:

Approach 1: HPLC workflow

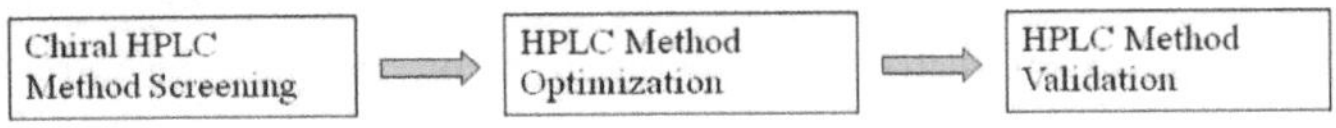

Approach 2: HPLC and SFC combined workflow

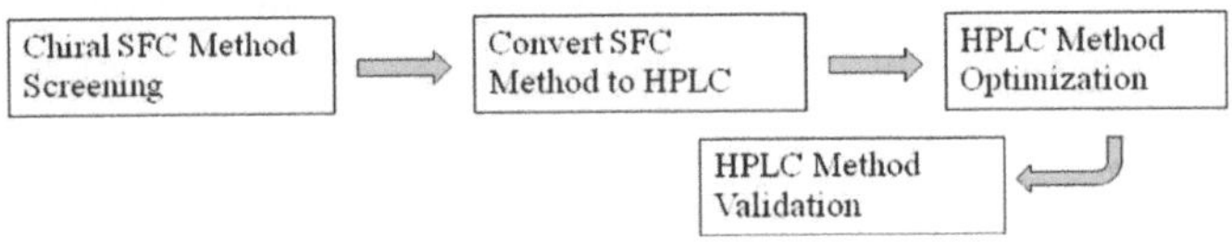

Approach 3: SFC workflow

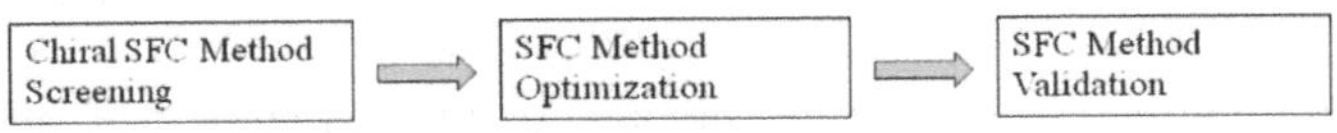

With the 3–5 times faster separation speed offered from the SFC technique, Approach 3 is the most efficient and effective workflow among the three, whose application feasibility needs to be proved for the clinical release and stability testing.

Pfizer has completed these feasibility studies in the past few years, which confirms that SFC can replace HPLC as the primary chiral purity tool for clinical API release and stability testing [23–24]. The detail of the studies will be described in Sections 9.3.2 and 9.3.3.

9.3.1 Method Validation Parameters

Analytical method validation is the process to confirm that the analytical procedure employed for a specific test is suitable for its intended use. Per GMP requirements, the analytical methods used in clinical release and stability testing need to be validated and performed on the qualified instruments. The analytical method validation is directed to the four most common types of analytical methods:

1. identification tests
2. quantitative tests for impurities content
3. limit tests for the control of impurities
4. quantitative tests of the active moiety in the samples of drug substance and drug product or other selected component(s) in the drug product

For chiral impurities analysis, the chromatographic method is required to pass the validation for impurity quantitative test prior to GMP use. The method validation parameters and acceptance criteria must ensure that the analytical method is appropriately validated for its intended use. The method validation protocol, detailed in Table 9.5, presents the suitability of the method for inclusion in proposed regulatory submissions and in accordance with current ICH guidance Q2(R1).

Table 9.5 Method validation protocol

Parameter	Acceptance criteria
Specificity	• Resolution (R_S) ≥1.5 from main band • $R_S \geq 1.0$ for separation of specified impurity from each other
Linearity/range	Range from reporting level (RL) to 2× specification: five levels • $r \geq 0.9950$ • y-Intercept ≤ 10.0% of response of maximum (2× specification) • Relative response factor of impurity standards across the range must be ±10% relative to that obtained at the expected specification level • For %area assay, response factor of RL level should be 90–110% of that obtained at nominal
Accuracy	3 replicates at 3 levels (LOQ, Spec and 2× of Spec) - Level (%) ≤0.15%, mean recovery 100 ± 15% - Level (%) >0.15%, mean recovery 100 ± 10%
System precision	6 injections of LOQ, RSD ≤ 10% for peak response
Repeatability	RSD ≤ 10% (use data obtained from accuracy validation)
LOQ	S/N at least 10

9.3.2 Feasibility Study on Chiral SFC Used for Clinical Release and Stability Testing

In 2010, analytical chemists at Pfizer conducted a feasibility study to determine if chiral SFC could be used for the clinical release and stability testing method by validating two NCE chiral APIs. The compound information is provided in Table 9.6.

Table 9.6 Information on selected APIs

	PF-00981823	PD-0348292
Current chiral spec	0.5%	0.1%
Target LOQ	0.1%	0.05%
Impurity profile complexity	**Low**, only API with its enantiomer PF-03507819	**High**, enantiomer, diastereoisomers, and other process impurities
HPLC release method availability	No	Yes

The chiral HPLC methods and alternative SFC methods were developed and validated on the following instrumentation for the feasibility studies:

1. Agilent HPLC 1100 system with DAD (G1315B) detector–Agilent 1100 system
2. Berger Analytical SFC equipped with Agilent 1100 DAD (G1315B) and MSD SL detectors—Chiral SFC method screening system
3. Berger Analytical SFC equipped with Agilent 1100 DAD detector (G1315B)-Berger SFC system
4. Aurora SFC Fusion^{TM} A5 module connected to Agilent 1100 HPLC system with DAD detector (G1315B)–Aurora-Agilent 1100 SFC system
5. Aurora SFC Fusion^{TM} A5 module connected to Agilent 1200 HPLC system with DAD detector (G1315B)–Aurora-Agilent 1200 SFC system
6. Waters Thar MethodStation SFC equipped with Water 2998 PDA detector–Thar MS SFC system

The chiral HPLC method development and validation was performed on the Agilent 1100 HPLC system (Instrument 1). The chiral SFC methods were developed from a chiral SFC method screening system (Instrument 2), then optimized and validated on the SFC analytical systems (Instruments 3–6). Both HPLC and SFC method were validated to the level required for compound registration. The key validation parameters were compared to demonstrate the following:

1. Chiral SFC methods can meet the similar requirements for HPLC methods and thus can be implemented as the primary method to replace conventional HPLC methods for chiral purity analysis in clinical release and stability testing.
2. Chiral SFC can improve business efficiency by shortening method development time without compromising on method performance.

9.3.2.1 Method validation results and discussions for PF-00981823

PF-0098123, a NCE in a phase I study, contains only one chiral center. The chiral separation is simple because only one enantiomer impurity (PF-03507819) needs to be separated. Method quantitation limit was determined at 0.1%, sufficient for the specification of ≤0.5%. Both a normal phase chiral HPLC method and a chiral SFC method were developed using the same Chiralpak AD-H, 5 μm, 4.6 × 250 mm column. Hexanes/isopropyl alcohol (IPrOH) was used as the HPLC method mobile phase, while isopropyl alcohol was the modifier in SFC method. HPLC run is 8 min, while SFC run is 3 min, ~3-fold faster. The chromatographic method conditions are summarized in Table 9.7 and the chromatograms of nominal concentration standard are shown in Figures 9.3–9.5.

Table 9.7 Chiral HPLC and SFC method conditions

	NP-HPLC (Agilent 1100)	**SFC (Berger and Aurora-Agilent 1100 SFC systems)**
Column	Chiralpak AD-H, 5 μm, 4.6 × 250 mm	
Detection	240 nm	
Sample conc.	0.7 mg/mL	
Column temp.	25°C	40°C
Mobile phase	Heptane:IPrOH (60:40)	CO_2 with 9% IPrOH modifier
Flow rate	0.8 mL/min	4.0 mL/min; (Pressure: 120 bar)
Injection volume	5 μL	10 μL
Isocratic run time	8 min	3 min

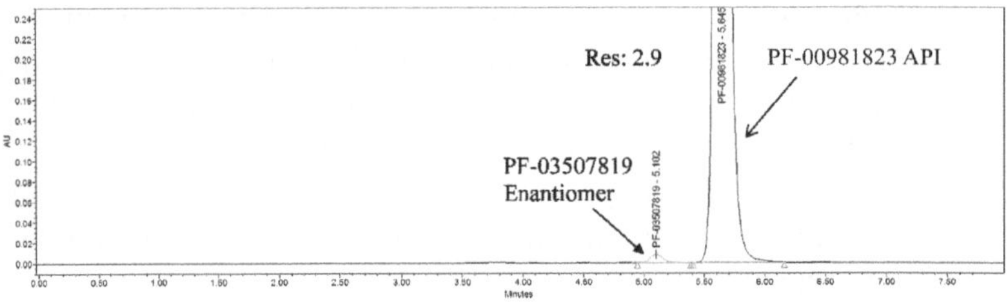

Figure 9.3 Chiral HPLC Chromatogram (Agilent 1100 HPLC System).

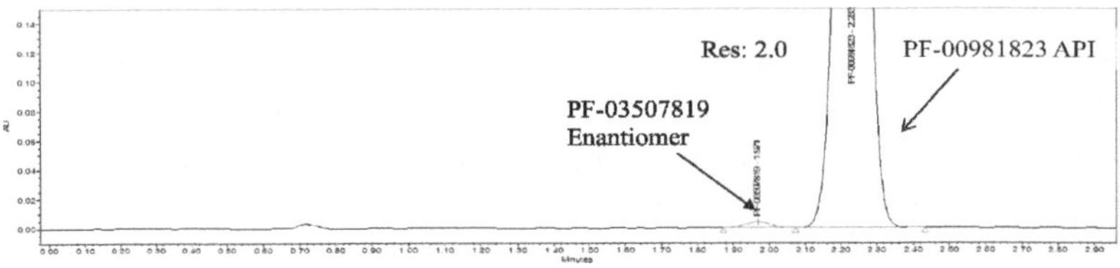

Figure 9.4 Chiral SFC Chromatogram of (Berger SFC System).

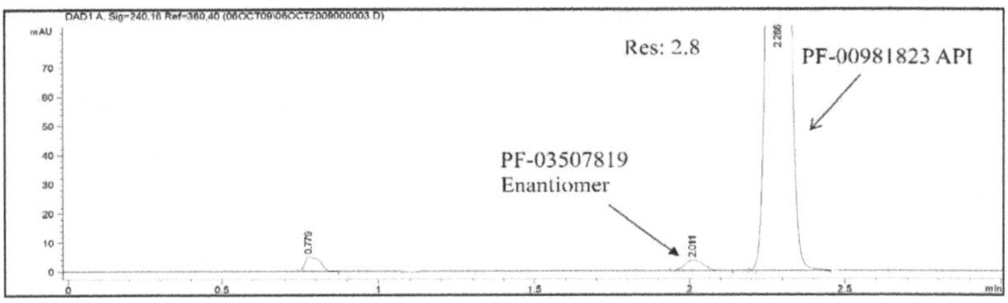

Figure 9.5 Chiral SFC Chromatogram (Aurora Fusion A5-Agilent 1100 SFC System).

The method validation of HPLC and SFC were performed under the same validation protocol (Table 9.5), with the validation results presented in Table 9.8.

Table 9.8 Method validation parameters and results

Parameter	NP-HPLC-Agilent 1100	SFC–Aurora/Agilent 1100	SFC–Berger
Specificity	Res = 2.9	Res = 2.8	Res = 2.0
Linearity (0.1–1.0%)	r = 0.9999 y-Int% = 0.79	r = 0.9999 y-Int% = 1.0	r = 0.9985 y-Int% = 0.97

Parameter		NP-HPLC-Agilent 1100	SFC–Aurora/Agilent 1100	SFC–Berger
Accuracy (*n* = 3 for each conc. level)	0.1%	105%	111%	114%
	0.5%	106%	109%	97%
	1.0%	106%	107%	99%
System precision at LOQ		2%	4%	7%
Repeatability		2%	2%	9%
LOQ @ 0.1%		S/N = 180	S/N = 20	S/N = 20

Experiments and results discussion are presented in the following text.

9.3.2.1.1 *Specificity*

The resolutions between API (PF-00981823) and its enantiomer (PF-03507819) all met the acceptance criteria of ≥1.5 for all three methods (Agilent 1100 HPLC system, Berger SFC and Aurora Fusion™ A5-Agilent 1100 SFC system). The resolution achieved from Aurora Fusion™ A5-Agilent 1100 SFC system (Rs = 2.8) is the same as that obtained on Agilent 1100 HPLC system (Rs = 2.9), while the Berger SFC system presented the lowest but acceptable resolution (Rs = 2.0).

9.3.2.1.2 *Linearity*

Five impurity level linearity samples were prepared at 0.1–1% of the nominal concentration. Comparable linearity results were obtained from all three chromatographic systems, which all passed the validation specification. The Berger SFC system performed inferiorly but still acceptable, while the Aurora Fusion™ A5-Agilent 1100 SFC system and Agilent 1100 HPLC system performed similarly.

9.3.2.1.3 *Accuracy*

Three replicate sample preparations each at three concentration levels of LOQ (0.1%), specification (0.5%) and 2× specification (1.0%) were prepared. The mean of the obtained recoveries from three samples at each level were calculated. All three methods passed the acceptance criteria that the mean recovery should be 100 ± 15% for LOQ (0.1%) level and 100 ± 10% for the remaining two levels.

9.3.2.1.4 *System precision*

The system precision was obtained from six injections of a LOQ sample; %RSD of the peak areas from the six injections met the acceptance criteria of RSD ≤10%, with HPLC method presenting the best result compared to the other two SFC methods.

9.3.2.1.5 *Repeatability*

The method repeatability was assessed from the accuracy data (a total of nine sample preparations). Both HPLC and SFC methods met the acceptance criteria of RSD ≤10%. Aurora-Agilent 1100 SFC system has the same repeatability result as the HPLC system.

9.3.2.1.6 *Limit of quantitation*

SFC method can achieve the target LOQ level at 0.1% with sufficient sensitivity (S/N = 20). However, it is not surprising that HPLC still presents a significantly higher sensitivity than those from the SFC systems due fewer sources contributed to the generation of noise for the UV detector used in HPLC systems.

In summary, both HPLC and SFC methods passed validation acceptance criteria and suitable for intended use. As expected, NP-HPLC still showed better sensitivity than the SFCs. However, the current advanced SFC system already achieved sufficient sensitivity for intended clinical release and stability testing. As discussed above, Aurora-Agilent 1100 SFC system demonstrated comparable performance to NP-HPLC. The successful validation of the chiral SFC method provides a streamline workflow to employ the same SFC method from the drug discovery phase as a screening method to the later phases as a clinical release testing method.

9.3.2.2 SFC method development and validation for PD-0348292

9.3.2.2.1 *PD-0348292 SFC Method Development*

With the above SFC method validation success, an extended feasibility study was conducted on a higher complexity chiral compound PD-0348292 to demonstrate the suitability of SFC in chiral separation compared to the available validated HPLC as both methods can now meet the clinical release testing requirements.

PD-0348292 contains two chiral centers resulting in 3 chiral impurities (an enantiomer and two diastereomers). Besides the chiral impurities, there are several other in-process impurities observed in the API that require separation. All the impurities are shown in Fig. 9.6.

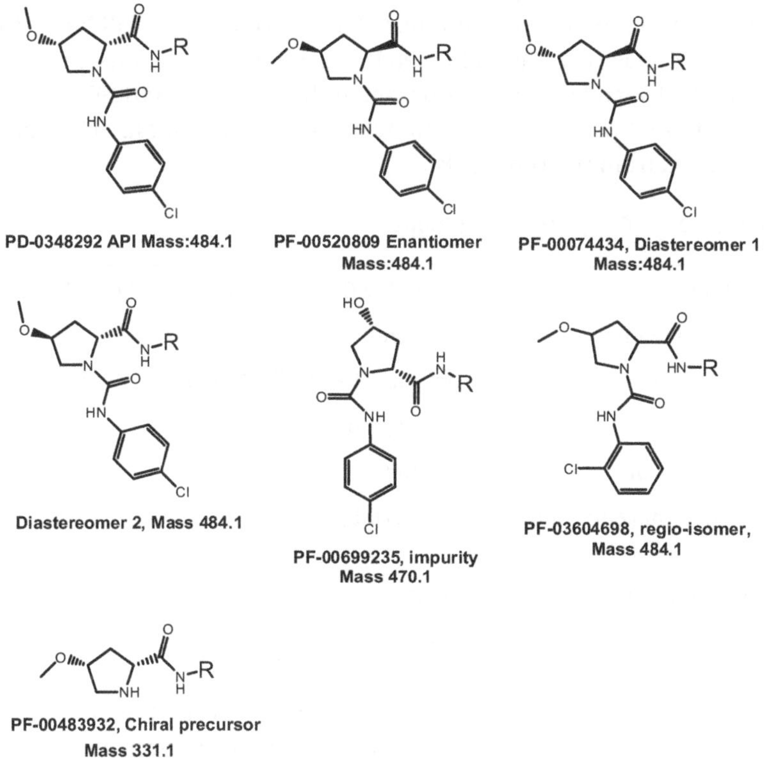

Figure 9.6 Chemical structures of PD-0348292 and its impurities.

PF-00483932 is the chiral precursor used the final synthesis step. PF-00520809 is the enantiomer. PF-00074434 and diastereomer 2 are the two diastereomers. PF-00699235 is a process-related impurity presented in the API batches, and PF-03604698 is a potential region-isomer impurity. For all these five impurities, authentic materials of PF-00520809 (enantiomer), PF-00074434 (diastereomer 1) and PF-00699235 (process-related impurity) were available for analytical method development, with the other two, diastereomer 2 and PF-03604698 (regio-isomer) not available.

During HPLC method development, it was found that the diastereomers and the process-related impurities (listed in Fig. 9.6) would interfere the separation and quantitation of the enantiomer (PF-00520809). It has been a challenge to develop an HPLC method for such a complex chiral separation. After extensive development work, a column coupling technique was utilized to achieve the quantitation of the enantiomer. The method was validated for clinical release use. The validated HPLC method conditions are listed in Table 9.9, and the chromatogram of the API (PD-0348292) spiked all potential process-related impurities is shown in Fig. 9.7, which was cited from reference [28].

Table 9.9 Chiral HPLC conditions

Column	2 coupled Chiralcel OD, 10 μm, 4.6 × 250 mm
Column temperature	Ambient
Mobile phase	Hexanes/MeOH/EtOH 880/90/30 (v/v/v)
Flow rate	1.2 mL/min
UV detection	245 nm
Sample concentration (nominal)	3.0 mg/mL
Injection volume	20 μL
Run time	110 min

An in-process impurity H-107654-001 (≤0.1%) was present in most batches, but not identified. As reported, even employing the exotic column coupling technique, the resolution between the critical pair, the enantiomer and the H-107654-001, was only 1.0, barely passing the minimum resolution requirement and causing the enantiomer peak integration and quantitation to be difficult. In addition, the precursor PF-00483932 co-eluted with API PD-0348292 in this method, making the %area quantitation not feasible. The enantiomer has to be quantitated by external enantiomer standard. Additionally, the HPLC run time is 110 min per injection.

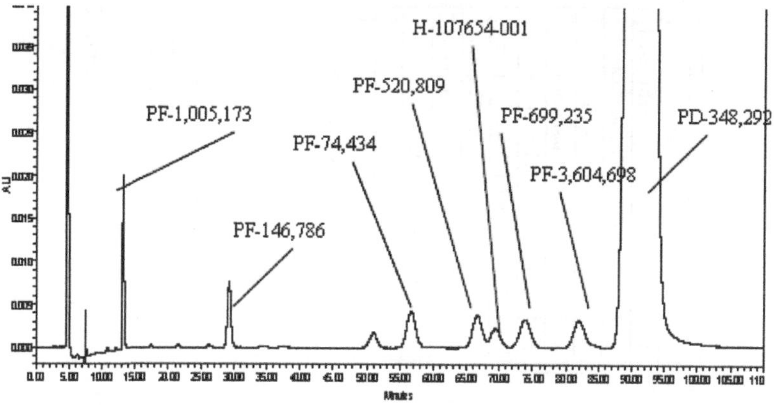

Figure 9.7 Chromatogram of PD-0348292 spiked with all potential impurities.

An SFC method was developed starting from the chiral method screening system (Berger analytical SFC equipped with Agilent MSD SL) connecting with a MS detector for peak tracking and identification. After the best separation method was obtained from the screening system, it was optimized with the final method conditions listed in Table 9.10, and the chromatogram shown in Fig. 9.8.

Table 9.10 Chiral SFC conditions

Column	Chiralcel OJ-H, 5 µm, 4.6 × 250 mm
Column temperature	40°C
Mobile phase	84% CO_2 with 16% modifier (0.1% IPAm in MeOH)
Flow rate	4.0 mL/min
Pressure	150 bar
UV detection	245 nm
Sample concentration (nominal)	3.0 mg/mL
Injection volume	20 µL
Run time	15 min

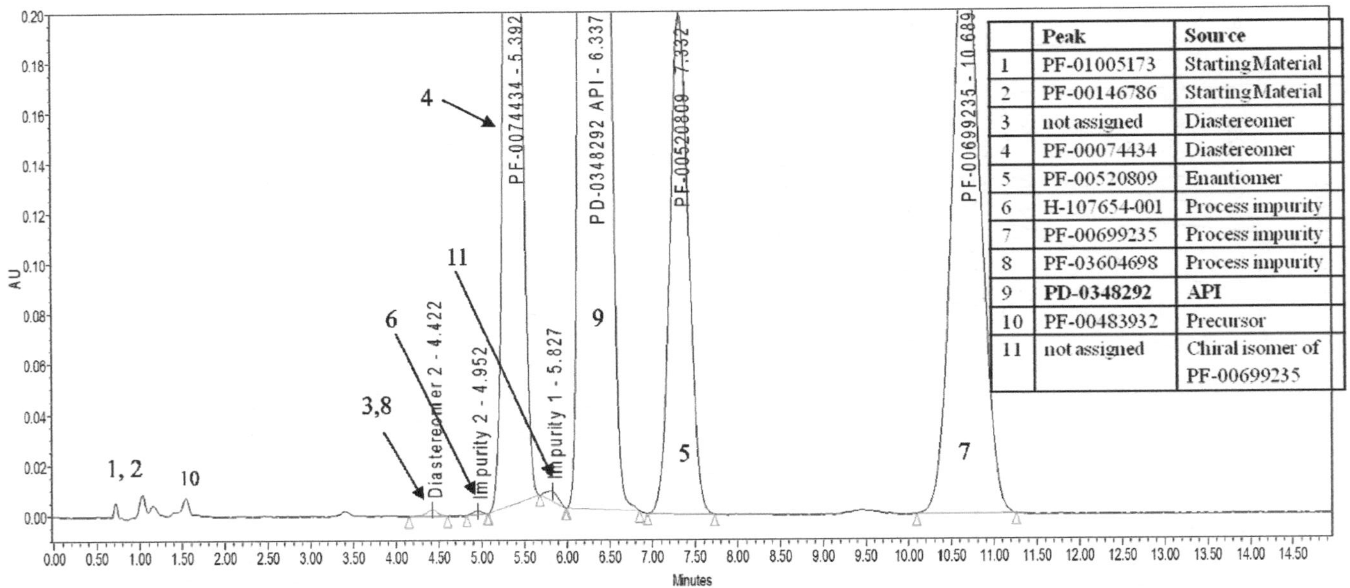

	Peak	Source
1	PF-01005173	StartingMaterial
2	PF-00146786	StartingMaterial
3	not assigned	Diastereomer
4	PF-00074434	Diastereomer
5	PF-00520809	Enantiomer
6	H-107654-001	Process impurity
7	PF-00699235	Process impurity
8	PF-03604698	Process impurity
9	**PD-0348292**	**API**
10	PF-00483932	Precursor
11	not assigned	Chiral isomer of PF-00699235

Figure 9.8 SFC Chromatogram of PD-0348292 spiked PF-00520809 (enantiomer), PF-00074434 (diastereomer) and PF-00699235 (impurity).

The starting materials of PF-01005173 (Peak 1) and PF-00146786 (Peak 2), as well as the chiral precursor of API, PF-000483932 (Peak 10) were eluted earlier in the chromatogram (<2 min). The API (PD-0348292, Peak 9), known enantiomer (PF-00520809, Peak 5), diastereomer 1 (PF-00074434, Peak 7) and impurity (PF-00699235) were baseline separated from each other (Rs > 1.5), with a total run time only 15 min (i.e., 1/7 of that from previous HPLC method).

Another advantage of SFC is its compatibility with mass spectrometry, a key part for its becoming a convenient and powerful identification tool for rapid method development. In this method development case, because not all authentic materials for known impurities were available, MS analysis was essential to accurately identify and assign the peaks for the impurities without available authentic material. As shown in SFC chromatogram (Fig. 9.8), only PF-00520809, PF-00074434 and PF-00699235 could be identified by the available authentic materials, the rest peaks were assigned by the assistance of MS analysis.

The MS total ion chromatogram and SIM at *m/z* 544 [M + 60 (isopropylamine)] to extract API and its chiral isomers are shown in Figs. 9.9 and 9.10.

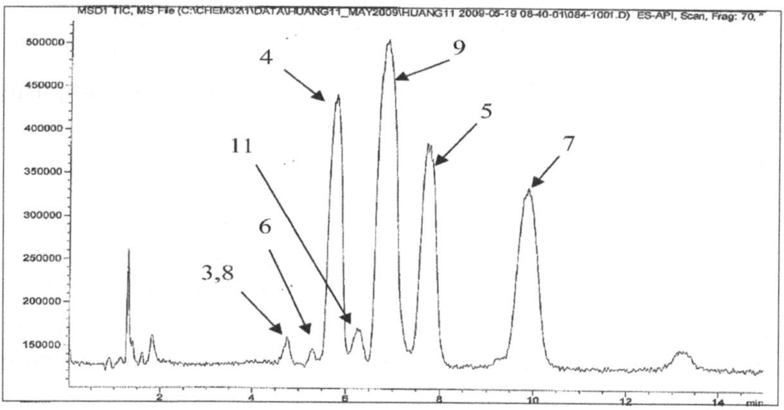

Figure 9.9 MS Total Ion Chromatogram of PD-0348292 spiked PF-00520809 (enantiomer), PF-00074434 (diastereomer) and PF-00699235(impurity), peak assignment (see Fig. 9.8).

Four peaks were obtained on the chromatogram of SIM at *m/z* 544. Compared to the SFC chromatogram of PD-0348292 (API, peak 9) spiked with PF-00520809 (enantiomer, peak 5) and PF-00074434 (diastereomer, peak 4), these three peaks were identified easily. The fourth peak at RRT 0.70 was thus identified to be either the second diastereomer (peak 3) or PF-03604698 (regio-isomer, peak 8), which has the same mass as the API.

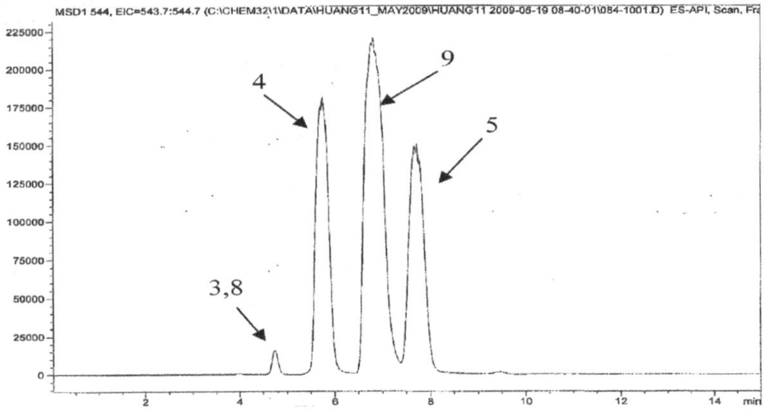

Figure 9.10 SIM at m/z 544 [M + 60 (isopropylamine)] to extract API and its Chiral Isomers peak assignment (see Fig. 9.8).

MS analysis also confirmed that API and enantiomer peaks are pure and well separated from the precursor PF-00483932. In addition, SFC/MS identified a new peak (peak 11) not observed in the HPLC method, which has the same mass as in-process impurity of PF-00699235, thus possibly its chiral isomer. With all other peaks identified, the peak 6 is most likely H-107645-001 (the critical pair with enantiomer) in the HPLC method.

Table 9.11 summarizes the peak assignment and specificity of SFC method. The starting materials of PF-01005173 (Peak 1) and PF-00146786 (Peak 2), as well as the chiral precursor of API, PF-000483932 (Peak 10) were eluted in the front of chromatogram, and the resolution among these peaks were not calculated and listed in the table.

In summary, this SFC method can successfully achieve the desired separation on a single column and is able to more accurately quantify the enantiomer (PF-00520809) and thus to simplify the quantitation by using %area instead of having to use the external enantiomer standard. The enantiomer is well separated from the API and other impurities with Rs = 2.6, a dramatic improvement from the previous HPLC method (Rs = 1.0), which improves the peak integration and quantitation accuracy. In addition, that no peak co-eluted with API makes quantitation by %area feasible, which improves work efficiency by eliminating the external standard preparation. The run time was significantly shortened from 110 min to 15 min (Berger SFC) and 10 min (Aurora-Agilent 1100/1200 SFC systems), as shown in Fig. 9.11. Compared to the available HPLC method, the SFC method is more accurate, efficient, and time saving; thus, it was validated to replace the HPLC method for clinical release and stability testing use.

Table 9.11 Peak Assignment and Resolution of SFC Method

Peak #	RRT	Compound	Description	ID method	Resolution[b]
3 or 8	0.70	Diastereomer or PF-030604698	Diastereomer of API or process-related impurity	SFC-MS	NA
6	0.78	H-107654-001	Process impurity	Note 1	2.2
4	0.85	PF-00074434	Diastereomer	AM[a]/ SFC-MS	1.7
11	0.92	Chiral isomer of PF-00699235	Process impurity	SFC-MS	1.4
9	1.0	PD-0348292	API	—	1.5
5	1.16	PF-00520809	Enantiomer	AM/ SFC-MS	2.6
7	1.69	PF-00699235	Process impurity	AM/ SFC-MS	6.5

Note: As no authentic material and mass data are available, this peak is assigned as the last peak after all other peaks have been positively identified.

[a]AM: Authentic material.

[b]Resolution between the closest peak.

9.3.2.2.2 *PD-0348292 SFC method validation results and discussions*

The above SFC method was then validated in order to replace the current effective HPLC method. As the specification for the enantiomer (PF-00520809) was ≤0.1%, LOQ needed to be at least 0.05%. The method validation was initially performed on Berger SFC system followed the similar validation protocol from the validated HPLC method (Table 9.5). The nominal concentration of 3.0 mg/mL was directly adopted from the HPLC method. The HPLC method validation data was cited from its relevant validation report. The validation results on both chiral normal phase (NP) HPLC and SFC methods are listed in Table 9.12.

Table 9.12 Method validation results of NP-HPLC and SFC methods

Parameter		NP-HPLC [a]20 μL injection)	SFC (Berger) (10 μL injection)
Specificity[b]		Res =1.0	Res = 2.6
Range		0.05–0.4% of nominal conc.	0.05–0.5% of nominal conc.
Linearity[c]		R = 0.9995 y-Intercept = 0.90% of 2× Spec response (0.2% concentration)	r = 0.9999 y-Intercept = 0.80% of 2× Spec response (0.2% concentration)
Accuracy (recovery)[d]	0.05% (LOQ)	93% to 107%	99%
	0.1% (Spec)	93% to 107%	107%
	0.2% (2× Spec)	93% to 102%	96%
System precision[e]		2.2	6.1
Repeatability[f]		4.8	6.7
LOQ (0.05%)		S/N = 100	S/N = 12

[a] Validation data were cited from the Pfizer internal validation report.
[b] Resolution between PF-00520809 (enantiomer) and its closet peak.
[c] Linearity was assessed from five concentrations covered the range.
[d] Recovery data calculated by external standard of 0.1% specification, triplicate preparations at each level.
[e] %RSD for the 6 injections of LOQ solution.
[f] %RSD of the nine recovery data from accuracy/recovery experiment.

The SFC validation results of specificity, linearity, accuracy, system precision at LOQ, repeatability and LOQ met all the acceptance criteria listed in Table 9.5. Thus, the SFC method is suitable for clinical release testing use. NP-HPLC showed better sensitivity, system precision and repeatability while SFC is superior in the specificity (resolution) and analysis time (110 min run for HPLC, 15 min run for Berger SFC and 10 min run for Aurora-Agilent SFC).

9.3.3 Method Transferability

9.3.3.1 Study design

Analytical methods need to be transferred to other sites or labs as a project progresses through clinical development and commercialization. Thus, method transferability was studied in order to integrate the chiral SFC platform within all analytical communities from early phase to late phases and commercialization. Three analytical SFC systems in different analytical laboratories were selected for the transferability study as listed in Table 9.13. PD-0348292 chiral SFC method was used as the testing example. Transferability would be demonstrated by validating the same method in three different labs and instruments.

Table 9.13 Analytical SFC instruments for transferability study

Instruments	Detector	Function area
Aurora-Agilent 1100 SFC System	Agilent DAD G1315B	Early phase analytical lab (support up to phase 2b)
Aurora-Agilent 1200 SFC System	Agilent DAD G1315B	Late phase analytical lab (support phase 2b and beyond)
Waters Thar MethodStation SFC	Waters 2998	GMP lab (support all phases for release and stability testing)

These SFC systems were newly installed and the UV detector sensitivity was higher compared to the Berger SFC system involved in the previous method validation. Thus, the nominal concentration was reduced from 3.0 mg/mL to 2.0 mg/mL.

9.3.3.2 Results and discussions

The method transferability study was conducted on three analytical SFC systems in three different analytical laboratories. Table 9.14 summarizes the validation results together with the original data from Berger system for comparison.

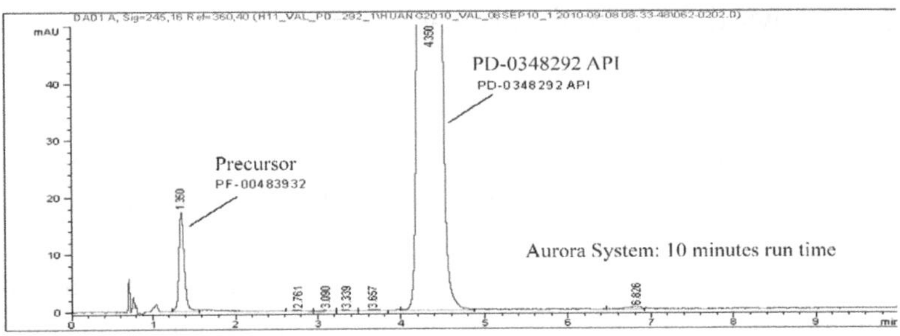

Figure 9.11 Chromatogram of PD-0348292 spiked with Precursor, PF-00483932.

Table 9.14 Validation results - the external standard (w/w) calculation

Parameter		Berger	Thar	Aurora 1100	Aurora 1200
Nominal conc.		3.0 mg/mL	2.0 mg/mL	2.0 mg/mL	2.0 mg/mL
Specificity[a]		Res = 2.6	Res = 2.0	Res = 2.3	Res = 2.5
Linearity (0.05–1.0%)		r = 0.9999 y-Int% = 0.8	r = 0.99994 y-Int% = 0.4	r = 0.99995 y-Int% = 0.7	r = 0.99993 y-Int% = 0.5
%Recovery/ accuracy (mean of 3 reps)[b]	0.05%	99%	96.50%	100.10%	99.50%
	0.10%	107%	97.80%	99.90%	98.70%
	0.20%	96%	98.60%	99.00%	97.30%
Repeatability[d]		6.70%	4.00%	0.70%	1.30%
Sys. precision[c]		6.10%	5.00%	0.87%	0.95%
S/N@LOQ		12	12	23	14

[a]Resolution: peak of interest (API and the enantiomer) with their closest peaks.

[b]%Recovery: Calculated by external standard of 0.1% spec.

[c]Six injections of LOQ@0.05%.

[d]Repeatability: calculated from nine preps of recovery data.

As demonstrated in the results table, the chiral SFC method has been successfully validated in the additional three SFC systems. No issue was foreseen for method transfer from Berger SFC system to Thar MethodStation SFC system as well as Aurora-Agilent 1100 and 1200 SFC systems.

After the method transferability study using different instruments, the method was further simplified to use %area normalization instead of external standard quantitation (w/w%). Additional validation included the system precision and linearity.

System precision for the three instruments was fully evaluated at nominal concentration, 0.05% (LOQ), and 0.1% (Spec) of nominal concentrations as shown in Table 9.15. Aurora systems (-Agilent 1100 or -1200) showed better precision than Thar SFC system for the 0.1 and 0.05% levels.

Table 9.15 System precision evaluation for Aurora-Agilent 1100/1200 SFC and Thar SFC

System precision solution used	Thar MethodStation	Aurora/1100	Aurora/1200
Nominal (2.0 mg/mL)	0.10%	0.29%	0.76%
0.1% of nominal (0.002 mg/mL)	3.8%	0.74%	0.64%
0.05% of nominal (LOQ, 0.001 mg/mL	5.0%	0.87%	0.95%

Additional linearity validation was performed with a five-point high-level linearity at the range of 80–120% of nominal, and a five-point low-level linearity at the range 0.05–1.0% of nominal. The linearity results obtained from Aurora-Agilent 1200 SFC system are presented in Table 9.16, and Figs. 9.12 and 9.13.

Excellent linear responses were achieved from an advanced analytical SFC system (such as Aurora-Agilent 1100 and 1200 systems) through high assay level to low impurity level, which enabled the quantitation of impurity by %area feasible. The additional validation results by using %area assay are presented in Table 9.17. All acceptance criteria are met.

Table 9.16 Additional linearity validation on Aurora-Agilent 1200 SFC system

% Nominal	RF	%RF*	Linearity
80.0%	6296.06	99.26	**High level**: 80–120% r = 0.99987 y-Int.: 0.5% of the nominal response Slope: 6353
90.0%	6325.11	99.70	
100.0%	6344.42	100.00	
107.0%	6319.27	99.60	
120.0%	6317.88	99.58	
0.05%	6427.43	101.43	Low level: 0.05–1.0% (spec: 0.1%) r = 0.99993 y-Int.: 0.5% of 0.5% level Slope: 6437 (99% of high-level slope) %RF of 0.05% (LOQ) = 101.43 (acceptance criteria within 90–110 of nominal)
0.1%	6430.60	101.46	
0.2%	6362.05	100.38	
0.5%	6309.53	99.55	
1.0%	6445.97	101.60	

* %RF: Relative to 100% nominal response.

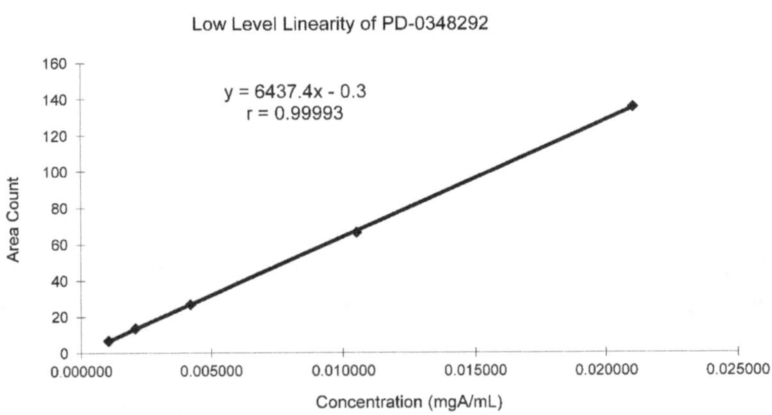

y-intercept of the 0.5 % nominal assay response: 0.5%

Figure 9.12 Low-level linearity (0.05%–1.0% of nominal conc., 2.0 mg/mL).

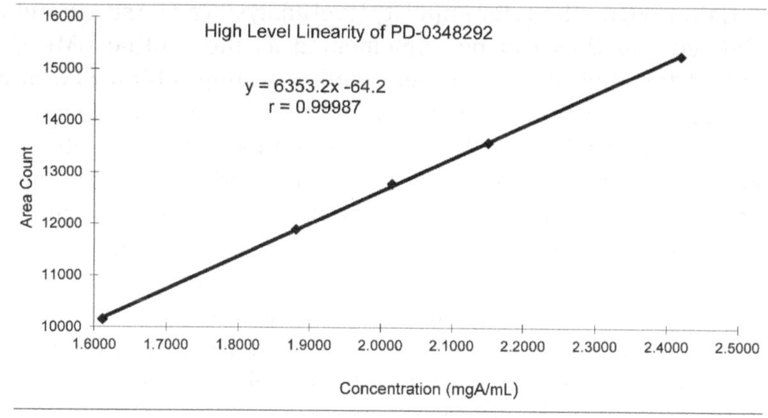

y-intercept of the nominal assay response: 0.5%,
Slope (6353) agrees with low level slope (6437) within 1.3%, confirming the good linearity across low to high level.

Figure 9.13 High-level linearity (80%–120% of nominal conc., 2.0 mg/mL).

Table 9.17 Additional Validation Results –Quantitated by %Area

Parameter		**Thar**	**Aurora 1100**	**Aurora 1200**
Nominal conc.		2.0 mg/mL	2.0 mg/mL	2.0 mg/mL
Specificity[a]		Res = 2.0	Res = 2.3	Res = 2.5
Linearity (low)		*r* = 0.99994 *y*-Int% = 0.4	*r* = 0.99995 *y*-Int% = 0.7	*r* = 0.99993 *y*-Int% = 0.5
%Response factor of LOQ to nominal		97	102	101
%Rec/accuracy (mean = 3)[b]	0.05%	101%	103%	102%
	0.10%	103%	102%	102%
	0.20%	104%	102%	100%
Repeatability[d]		2.8%	0.7%	1.3%
Sys. precision[c]		5.0%	0.87%	0.95%
S/N@LOQ		12	23	14

[a]Resolution: peak of interest (API and the enantiomer) with their closest peaks.

[b]%Recovery: Calculated vs. 100% nominal.

[c]Six injections of LOQ@0.05%.

[d]Repeatability: Calculated from nine preps of recovery data.

Chiral SFC methods for impurity level analysis can be successfully validated and thus can be implemented as the routine GMP QC method for clinical release and stability testing. Chiral SFC also showed superior separation power and analysis speed compared to HPLC methods when working on very complex analysis. Different SFC systems were also studied for the method transferability between Berger SFC, Waters Thar MethodStation SFC system and Aurora-Agilent 1100 and 1200 SFC systems. As all the systems passed method validation criteria, no issues were expected to be encountered when transferring the method among different systems in different labs. This demonstrates that SFC methods for chiral impurity analysis could be suitable for later phases, registration and commercial use in the future.

9.4 Conclusions

Recent advanced analytical SFC systems have led to the overall improvement on the detection sensitivity (due to noise reduction) and system robustness, without which the implementation of SFC as an alternative to conventional HPLC method in chiral impurity analysis would be never possible. The dramatic enhancement on the UV detection sensitivity can help achieve LOQ at 0.1% or even lower (e.g., 0.05%) of nominal concentration, essential for an impurity method. With a couple of feasibility study examples, the SFC method development and validation workflow promotes a consistent and efficient approach for chiral analysis. Chiral SFC demonstrated to be an advanced technique to potentially replace conventional HPLC as the primary chiral analysis tool for clinical release and stability testing throughout all phases of drug development and commercialization. The implementation of SFC in the regulated analytical laboratories would significantly shorten the method development, validation and release testing analytical time, and bring a more efficient and effective streamline practice to analytical chemists in the pharmaceutical industry.

Along with the continuous improvement on SFC instrumentation and further opportunities to explore how to better apply the technology in pharmaceutical analysis, SFC will become increasingly accepted as a fast and efficient separation tools for pharmaceutical analysis. In the future, SFC will not only continue its role as a GMP chiral analysis tool, but also will be utilized as a complementary analytical tool orthogonal to RP-HPLC to solve some trace-level impurity analysis challenges and ensure quality and safety of drug product. SFC also can be potentially used as an alternative achiral analytical tool for those challenging analytes in RP-HPLC separations for pharmaceutical analysis.

Acknowledgments

The author would like to thank Pfizer PharmaTherapeutics Pharmaceutical Sciences supporing scientific initiatives and, in particular, the support obtained from Karen M. Alsante, Michael Coutant, Richard L. Iwrin, and Michael Lovdahl who provided necessary management guideline and support for implementation of SFC in Pfizer ARD and GMP Analytics (QC) laboratories for GMP use. Thanks are extended to Xiaohua Zhang of Merck for her generous review and helpful discussion, as well as to Agilent, Aurora SFC Inc., and Waters for their cooperation on Pfizer's SFC instrument qualifications. Appreciation also goes to Gregory K. Webster of AbbVie, Inc. and the publisher for the invitation to contribute to the book.

References

1. International Conference on Harmonisation, Harmonised Tripartite Guideline, Impurities In New Drug Substances, Q3A(R2).
2. International Conference on Harmonisation, Harmonised Tripartite Guideline, Validation of Analytical Procedures: Text and Methodology, Q2(R1).
3. Swartz, E.M. (2005) UPLC: An introduction and review, *J. Liquid Chromatogr. Relat. Technol.*, 28, 1253–1263.
4. Nováková, L. (2006) Advantages of application of UPLC in Pharmaceutical analysis, *Talanta*, 68, 908–918.
5. Gōrōg, S. (2007) The changing face of pharmaceutical analysis, *Trends Anal. Chem.*, 26(1), 12–17.
6. Dong, M. (2007) Ultrahigh-pressure LC in pharmaceutical analysis: performance and practical issues, *LCGC North Am.*, 25(7), 656–666.
7. Srivastava, B., Sharma B.K. (2010) UPLC: A chromatography technique, *Int. J. Pharm. Qual. Assur.*, 2(1), 19–25.

8. White, C. (2005) Integration of supercritical fluid chromatography into drug discovery as a routine support tool: Part 1. Fast chiral screening and purification, *J. Chromatogr. A*, 1074, 163–173.
9. Maftouh, M., Granier-Loyaux, E. (2005) Screening approach for chiral separation of pharmaceuticals: Part III. supercritical fluid chromatography for analysis and purification in drug discovery, *J. Chromatogr., A*, 1088, 67–81.
10. Zhang, Y., Watts, W. (2004) Rapid method development for chiral separation in drug discovery using multi-column parallel screening and circular dichroism signal pooling, *J. Chromatogr. A*, 1049, 75–84.
11. Abbott, E., Veenstra, T.D. (2008) Clinical and pharmaceutical applications of packed-column supercritical fluid chromatography, *J. Sep. Sci.*, 31, 1223–1230.
12. Taylor, L. (2010) Supercritical fluid chromatography, *Anal. Chem.*, 82, 4925–4935.
13. Anton, K., Siffrin, C. (1999) Packed-column SFC in the pharmaceutical industry: cGMP aspects, Dossier, *Analysis*, 27(8), 691–701.
14. Wang, Z., Zhang, H., Liu, O., Donovan, B. (2011) Development of an orthogonal method for mometasone furoate impurity analysis using supercritical fluid chromatography, *J. Chromatogr. A*, 1218, 2311–2319.
15. Brunelli, C. (October 2012) Latest advances and applications of SFC for analysis in drug development, paper presented at SFC 2012 International Conference, Brussels, Belgium.
16. Zeng, L. (October 2012) Advanced SFC/MS Technologies to support drug discovery and development, paper presented at SFC 2012 International Conference, Brussels, Belgium.
17. Secreast, S. (2007) Applying analytical SFC to process analytics, paper presented at Pfizer Internal Analytical Science Forum, Ann Arbor, MI, January 2007.
18. Helmy, R., Biba, M., Zang, J (2007) Improving sensitivity in chiral supercritical fluid chromatography for analysis of active pharmaceutical ingredients, *Chirality*, 19, 787–792.
19. Berger, T.A., Berger, B.K. (2010), Minimizing UV noise in supercritical fluid chromatography. I. Improving back pressure regulator pressure noise, *J. Chromatogr. A*, 1218(16), 2320–2326.
20. Rodwell, P., Davis, A., Huang, Y., Carrick, P. (19–23 June 2011) Analytical SFC system performance qualification and method validation in early development, paper presented at HPLC 2011, Budapest, Hungary.

21. Huang, Y. (September 2011) Analytical SFC for pharmaceutical impurities: Implementation and challenges in GMP labs, paper presented at Pfizer SFC symposium, Groton, CT USA.

22. Sun, J. (2011) Critical attributes and technical challenges of successful SFC instrument qualification, *Pharm. Rev.*, May/June, 79–83.

23. Huang, Y., Wang, J., Vuong, D., Hardink, M. (September 2010), A feasibility study for implementation of chiral SFC as the primary chiral purity tool for API release and stability testing through all clinical phases, paper presented at SFC 2010 International Conference, Stockholm, Sweden.

24. Huang, Y., Wang. J., Vuong, D., Hardink, M. (March 2011) Implementation of chiral SFC as the primary chiral purity tool for API release and stability testing—feasibility and transferability study, paper presented at Pittcon 2011, Atlanta, GA USA.

25. Huang, Y., Riley, F., Hardink, M., Vuong, D.V., Alsante, K. (October 2012) Method validation and transferability on advanced analytical SFC instruments: Update of chiral implementation in Pfizer GMP regulated laboratories, paper presented at SFC 2012 international Conference, Brussels, Belgium.

26. U.S. Food and Drug Administration. U.S. Government Printing Office: Washington, D.C., 1987, General Principles of Validation.

27. UPS general chapter <1058>, Analytical Instrument Qualification.

28. Xiang, Y., Sluggett, G.W., Jiang, H. (2008) Coupling two chiral columns for resolving process impurity interference in the PD-0348292 enantiomeric separation, Analytical Matters (Pfizer internal analytical publication), Vol., 3, 163–167.

Chapter 10

Supercritical Fluid Chromatography–Mass Spectrometry

Laila Kott
Millennium Pharmaceuticals (The Takeda Oncology Company), 40 Landsdowne St., Cambridge, MA, USA
Laila.Kott@mpi.com

10.1 Introduction

When Terry Berger's reference text on supercritical fluid chromatography (SFC) was published in 1995, mass spectral detection was mentioned only in four separate sentences, in the entire book [1]. Currently, a literature search yields about 150 references to supercritical fluid chromatography–mass spectrometry (SFC-MS), where about one quarter of them show some application to the pharmaceutical industry, under the general categories of medicinal/discovery chemistry [2,3], chiral separations [4], high-throughput screening [5–7], and biologics (lipids, membrane proteins, peptides) [8,9]. However, the most prevalent use of SFC-MS is its use for mass-directed isolations and purifications [3,10].

Supercritical Fluid Chromatography: Advances and Applications in Pharmaceutical Analysis
Edited by Gregory K. Webster

ISBN 978-981-4463-00-3 (Hardcover), 978-981-4463-01-0 (eBook)
www.panstanford.com

Prior to coupling SFC to mass spectrometers, the SFC system used both UV and flame ionization detectors (FID). Initially this may not seem intuitive; however, when one considers that SFC is essentially a hybrid of gas chromatography and liquid chromatography (supercritical CO_2 is gaseous carbon dioxide molecules pressed as closely together as they can be, without forming a liquid), the use of both these detectors makes sense. A typical SFC-FID setup is shown in Fig. 10.1 [11].

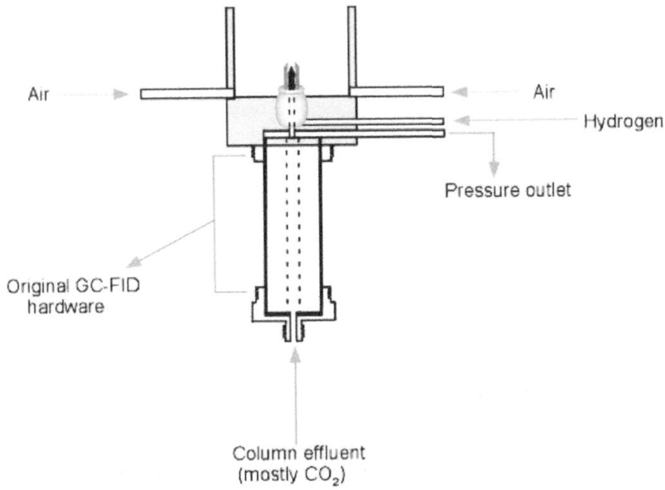

Figure 10.1 Schematic of a flame ionization detector (FID) modified for use with SFC.

The SFC-MS can also be roughly divided into two parts: use of non-atmospheric ionization sources and interfaces (i.e., under vacuum) with capillary columns and the era of atmospheric pressure ionization (API) sources using packed columns.

10.2 Sources

10.2.1 Vacuum Sources

The low capillary column flow rates, coupled well to the vacuum conditions required by mass spectrometers. Following that, the properties of CO_2 allowed for ion–molecule reactions, which lead to ionization techniques such as charge exchange ionization. Such techniques, however, proved problematic since the CO_2 pressure gradient at the column inlet would cause large variations in the mass spectrometer source pressure. The result was that the efficiency of the charge exchange ionization was low [12]. Some of the vacuum interfaces and sources used with SFC-MS include direct introduction, thermospray, and particle beam [13].

10.2.1.1 Direct introduction

The first type of vacuum interface used was plumbed directly to the SFC so that the column effluent flowed directly into a chemical ionization or electron impact source (similar to those found on gas chromatographs). There were, however, problems with this interface including [13]

- the need to heat the effluent such that it expanded rapidly into the gas phase, to ensure transfer of high mass components to the mass spectrometer;
- the need for high fluid pressures, which resulted in pressure gradients in the ion source that could affect ionization and tuning

To compensate for these problems, extra cryopumping in the analyzer and ion source have been evaluated for the enhancement of performance [14].

10.2.1.2 Thermospray interface

The thermospray interface can be considered a derivative of direct fluid introduction. It is comprised of a jet of vapor generated out of a heated vaporizer tube and a mechanical pump attached to the ion source, which can allow for high liquid flow rates. This setup is also more easily coupled to packed columns with smaller diameters (ranging from 3 to 5 mm). Downstream from the interface is the ionization source where reagent gas can be generated either by a filament (filament-on) or by using a discharge electrode (discharge-on) [15, 16]. In the presence of carbon dioxide, the result is charge exchange ionization; however, if proton-donating modifiers are used, proton transfer chemical ionization can occur.

In the charge exchange process sample constituents are ionized by collisions with a reagent gas that does not contain protons, i.e., CO_2, N_2, air, and He. The process of charge exchange ionization is [17]

$$N_2^{+} + M_{analyte} \rightarrow N_2 + M^{+}_{analyte} \quad (10.1)$$

More fragmentation occurs during charge exchange ionization than chemical ionization, due the high amount of energy being transferred. This leads to spectra similar to that produced by electron impact (EI), which allows for these spectra generated by SFC-MS to be compared to the EI spectral libraries [18].

10.2.1.3 Particle beam interface

In the particle beam interface, the column effluent is nebulized in a chamber connected to a momentum separator. This separator preferentially conveys the high mass analytes to the mass spectrometer ion source, which most often is an electron impact or chemical ionization [19].

10.2.2 Atmospheric Sources

The current focus of SFC-MS sources are the atmospheric pressure ionization (API) sources, used in tandem with packed columns. Most of the columns commonly used for normal phase chromatography, both chiral and achiral, can be used with SFC [5,20]. Integrating SFC and a mass analyzer is fairly straightforward and is easier than LC-MS, as CO_2 is highly volatile, thus making volatilization during the ionization process easier.

The earlier SFC-MS ionization sources followed the evolution of the ionization interfaces for LC-MS and GC-MS, and as these interfaces continue to improve, they too are applied to SFC [21,22].

The most popular ionization sources for SFC-MS are atmospheric pressure chemical ionization (APCI) and electrospray ionization (ESI) as they allow direct infusion of the chromatographic effluents into a mass analyzer. Both ESI and APCI are used in the pharmaceutical industry, with ESI being more prevalent for LC-MS applications. Their use in LC-MS is based on ionizability in solution, molecular weight, volatility, polarity and flow rate, if high flows are required (see Fig. 10.2 and 10.3). It is possible to use these sources without further modification for SFC-MS applications [23]; however, some modifications may be required to enhance ionization and achieve a stable baseline [24].

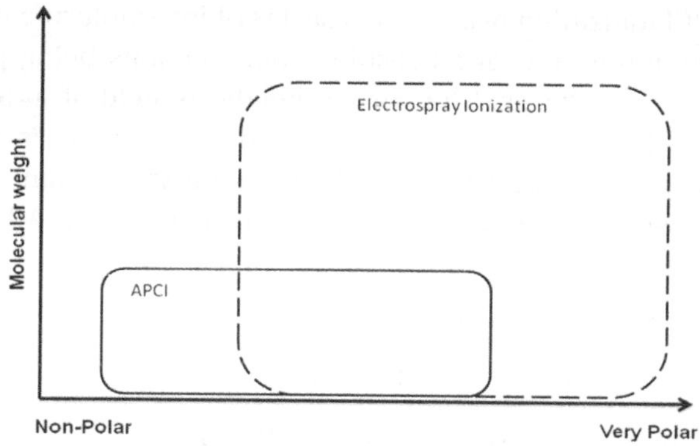

Figure 10.2 Ranges of use for API sources in LCMS that can be applied to SFC-MS. Adapted from [25].

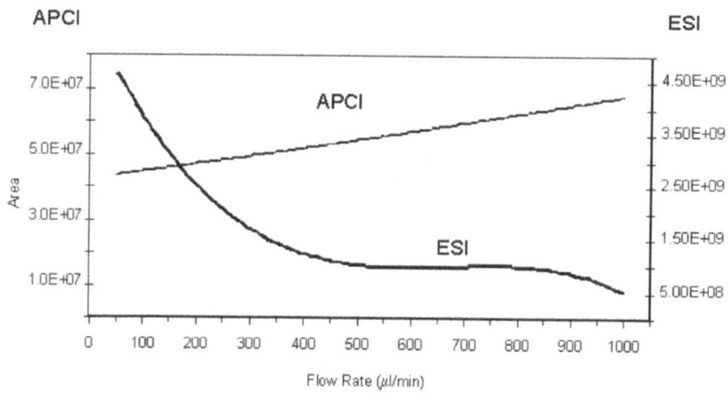

Figure 10.3 Change in sensitivity versus flow rate for APCI and ESI. Reprinted with permission from *Anal. Chem.* 75, 973, 2003. Copyright 2003 American Chemical Society [26].

10.2.2.1 Atmospheric pressure chemical ionization

Atmospheric pressure chemical ionization is probably the most common source currently used in SFC-MS. It is easily adaptable to SFC systems and can handle high flow rates. It is a mass-dependent source, which means the greater the amount sample, the better the sensitivity. It is less affected by ion pairing agents, than is ESI. This also translates into being less affected by ion suppression than ESI. The key parameter in APCI is the probe temperature, where one has to find the temperature where there is the maximum sensitivity coupled with minimum decomposition.

In APCI ionization occurs via a series of ion–molecule reactions, where the initial ionization results from elections being produced by a corona discharge. The corona needle is held at 1–5 kV (see Fig. 10.4). The mechanism of ionization is *vaporization > solvent ionization > charge transfer to analyte*. The most common ionization reactions, starting from atmospheric nitrogen, in the APCI source are [27]

$$N_2 + e^- \rightarrow N_2^{+\bullet} + 2e^- \tag{10.2}$$

$$N_2^{+\bullet} + X-OH \rightarrow X-OH^{+\bullet} + 2N_2 \tag{10.3}$$

$$X-OH^{+\bullet} + X-OH \rightarrow [X-OH+H]^+ + OX^\bullet \tag{10.4}$$

$$[X-OH+H]^+ + M_{analyte} \rightarrow [M_{analyte}+H]^+ + X-OH \tag{10.5}$$

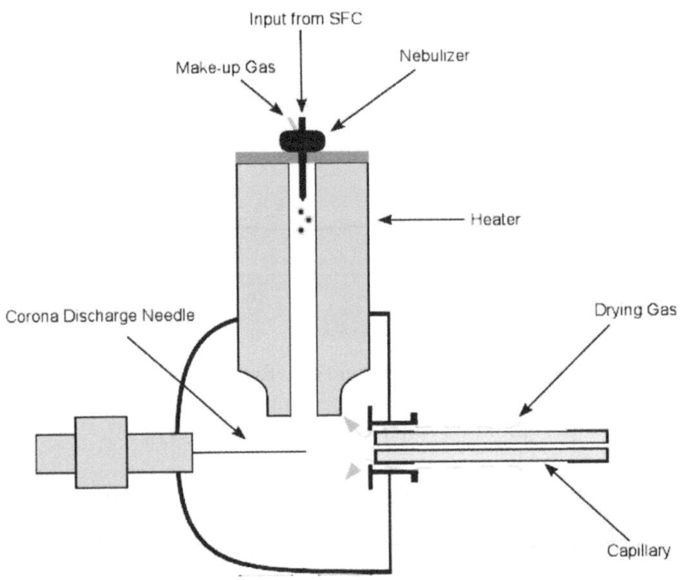

Figure 10.4 Schematic of an atmospheric pressure chemical ionization (APCI) mass spectrometer source.

Since carbon dioxide does not have a proton to donate, use of APCI usually requires the addition of hydrogen donors (X-OH), by either incorporating them as part of the mobile phase or by post-column effluent addition, prior to the ionization source. The proton transfer reaction occurs when the proton affinity of the analyte molecules is greater than that of the reagent gas/mobile phase component. Some proton affinities of SFC reagents are shown in Table 10.1.

Table 10.1 Proton affinities of some SFC additives

Mobile phase component	Proton affinity (kJ/mol) [28]
Hydrogen	422.3
Carbon dioxide	540.5
Methane	543.5
Water	691
Trifluoracetic acid	711.7
Formic acid	742
Methanol	754.3
Ethanol	776.4
Acetonitrile	779.2
Acetic acid	783.7
IPA	793.0
Ethanoic acid	797.2
Ammonia	853.6
Methylamine	899.0
Ethylamine	912.0
Diethylamine	952.4
Dimethyl ethylamine	960.1
Triethylamine	981.8

An APCI source can tolerate the generally higher flow rates of SFC, and since it is coupled with a highly volatile mobile phase, it does not suffer from reduced ionization efficiency as may be experienced in LC-MS at higher flow rates. The source is durable and easily cleaned.

Morgan et al. [23] investigated which parameters are and are not important when optimizing an APCI source for use in SFC-MS. They report that the APCI source could be used successfully, even in the absence of a proton donor as all their experiments were run using only 100% CO_2 as the mobile phase. The source of protons was believed to be either ambient moisture in the source or small amounts of it in the supercritical CO_2. They report that the transfer line, be it PEEK or stainless steel, need not be heated and that neither the internal diameter (I.D.) nor the length of the transfer line were important parameters as they show retention of peak shape I.D.s from 125 μm to 1 mm and transfer line lengths ranging from 75 to 300 cm. They did find that the internal diameter of the inlet capillary of the APCI source to have a profound effect on peak shape. They tested capillaries of I.D.s ranging from 50 to 180 μm and found that 75 μm was the optimum inlet capillary diameter for the best peak shape.

10.2.2.2 Electrospray ionization

Electrospray ionization occurs as the solvent is nebulized, followed by evaporation to leave ionized particles in the gas phase. A high electrical field (~3 kV) is applied between the needle and a nearby counter electrode, during nebulization. The resulting charged droplets shrink via evaporation, leaving small highly charged particles that are propelled toward the mass analyzer. ESI, in contrast to APCI, produces fewer ion fragments for most polar compounds [29]. Figure 10.5 shows the ionization process in ESI and Fig. 10.6 shows a schematic of an electrospray source.

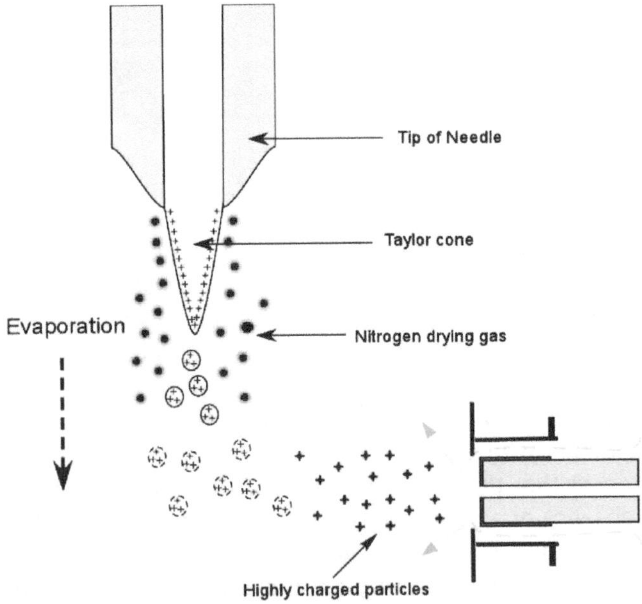

Figure 10.5 Process of ionization in electrospray sources.

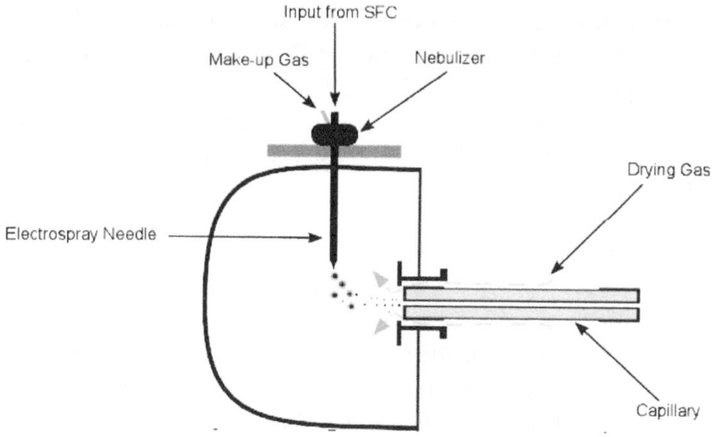

Figure 10.6 Schematic of an electrospray ionization mass spectrometer source.

> Both ionization sources discussed above are usually used as they would be in LC-MS, i.e., having optimized source temperatures, gas flow rates and high voltages. In their drive to understand ionization mechanisms Thitle et al. [30] observed several compounds (diphenhydramine, terfenadine, oxybuytinin, reserpine, mebendazole, nicotinanilide, erythromycin) under normal ESI and APCI ionization conditions and ionization when these sources have the high voltages turned off. They showed some unexpected ionization. In some cases there was an increase in sensitivity when the high voltage was turned off, using CO_2 and methanol as the mobile phase.

10.2.2.3 Atmospheric pressure photoionization

Atmospheric pressure photoionization (APPI) is used less than APCI or ESI but can be useful, especially with polyaromatic hydrocarbons or steroids. APPI uses a lamp (usually krypton) that has energy of 10.6 eV. Ionization occurs when the ionization potential of the lamp exceeds that of the analyte. Solvents that have ionization potentials (IP) that are higher than the lamp absorb the energy and little to no ionization occurs as the energy is dissipated in the solvent. Some common solvent ionization potentials are [31, 32]:

- water (12.6 eV);
- acetonitrile (12.3 eV);
- methanol (10.9 eV).

APPI works well in solvents like isopropyl alcohol (10.4 eV), hexane (10.0 eV) and heptane (9.83 eV), more normal phase chromatographic conditions [31, 33].

Ionization occurs when a photon interacts with an analyte ($M_{analyte}$) to create a radical ion, which can then be protonated by protic solvent molecules:

$$M_{analyte} + h\nu \rightarrow M_{analyte}{}^{+\bullet} + e^{-} \quad (10.6)$$

$$M_{analyte}{}^{+\bullet} + X-H \rightarrow [X]^{-\bullet} + \left[M_{analyte} + H\right]^{+} \quad (10.7)$$

Depending on the protic solvent it is possible to have quantities of both $M_{analyte}{}^{+\bullet}$ and $[M_{analyte} + H]^{+}$. If there is no source of protons, dopants (D) with low ionization potentials may be added, post separation, pre-source to enable protonation of the analyte.

$$D + h\nu \rightarrow D^{+\bullet} + e^{-} \quad (10.8)$$

$$D^{+\bullet} + D \rightarrow \left[D+H\right]^{+} + D^{\bullet} \quad (10.9)$$

$$\left[D+H\right]^{+} + M_{analyte} \rightarrow \left[M_{analyte} + H\right]^{+} + D \quad (10.10)$$

Examples of dopants include acetone with an IP of 9.72 eV [34] and toluene with an IP of 8.82 eV [35]. Figure 10.7 shows a schematic of the APPI source.

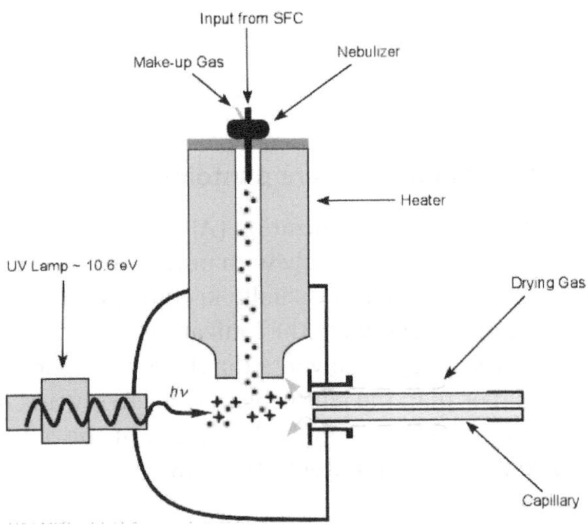

Figure 10.7 Schematic of an atmospheric photoionization mass spectrometer source.

It has been suggested that APPI is less susceptible to sample matrix effects [36], making it attractive for the analysis of drugs or metabolites in tissues, blood or plasma and works well for very non-polar molecules [37]. Figure 10.8 shows where the APPI source applicability fits in with APCI and ESI.

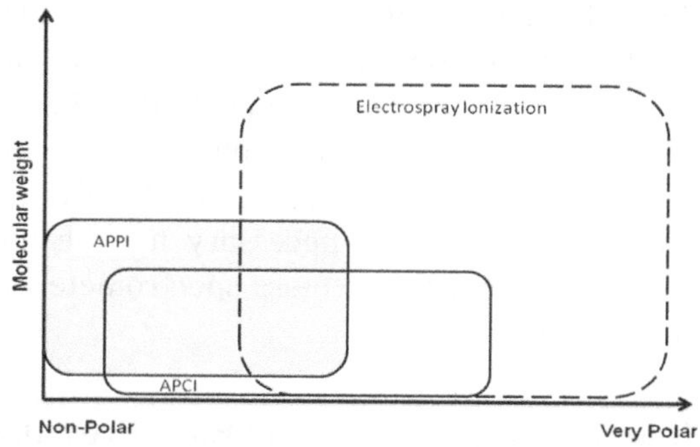

Figure 10.8 Ranges of use for APPI, APCI and ESI sources.

APPI can be more sensitive than APCI or ESI, as the common solvents used in SFC do not ionize well under APPI conditions. This leads to lower background ionization and noise [37]. (i.e., IP of CO_2 is 13.8 eV [31] and methanol is 10.9 eV) An example of this is shown in Fig. 10.9 [37].

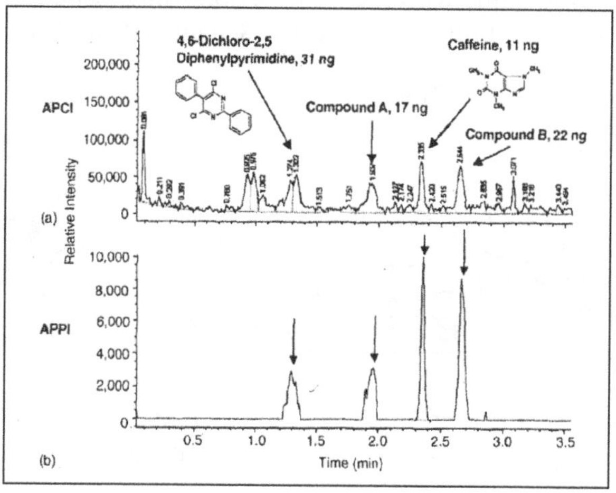

Figure 10.9 SFC-MS analysis of a standard mixture using (a) APCI and (b) APPI, where all conditions were held constant except the ionization source. Reprinted with permission from *Int. J. Mass. Spectrom.*, 238, 85, 2004.

10.3 Source and Mass Analyzer Interfaces

The ionization source is only a part of the setup in an SFC-MS system. How the source is coupled to the mass analyzer is also important. There are several options, some of which are more universal or user friendly and others that are more sensitive. This is more straightforward in LC-MS, whereas the phase behavior of the CO_2 makes these interfaces more challenging. As the SFC mobile phase elutes from the column, the pressure on the gas is released, the density decreases, the CO_2 can boil off and any dissolved components can precipitate. It is, therefore, necessary for the interface that links the ionization source to the mass spectrometer to control or counteract these problems.

10.3.1 Flow Splitting Prior to the Back Pressure Regulator

Since the mobile phase in SFC is predominantly a gas (CO_2), the pressure downstream from the column must be controlled. This is often done by means of a back pressure regulator (BPR). There are several options on where to place the BPR (if one is used at all).

The configuration used most often by instrument vendors is to place the BPR in line with the UV detector, which would be after the flow has been split off to MS detector. See Fig. 10.10 [24].

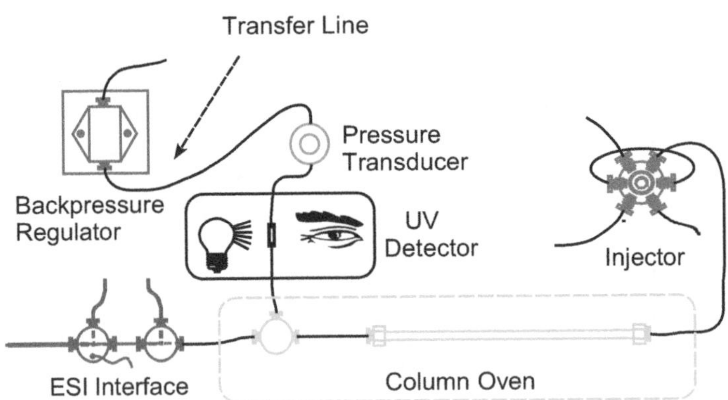

Figure 10.10 Schematic of an SFC-MS interface showing flow splitting prior to the BPR. Reprinted with permission from *Eur. J. Mass Spectrom.* 11, 189, 2005. Copyright (2005) IM Publications LLP.

In this arrangement, the proportions of the transfer line are chosen to ensure that most of the mobile phase carries on to the BPR and only a small portion of the column eluent is diverted to the MS detector by means of a zero dead volume tee. According to Pinkston, for a CO_2/methanol mobile phase system running at 2 mL/min, 200 bar and room temperature, a good starting configuration would be a 125 μm I.D., 1 m-long transfer line [24].

The approach is easily controlled by the instrument software and generally maintains good chromatographic fidelity [38]. There are, however, a few disadvantages, one being only a portion of the effluent is diverted to the source, which can affect sensitivity when using mass-dependent ionization sources. When using a fixed tee, the split ratios can vary with mobile phase pressure (i.e., during a gradient elution) and any change in the transfer line length will also affect this ratio. This could potentially make a method difficult to transfer from lab to lab and may make a system more difficult to qualify in a GxP environment. Both are requirements in the pharmaceutical industry. Method validation and transfer exercises will highlight whether the pre-BPR splitting approach will work.

A variation of this interface was reported by Crowther and Henion [39], where they placed the mass spectrometer between the UV effluent and the back pressure regulator. The split to the BPR was affected by a diaphragm in the liquid injection probe, that essentially split the flow in the probe and was therefore regulated by the both the geometry of the diaphragm and the BPR. Although the approach worked well, the diaphragms wore out quickly making this a difficult setup to have in a regulated environment.

10.3.2 Total Flow Using a Pressure Regulating Fluid Interface

To counteract the variability of the split flow approach described in Section 10.3.1, the total eluent flow can be introduced into the mass spectrometer. There are several ways to achieve this, the first of which is illustrated in Fig. 10.11 [24].

In this case, the BPR is replaced by a zero dead volume tee, where the column effluent and liquid that is miscible with CO_2 are mixed. This miscible liquid works as the pressure regulator and as can be seen by the diagram, that all the effluent is then pumped into the ionization source [40].

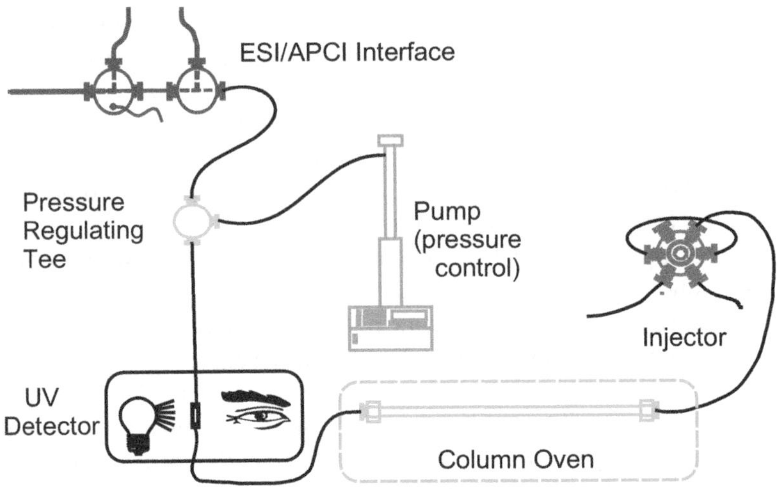

Figure 10.11 Schematic of an SFC-MS interface showing total flow introduction by using a regulating fluid. Reprinted with permission from *Eur. J. Mass Spectrom.* 11, 189, 2005. Copyright (2005) IM Publications LLP.

This interface has several advantages:

- Pressure-regulating fluid can be chosen to enhance ionization, especially when using ESI, as CO_2 alone will not form charged droplets.
- All column effluent is introduced into the mass spectrometer.
- The tee can be located close to the mass spectrometer, reducing the need for long transfer lines.
- Maintains good chromatographic fidelity.

The main disadvantages of this interface are as follows:

- Control of the downstream pressure is via an independent pump that is not an integral part of the existing chromatography system, nor is it necessarily controlled by the instrument software.
- High solvent load to the ionization source.
- Requires a pulse-free high-pressure pump.
- Usable flow rates are dictated by the transfer line from the tee to the mass spectrometer—possible difficulty for labs in the pharmaceutical industry requiring GxP controls.

10.3.3 Total Flow Using a BPR

To avoid some of the disadvantages of using a pressure-regulating fluid interface, total eluent flow can be introduced into the mass spectrometer using a mechanical back pressure regulator (BPR). Since commercial BPRs are not necessarily able to maintain good chromatographic fidelity, all large-volume components must be removed from the flow path. Such a setup is shown in Fig. 10. 12 [24].

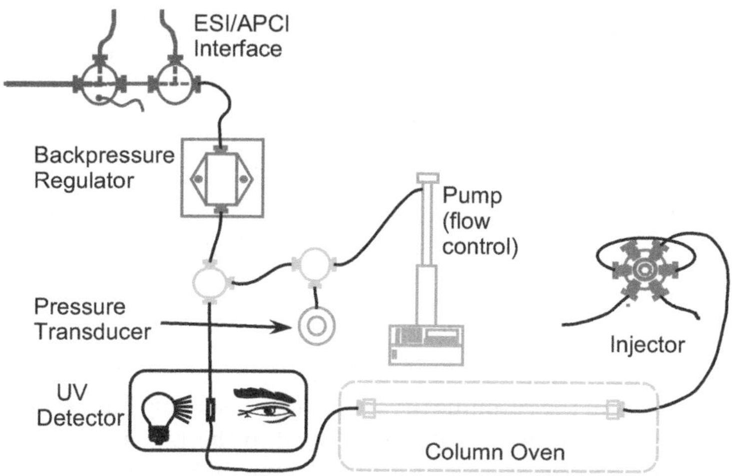

Figure 10.12 Schematic of an SFC-MS interface showing total flow introduction by using a BPR. Reprinted with permission from *Eur. J. Mass Spectrom.* 11, 189, 2005. Copyright (2005) IM Publications LLP.

The main advantages of this setup are that the total effluent flow goes to the mass spectrometer and that the instrumental parameters are more easily controlled by the system's software. As long as the dead volume in the BPR and transfer lines is minimized, the chromatographic separations remain intact. If the dead volumes become excessive, they can result in peak broadening. If there is a pressure drop, the CO_2 may boil off, which can cause precipitation of the analyte in the source and loss of sensitivity. With careful control over the dead volume of the system, this setup fairly easy to use. Again, due to several instrumental variables, such systems would be hard to qualify in a regulated environment.

10.3.4 Total Flow Using a Passive BPR

Perhaps the simplest interface to the mass spectrometer is one in which the total effluent is directed toward the mass spectrometer; however, the back pressure is controlled with a single piece of restrictor tubing. A schematic of such a setup is depicted in Fig. 10.13 [24].

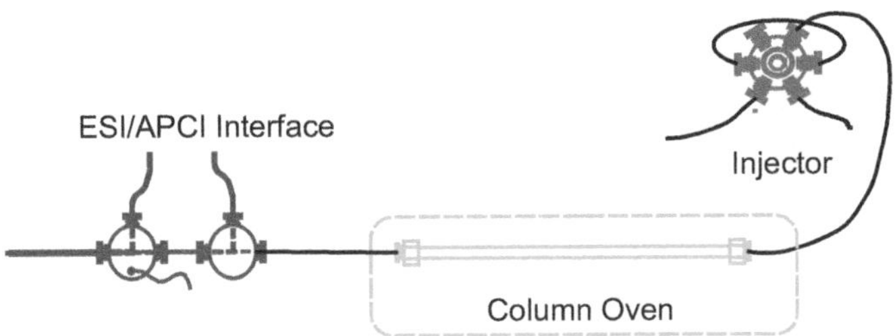

Figure 10.13 Schematic of an SFC-MS interface showing total flow introduction by using a passive BPR. Reprinted with permission from *Eur. J. Mass Spectrom.* 11, 189, 2005. Copyright (2005) IM Publications LLP).

This interface works best for bioanalytical applications, where flow rates and the resulting pressures are reasonably high (1–7 mL/min), the column temperatures are held in the range of 40– 60°C, and there is a higher organic content in the mobile phases, thereby retarding any phase transitions, which could result in analyte precipitation.

10.4 Mass Analyzers

All the above-mentioned sources and interfaces need next to be connected to a mass analyzer. Similar to LC-MS, the choice of mass analyzer depends on the application.

The workhorse mass analyzer for SFC-MS in the pharmaceutical industry is the single-quadrupole mass spectrometer. This type of mass spectrometer is ubiquitous in research and development laboratories, easily controlled by existing software, and easily adaptable to the sources and interfaces mentioned in Sections 10.2 and 10.3. Single-quadrupole mass spectrometers are particularly useful for both achiral [6] and chiral [41] high-throughput screening.

One of the more prevalent mass analyzers used in the analysis of trace components in complicated matrices is the triple-quadrupole mass analyzer. Triple-quadrupole analyzers consist of an initial quadrupole (which is essentially the same as any single-quadrupole instrument) set to scan a single ion mass, followed by a collision cell (this is the second quadrupole) and followed by the third quadrupole, which is attenuated to a specific mass of a daughter product. The quadrupoles filter out any interference or matrix effects, as interfering components would have to contain the same initial ion filtered by the first quadrupole and the same daughter product selected by the third quadrupole. As an example, Hoke II et al. successfully developed and validated a method for the quantitation of dextromethorphan in human plasma in the range of 0.256 to 65.6 ng/mL using packed column SFC and a triple-quadrupole mass analyzer [42].

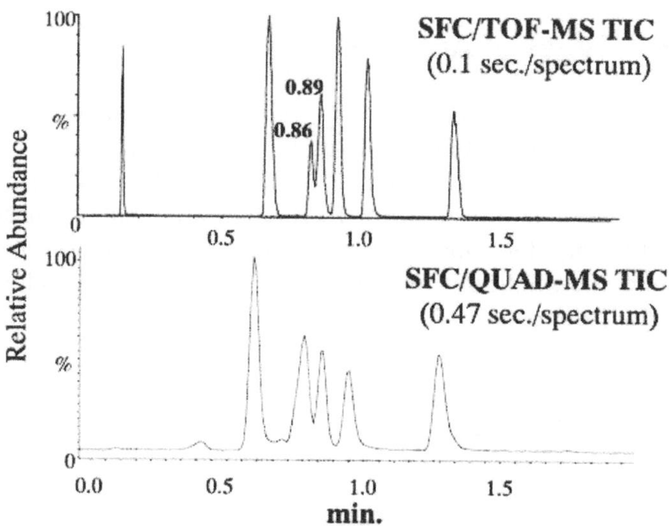

Figure 10.14 Ion chromatograms showing the same separation collected by a TOF mass spectrometer (top) and a single quadrupole mass spectrometer (bottom). Resolution between the peaks labeled 0.86 and 0.89 is lost with the slower acquisition rate of the single quadrupole instrument. Reprinted with permission from *J. Comb. Chem.* 5, 451, 2003. Copyright 2003 American Chemical Society [43].

Time-of-flight (TOF) mass spectrometers couple well with high-speed separation techniques, like SFC, due to a high rate of data acquisition. Bolaños et al. [43] showed this to be true when they compared ion chromatograms of a mixture of six components (steroids and a polar molecule), collected from both a single-quadrupole and a TOF mass spectrometer. Because of the data acquisition rate, the TOF instrument gave better peak resolution (see Fig. 10.14).

Other examples of mass analyzers combined with SFC include Q-TOF [44], Fourier transform [45], ion trap [46], magnetic sector [47], and double focusing [48,49] mass spectrometers. However, with ultra-high-pressure chromatography (UHPLC) maturing, new single-quadrupole and other mass analyzers capable of higher scan speeds are being developed to be coupled to the ultra-high-pressure systems. In the future, these newer systems could also be coupled to SFC systems.

10.5 Types of Analyses

SFC-MS analyses can be used for a wide range of analyses ranging from normal phase type analyses to chromatography that more emulate reverse phase chromatography by using small amounts of water as an additive. Figure 10.15 shows the mobile phase and polarity range for supercritical fluid chromatography. Table 10.2 presents the conditions for some representative chiral and achiral SFC-MS analyses.

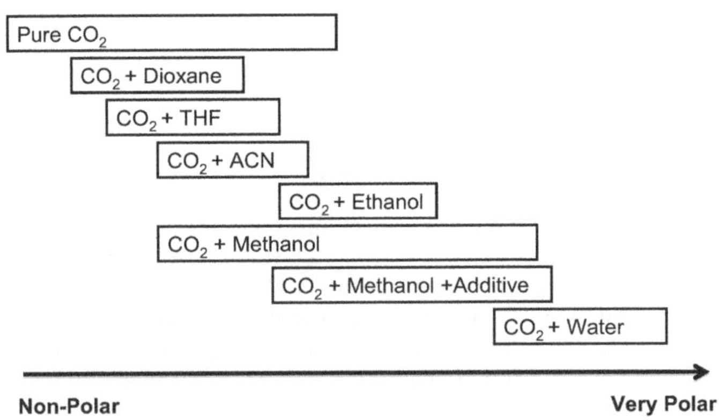

Figure 10.15 Schematic showing polarity ranges of mobile phase combinations in SFC-MS.

Table 10.2 Some representative chiral and achiral SFC-MS analyses

Compound	MS Interface	Ion mode	Column	Mobile phase	Mass spectrometer	Ref.
Clenbuterol, salbutamol, tebutaline, fenoterol	APCI	+	Cyano	CO_2/MeOH/TFA/DEA	Single Quad	[50]
Androstenone	APCI	+	Cyano/C18	CO_2	Triple Quad	[51]
(R,S)-Propranolol, (+) pindolol	APCI	+	Chiralcel OD-H	CO_2/MeOH	Triple Quad	[52]
Dextromethrophan	ESI	+	Cyano	CO_2/MeOH	Triple Quad	[53]
Hydrocortisone, testosterone, troleandomycin, picromycin, oleandomycin	APCI	+	Cyano	CO_2/MeOH/IPA	Q-TOF	[44]
(R,S)-Ketoprofen	ESI	+	Chirex 3005	CO_2/MeOH	Triple Quad	[42]
(R,S)-Warfarin	APCI	+	Chiralpak AD	CO_2/MeOH	Triple Quad	[54]
Estrogen metabolites	APCI	+	Diol/cyanopropyl	CO_2/MeOH	Triple Quad	[55]

(*Continued*)

Table 10.2 (*Continued*)

Compound	MS Interface	Ion mode	Column	Mobile phase	Mass spectrometer	Ref.
Cytarabine, clofazimine	APCI	+	Silica	CO_2/MeOH/ ammonium acetate	Triple Quad	[56]
Polypeptides	ESI	+	Ethylpyridine/ aminopropyl	CO_2/MeOH/ TFA	Single Quad	[8]
Pyrimidine, caffeine, sulfanilamide	APCI	+	Zymor Zyrosil-hybrid	CO_2/MeOH	Single Quad/ TOF	[43]
Artemisinin	APCI	+	Cyano	CO_2/MeOH	Single Quad	[57]
Sulfanilamides	APCI	+	Cyanopropyl	CO_2/MeOH	Single Quad	[58]
Pesticides	APCI	+	Cyano	CO_2/MeOH	Single Quad	[59]
Primaquine	ESI	+	Chiralpak AD-H	CO_2/EtOH/ DEA	Triple Quad	[60]

Source: Reproduced from [29] with permission.

10.5.1 Chiral SFC-MS

Chiral SFC has many advantages over normal phase liquid chiral chromatography. The first advantage is its speed; due to the low viscosity of CO_2, SFC can be run much faster flow rates than normal phase HPLC, even when employing columns of the same dimensions. This leads to faster equilibration, even when changing solvents. This is particularly useful when using a multi-solvent screening system. The second advantage is that when SFC chromatography is scaled up to the preparatory level, the recovery of enantiomers is much simpler as the amount of solvent that needs to be removed post SFC separations is minimal by comparison. This is especially useful during the early phases of drug discovery. Unlike normal phase chromatography, however, where the most common modifier is isopropyl alcohol (IPA), the preferred starting point for method development would be with methanol as SFC-MS modifier. Owing to its lower surface tension, it gives better ionization efficiency and therefore sensitivity in the mass spectrometer over other alcohols, such as ethanol and IPA [38].

In comparison with chiral LC-MS, chiral SFC-MS is similar with respect to specificity, linearity, accuracy and robustness [29]. In their search for a fast bioanalytical method for the determination of warfarin enantiomers in human plasma, Coe et al. [54] optimized many aspects of their SFC-MS system and compared the results to work done by LC-MS. They tested an atmospheric pressure photoionization source (APPI) and ESI, finally settling on APCI. In order to improve sensitivity, they tried several different mass analyzers and settled on the API 4000, citing that the ceramic tube, as opposed to the quartz tube in other instruments, and the dual-heater in the source resulted in less tailing and greater ionization efficiency, than the other systems studied. Additionally, as discussed above, plumbing and location of the back pressure regulator were investigated. It was found that to reduce the major source of carryover, it was best to simply replace the BPR with narrow bore PEEK tubing, as was discussed in Section 6.3.4. After optimization of the internal standard, the method was found to be precise, linear and accurate, for both enantiomers. Selectivity over matrix components and thalidomide, which was co-administered, was shown. Overall, the resulting method was faster and showed higher peak resolution power [54].

Many pharmaceutical investigators are looking at using SFC-MS for screening or automating chiral separations. Zhao et al. [38] described a system using the available normal phase chiral stationary phases, coupled with the sensitivity of mass spectrometry and sample pooling which allowed them to scan samples containing various enantiomeric pairs (providing they had unique masses) in a single chromatographic run, increasing throughput. Zeng et al. [41] have automated a parallel four-column system for chiral screening and optimization. By running the screens in parallel, the overall time required was less making this type of system good for laboratories requiring high throughput, as in early drug discovery labs.

Similar to the developments in liquid chromatography [61], Zeng et al. [4] have applied the concept of SFC/SFC/MS (two-dimensional chromatography or 2DSFCMS) to the purification of pharmaceutical samples. In their system, the first dimension is an achiral separation that separates the target from all other impurities; the second dimension performs the chiral separation. In liquid chromatography, it is often the case that the achiral and chiral methods use different mobile phase conditions (reverse vs. normal phase). This is not a problem when applying 2D chromatography to SFC. Figure 10.16 shows the resulting chromatogram from the reported 2DSFCMS system.

10.5.2 Achiral SFC-MS

Achiral SFC-MS can cover a very broad range of compounds as shown in Table 10.2. Crowther and Henion [39] have reported fast analyses for the detection of codeine in urine and other alkaloids, as well as phenylbutazone and its metabolites, also from urine.

In the pharmaceutical industry, SFC-MS systems have been optimized for high-throughput using 96-well plates, rapid switching between positive and negative ionization modes and the ability to run up to 1000 samples in a 24 h period [6,7]. Using more polar combinations of mobile phases, Agrawal et al. [62] developed and validated a fast SFC-MS method that was able to separate two antidiabetic drugs for type-2 diabetes. Using an 80:20 mixture of acetonitrile and water, they successfully separated and monitored metformin and gliclazide in blood plasma samples on a C_{18} column in less than 5 min. De la Puente et al. [63] developed a comprehensive achiral screening system, using a variety of stationary phases, modifiers and elution gradients. They report an 85% success rate having applied their system on a variety of structurally diverse compounds.

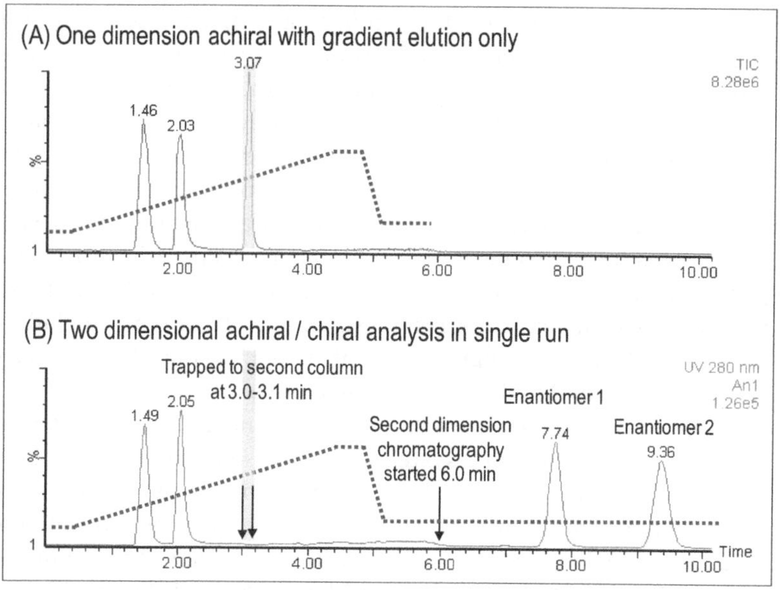

Figure 10.16 2D SFC/SFC/MS system showing a three-component mixture. (A) using the total ion chromatogram (TIC) to define the peak and (B) showing where the peak was cut, loaded on to the chiral column and the resulting separation. Reprinted with permission from *J. Chromatogr. A.* 1218, 3080, 2011. Copyright 2011 Elsevier B.V. [4].

10.5.3 Prep SFC-MS

Wang et al. [10] used mass-directed fraction collecting with their prep SFC system, to separate the enantiomers of several pharmaceutically relevant compounds including warfarin, metoprolol, promethazine, cortisterone, tolbutamide, coumarin, atenolol, imipramine, alprenolol, erythromycin, and verapamil, showing that prep SFC-MS can be used instead of prep SFC-UV for the isolation and purification of enantiomers. By resolving some of the earlier problems, such as the proper integration of the software, the interface between the mass spectrometer and the SFC, and loss of sample at the fraction collector due to aerosol formation, Zhang et al. [64] have also successfully implemented a mass-directed SFC-MS purification system for pharmaceuticals and use it regularly in the purification of their library compounds.

Owing to the need for pure active pharmaceutical ingredients (chirally and achirally pure), there has been more interest in prep SFC-MS gradient systems. Evaluation of such systems currently shows that high recoveries are possible and that the instrumentation produces good gradients and reasonable flow rate and is capable of mass-directed fractionation [65]. As this technology matures and is used more widely, transferability, robustness and other system suitability parameters will have to be evaluated. Prep SFC-MS systems have also been applied to achiral separations of active pharmaceutical ingredients (API). The approach was desirable for the same reasons as using prep SFC-MS for chiral work including different selectivities, highly concentrated fractions due the easy removal of CO_2 and high recovery rates [66].

Zeng et al. [4] combined both achiral and chiral purifications in their 2D SFC/SFC/MS system. Their system affords two-step purifications in one process. The first dimension consists of an achiral separation to remove impurities from the chiral compounds and the second dimension completes the chiral separations. The evolution of such 2D SFC/SFC/MS systems could prove to be the most efficient way to purify chiral compounds.

10.5.4 Structure Elucidation

Since, in most cases, the bulk of the mobile phase in SFC-MS is CO_2 and only small volumes of additives are used, the use of deuterated solvents (i.e., d_4-MeOD) becomes a very real and not very expensive way of determining the location of exchangeable protons or studying H-D exchange reactions. This is an area of SFC-MS that is expected to develop as more analytical SFC-MS systems come online.

10.6 Traditional Problems

Traditionally SFC-UV and to a certain extent SFC-MS have suffered from a lack of sensitivity. Many new advancements in the past few years have increased the sensitivity and stabilized the baseline in SFC-MS; however, it still remains that often the sample concentrations have to be 10× larger to get the same limits of detection as seen by LCMS. There is science in this field, which will help to further increase the use of this technique.

SFC-MS is still behind, in use, to LC-MS or even perhaps, GC-MS, but as the competition in the arena of drug discovery gets more heated, the analytical and process chemists will be looking for tools like SFC-MS, that speed up their processes and perhaps, solve problems that are not answerable by other means.

Acknowledgments

The author would like to thank Lu Zeng (Takeda California) for his help in reviewing this work.

References

1. Berger, T.A. (1995). *Packed Column SFC* (The Royal Society of Chemistry, Cambridge, UK).
2. Lazarescu, V., Mulvihill, M.J., and Ma, L. (2011). A generic workflow for achiral SFC purification of complex pharmaceutical mixtures. *LCGC North Am.*, **29**, pp. 438–444.
3. Aurigemma, C., and Farrell, W. (2010). FastTrack to supercritical fluid chromatographic purifications: Implementation of a walk-up analytical supercritical fluid chromatography/mass spectrometry screening system in the medicinal chemistry laboratory, *J. Chromatogr. A*, **1217**, pp. 6110–6114.
4. Zeng, L., Xu, R., Zhang, Y., and Kassel, D.B. (2011). Two-dimensional supercritical fluid chromatography/mass spectrometry for the enantiomeric analysis and purification of pharmaceutical samples, *J. Chromatogr. A*, **1218**, pp. 3080–3088.
5. Zhao, Y., Sandra, P., Woo, G., Thomas, S., Gahm, K., and Semin, D. (2005). High-throughput drug discovery applications for supercritical fluid chromatography-mass spectrometry, *Pharm. Discov.*, **5**, pp. 30–41.
6. Ventura, M.C., Farrell, W.P., Aurigemma, C.M., and Grieg, M.J. (1999). Packed column supercritical fluid chromatography/mass spectrometry for high-throughput analysis, *Anal. Chem.*, **71**, pp. 2410–2416.
7. Ventura, M.C., Farrell, W.P., Aurigemma, C.M., and Grieg, M.J. (1999). Packed column supercritical fluid chromatography/mass spectrometry for high-throughput analysis. Part 2, *Anal. Chem.*, **71**, pp. 4223–4231.

8. Zheng, J., Pinkston, J.D., Zoutendam, P.H., and Taylor, L.T. (2006). Feasibility of supercritical fluid chromatography/mass spectrometry of polypeptides with up to 40-mers, *Anal. Chem.*, **78**, pp. 1535–1545.
9. Bamba, T., Shimonishi, N., Matsubara, A., Hirata, K., Nakazawa, Y., Kobayashi, A., and Fukusaki, E. (2008). High throughput and exhaustive analysis of diverse lipids by using supercritical fluid chromatography-mass spectrometry for metabolomics, *J. Biosci. Bioeng.*, **105**, pp. 460–469.
10. Wang, T., Barber, M., Hardt, I., and Kassel, D.B. (2001). Mass-directed fractionation and isolation of pharmaceutical compounds by packed-column supercritical fluid chromatography/mass spectrometry, *Rapid Comm. Mass. Spec.*, **15**, pp. 2067–2075.
11. Rawdon, M.G. (1984). Modified flame ionization detector for supercritical fluid chromatography. *Anal. Chem.*, **56**, pp. 831–832.
12. Arpino, P., Sadoun, F., and Virelizer, H. (1996). Recent developments in combined supercritical fluid chromatography/mass spectrometry (SFC/MS), *Mass spectrometry in Biomolecular Sciences* (Kluwer Academic Publishers, Dordrecht, The Netherlands).
13. Niessen, W.M.A. (1996). Developments in interface technology for combined liquid chromatography, capillary electrophoresis and supercritical fluid chromatography-mass spectrometry, *Journal of Chromatography Library volume 59, Applications of LC-MS in Environmental Chemistry* (Elsevier Science B.V., The Netherlands).
14. Pinkston, J.D., and Bowling, D.J. (1993). Investigation of cryopumping for enhanced performance in supercritical fluid chromatography/mass spectrometry, *Anal. Chem.*, **65**, pp. 3534–3539.
15. Berry, A.J., Games, D.E., Mylcreest, I.C., Perkins, J.R., and Pleasance, S. (1988). Capillary and packed column SFC/MS, *J. High Resolut. Chromatogr.* **11**, pp. 61–64.
16. Chapman, J.R. (1988). Coupled supercritical fluid/mass spectrometry using a thermospray source, *Rapid Comm. Mass. Spec.*, **2**, pp. 6–7.
17. Hunt, D.F., and Gale, P.J. (1984). Simulation of an electron impact mass spectra by charge exchange in chemical ionization mass spectrometry, *Anal. Chem.*, **56**, pp. 1111–1114.
18. Houben, R.J., Leclercq, P.A., and Cramers, C.A. (1991). Ionization mechanisms in capillary supercritical fluid chromatography—chemical ionization mass spectrometry, *J. Chrom.*, **554**, pp. 351–358.
19. Jedrzejewski, P.T., and Taylor, L.T. (1995). Packed column supercritical fluid chromatography-mass spectrometry with particle beam interface

aided with particle forming solvent, *J. Chromatogr. A*, **703**, pp. 489–501.

20. Wang, Z., Li, S. Jonca, M., Lambros, T., Ferguson, S. Goodnow, R., and Ho, C. (2006). Comparison of supercritical fluid chromatography and liquid chromatography for the separation of urinary metabolites of nobiletin with chiral and non-chiral stationary phases, *Biomed. Chromatogr.*, **20**, pp. 1206–1215.
21. Zaugg, S.D., Deluc, S.J., Holzer, G.U., and Voorhees, K.J. (1987). Simplified interface for electron impact ionization supercritical fluid chromatography/mass spectrometry, *J. High Resolut. Chromatogr. Chromatogr. Commun.*, **10**, pp. 100–101.
22. Arpino, P.J., Dillettato, D., Nguyen, K., and Bruchet, A.J. (1990). Investigation of antioxidants and UV stabilizers from plastics. Part I: Comparison of HPLC and SFC; Preliminary SFC/MS study, *J. High Resolut. Chromatogr.*, **13**, pp. 5–12.
23. Morgan, D.G., Harbol, K.L., and Kitrinos, P.N., Jr. (1998). Optimization of a supercritical fluid chromatograph—atmospheric chemical ionization mass spectrometer interface using an ion trap and two quadrupole mass spectrometers, *J. Chromatogr. A*, **800**, pp. 39–49.
24. Pinkston, J.D. (2005). Advantages and drawbacks of popular supercritical fluid chromatography/mass spectrometry interfacing approaches—a user's perspective, *Eur. J. Mass. Spectrom.*, **11**, pp. 189–197.
25. Garcia, D.M., Huang, S.K., and Stansbury, W.F. (1996). Optimization of the atmospheric pressure chemical ionization liquid chromatography/mass spectrometry interface, *J. Am. Soc. Mass Spectrom.*, **7**, pp. 59–65.
26. Gallagher, R.T., Balogh, M.P., Davey, P., Jackson, M.R., Sinclair, I., and Southern, L.J. (2003). Combined electrospray ionization—atmospheric pressure chemical ionization source for use in high-throughput LC-MS applications, *Anal. Chem.*, **75**, pp. 973–977.
27. Niessen, W.M.A. (2006). *Liquid Chomatography—Mass Spectrometry*, 3rd ed. (Taylor & Francis, USA).
28. Hunter, E.P., and Lias, S.G. (1998). Evaluated gas phase basicities and proton affinities of molecules: An update, *J. Phys. Chem. Ref. Data*, **27**, pp. 413–656.
29. Li, F., and Hsieh, Y. (2008). Supercritical fluid chromatography-mass spectrometry for chemical analysis, *J. Sep Sci.*, **31**, pp. 1231–1237.
30. Thite, M.A., Boughtflower, R., Caldwell, J., Hitzle, L., Holyoak, C., Lane, S.J., Oakley, P., Pullen, F.S., Richardsen, S., and Langley, G.J. (2008).

Ionisation in the absence of high voltage using supercritical fluid chromatography: a possible route to increased signal, *Rapid Comm. Mass. Spec.,* **22**, pp. 3673–3682.

31. Benoit, F.M., and Harrison, A.G. (1976). Predictive value of proton affinity. Ionization energy correlations involving oxygenated molecules, *J. Am. Chem. Soc.,* **99**, pp. 3980–3984.
32. Franklin, J.L., Wada, Y., Natalis, P., and Hierl, P.M. (1966). Ion–molecule reactions in acetonitrile and propionitrile, *J. Phys. Chem.,* **70**, pp. 2353–2361.
33. Meot-Ner, M., Sieck, L.W., and Ausloos, P. (1981). Ionization of normal alkanes: Enthalpy, entropy, structural and isotope effects, *J. Am. Chem. Soc.,* **103**, pp. 5342–5348.
34. Brundle, C.R., Robin, M.B., Kuebler, N.A., and Basch, H. (1972). Perfluoro affect in photoelectron spectroscopy. I. Nonaromatic molecules, *J. Am. Chem. Soc.,* **94**, pp. 1454–1465.
35. Miller, L.L., Nordblom, G.D., and Mayeda, E.A. (1972). A simple, comprehensive correlation of organic oxidation and ionization potentials, *J. Org. Chem.,* **37**, pp. 916–918.
36. Hsieh, Y., Merkle, K., Wang, G., Brissen, J.-M., and Korfmacher, W.A. (2003). High-performance liquid chromatography-atmospheric pressure photoionization/tandem mass spectrometric analysis for small molecules in plasma, *Anal. Chem.,* **75**, pp. 3122–3127.
37. Bolaños, B., Grieg, M., Ventura, M., Farrell, W., Aurigemma, C.A., Li. H., Quezner, T.L., Tivel, K., Bylund, J.M.R., Tran, P., Pham, C., and Phillipson, D. (2004). SFC/MS in drug discovery at Pfizer, La Jolla, *Int. J. Mass. Spectrom.,* **238**, pp. 85–97.
38. Zhao, Y., Woo, G., Thomas, S., Semin, D., and Sandra, P. (2003). Rapid method development for chiral separation in drug discovery using sample pooling and supercritical fluid chromatography—mass spectrometry, *J. Chromatogr. A,* **1003**, pp. 157–166.
39. Crowther, J.B., and Henion, J.D. (1985). Supercritical fluid chromatography of polar drugs using small-particle packed columns with mass spectrometric detection, *Anal. Chem.,* **57**, pp. 2711–2716.
40. Baker, T.R., and Pinkston, J.D. (1997). Development and application of pack column supercritical fluid chromatography/pneumatically assisted electrospray mass spectrometry, *Am. Soc. Mass Spectrom.,* **9**, pp. 498–509.
41. Zeng, L., Xu, R., Laskar, D.B., and Kassel, D.B. (2007). Parallel supercritical fluid chromatography/mass spectrometry system for high-throughput

enantioselective optimization and separation, *J. Chrom. A*, **1169**, pp. 193–204.

42. Hoke, II, S.H., Tomlinson, J.A., Bolden, R.D., Morand, K.L., Pinkston, J.D., and Wehmeyer, K.R. (2001). Increasing bioanalytical throughput using pcSFC-MS/MS: 10 minutes per 96-well plate, *Anal. Chem.*, **73**, pp. 3083–3088.

43. Bolaños, B.J., Ventura, M.C., and Grieg, M.J. (2003). Preserving the chromatographic integrity of high-speed supercritical fluid chromatography separations using time of flight mass spectrometry, *J. Comb. Chem.*, **5**, pp. 451–455.

44. Garzotti, M., Rovatti, L., and Hamdan, M. (2001). Coupling of a supercritical fluid chromatography system to a hybrid (Q-TOF 2) mass spectrometer: on-line accurate mass measurements, *Rapid Comm. Mass. Spec.*, **15**, pp. 1187–1190.

45. Lee, E.D., Henion, J.D., Cody, R.B., and Kinsinger, J.A. (1987). Supercritical fluid chromatography/Fourier transform mass spectrometry, *Anal. Chem.*, **59**, pp. 1309–1312.

46. Todd, J.F.J., Mylchreest, I.C., Berry, A.J., Games, D.E., and Smith, R.D. (1988). Supercritical fluid chromatography/mass spectrometry with an ion trap detector, *Rapid Comm. Mass. Spec.*, **2**, pp. 55–58.

47. Kalinoski, H.T., Udseth, H.R., Chess, E.K., and Smith, R.D. (1987). Capillary supercritical fluid chromatography-mass spectrometry, *J. Chrom.*, **394**, pp. 3–14.

48. Ekeberg, D., and Jablonska-Jentoft, A.M. (2007). Ion-neutral reactions in the ion source of a mixture of CO_2/C H_3OH, CO_2/C_2H_5OH, and CO_2/2-C_3H_7OH by capillary supercritical fluid chromatography-mass spectrometry (SFC-MS), *J. Am. Soc. Mass Spectrom.*, **18**, pp. 2173–2179.

49. Reinhold, V.N., Sheeley, D.M., Kuei, J., and Her, G.-R. (1988). Analysis of high molecular weight samples on a double-focusing magnetic sector instrument by supercritical fluid mass spectrometry, *Anal. Chem.*, **60**, pp. 2719–2722.

50. Jones, D.C., Dost, K., Davidson, G., and George, M.W. (1999). The analysis of β-agonists by packed column supercritical fluid chromatography with ultra-violet and atmospheric pressure chemical ionization mass spectrometric detection, *Analyst*, **124**, pp. 827–831.

51. Tuomola, M., Hakala, M., and Manninen, P. (1998). Determination of androstenone in pig fat using packed column supercritical fluid—mass spectrometry, *J. Chromatogr. B*, **719**, pp. 25–30.

52. Chen, J., Hsieh, Y., Cook, J., Morrison, R., and Korfmacher, W.A. (2006). Supercritical fluid chromatography—tandem mass spectrometry for the enantioselective determination of propranolol and pindolol in mouse blood by serial sampling, *Anal. Chem.*, **78**, pp. 1212–1217.

53. Hoke, II, S.H., Pinkston, J.D., Bailey, R.E., Tanguay, S.L., and Eichhold, T.H. (2000). Comparison of packed column supercritical fluid chromatography—tandem mass spectrometry with liquid chromatography—tandem mass spectrometry for bioanalytical determination of (R) and (S) ketoprofen in human plasma following automated 96-well solid phase extraction, *Anal. Chem.*, **72**, pp. 4235–4241.

54. Coe, R.A., Rathe, J.O., and Lee, J.W. (2006). Supercritical fluid chromatography—tandem mass spectrometry for fast bioanalysis of R/S-warfarin in human plasma, *J. Pharm. Biomed. Anal.*, **42**, pp. 573–580.

55. Xu, X., Roman, J.M., Veenstra, T.D., Anda, J.V., Ziegler, R.G., and Issaq, H.J. (2006). Analysis of fifteen estrogen metabolites using packed column supercritical fluid chromatography-mass spectrometry, *Anal. Chem.*, **78**, pp. 1553–1558.

56. Hsieh, Y., Li, F., and Duncan, C.J.G. (2007). Supercritical fluid chromatography and high-performance liquid chromatography/tandem mass spectrometric methods for the determination of cytarabine in mouse plasma, *Anal. Chem.*, **79**, pp. 3856–3861.

57. Dost, K., and Davidson, G. (2003). Analysis of artemisinin by packed-column supercritical fluid chromatography-atmospheric pressure chemical ionization mass spectrometry technique, *Analyst*, **128**, pp. 1037–1042.

58. Dost, K., Jones, D.C., and Davidson, G. (2000). Determination of sulfonamides by supercritical fluid chromatography-atmospheric pressure chemical ionization mass spectrometric detection, *Analyst*, **125**, pp. 1243–1237.

59. Dost, K., Jones, D.C., Auerbach, R., and Davidson, G. (2000). Determination of pesticides in soil samples by supercritical fluid chromatography-atmospheric pressure chemical ionization mass spectrometric detection, *Analyst*, **125**, pp. 1751–1755.

60. Brondz, I., Ekeberg, D., Bell, D.S., Annino, A.R., Hustad, J.A., Svendsen, R., Vlachos, V., Oakley, P., Langley, G.J., Mohini, T., Amaury, C.-G., and Mikhalitsyn, F. (2007). Nature of the main contaminant in the drug primaquine diphosphate: SFC and SFC-MS methods of analysis, *J. Pharm. Biomed. Anal*, **43**, pp. 937–944.

61. Kammerer, B., Kahlich, R., Ufer, M., Laufer, S., and Gleitner, C.H. (2005). Achiral-chiral LC/LC-MS/MS coupling for determination of chiral discrimination effects in phenprocoumon metabolism, *Anal. Biochem.*, **339**, pp. 297–309.

62. Agrawal, Y.K., Gogoi, P.J., Manna, K. Bhatt, H.G., and Jain, V.K. (2010). A supercritical fluid chromatography/tandem mass spectrometry method for the simultaneous quantification of metformin and gliclazide in human plasma, *Indian J. Pharm. Sci.*, **72**, pp. 50–57.

63. de la Puente, M.L., López Soto-Yarritu, P., and Burnett, J. (2011). Supercritical fluid chromatography in research laboratories: Design, development and implementation of an efficient generic screening for exploiting this technique in an achiral environment, *J. Chromatogr. A*, **1218**, pp. 8551–8560.

64. Zhang, X., Towle, M.H., Felice, C.E., Flament, J.H., and Goetizinger, W.K. (2006). Development of a mass-directed preparative supercritical fluid chromatography purification system, *J. Comb. Chem.*, **8**, pp. 705–714.

65. Ebinger, K., Weller, H.N., Kiplinger, J., and Lefebvre, P. (2011). Evaluation of a new preparative supercritical fluid chromatography system for compound library purification: The TharSFC *SFC-MS Prep-100* system, *JALA*, **16**, pp. 241–249.

66. Van Anda, J. (2010). Use of SFC/MS in the purification of achiral pharmaceutical compounds, *Am. Pharm. Rev.*, **13**, pp. 111–115.

Chapter 11

Supercritical Fluid Chromatography of Natural Products

Ying Wang
Center for Proteomic Chemistry/Natural Product Unit,
Novartis Institutes for Biomedical Research,
CH-4002, Basel, Switzerland
ying.wang@novartis.com

11.1 Introduction

Natural products have historically been the most important source of drugs for treatment of various diseases, as well as of lead compounds that have provided the basis and inspiration for the semi-synthesis or total synthesis of effective new medicines. The analysis on the drug source of 1024 new chemical entities from 1981 to 2008 demonstrated that 38% of them are directly related to natural products, including 7% unmodified natural products, 27% modified natural products and 5% synthetic compounds with a natural product pharmacophore [1]. Although the new drug approvals have continued to fall in the last years, the contributions of natural products are still quite consolidated. For example, a total of

Supercritical Fluid Chromatography: Advances and Applications in Pharmaceutical Analysis
Edited by Gregory K. Webster

ISBN 978-981-4463-00-3 (Hardcover), 978-981-4463-01-0 (eBook)
www.panstanford.com

19 natural product–based drugs were approved from 2005 to 2010, including 7 natural products, 10 semi-synthetic natural products, and 2 natural product derivatives [2]. As the catalysts for innovation in recent strategic changes, natural products will be increasingly valued as source of both potential chemotherapeutic agents and promising leads in the new drug research and development [3, 4].

The term "natural products" generally refers to secondary metabolites produced by organisms from any terrestrial or marine source, such as plants, microorganisms and animals. As their response to needs and challenges of the natural environment, they have been apparently optimized through many centuries evolution. Therefore, natural products cover an unsurpassed chemical diversity with unique breadth of biological functions.

The chemical diversity of natural products certainly makes their separation and purification quite challenging, but it stimulates the outstanding development of innovative technologies, which have higher performance, selectivity and throughput in both analytical and preparative scales. These modern technologies not only enhance established natural product-based therapies to better meet patients' needs but also promote the integration of natural products into contemporary new drug research and development.

Various chromatographic techniques have been successfully applied in natural products separation, such as gas chromatography, centrifugal thin-layer chromatography, ion-exchange chromatography, flash chromatography, medium pressure chromatography, high-performance chromatography and counter-current chromatography [5, 6].

After the discovery of the critical phenomenon in 1869, the first supercritical chromatographic separation was reported in 1962. Considering its mobile phase density and diffusion coefficient of the solutes, the operating range of supercritical fluid is exactly between the areas of typical gas and liquid [7]. With its unique features and significant advantages, supercritical fluid chromatography (SFC) has been developed as a continuum between high-performance liquid chromatography and gas chromatography and is becoming an important complementary technique to other chromatographic methods in the field of natural products.

In this chapter, we attempt to give an overview on the broad chemical diversity of biologically active natural products with some representative structures for each selected skeleton classes, and the wide applicability of supercritical fluid chromatography for their separation with demonstrated analytical and preparative benefits.

11.2 Analytical Supercritical Fluid Chromatography of Plant Metabolites

11.2.1 Sesquiterpenes

Sesquiterpenes are a structurally highly diverse class of terpenoids with 15 carbon atoms skeleton and derived biosynthetically from farnesyl pyrophosphate. There are more than 70 different ring systems that have been identified [8]. Many of them show quite broad and significant biological activities.

Artemisinin is a member of the amorphane sub-class of cadinane sesquiterpenes with artemisinic acid as its major bioprecursor (Fig. 11.1). It is a sesquiterpene lactone with an endoperoxide oxygen bridge across the seven-member ring. As the most promising compound with outstanding anti-malarial activity at low toxicity, it was found from the aerial parts of the plant *Artemisia annua* L., which were used for many centuries in traditional Chinese medicine for the treatment of high fever and malaria.

Artemisinin is effective against chloroquine resistant strains of *Plasmodium falciparum* and *P. vivax*, as well as against cerebral malaria. An ether-substituted derivative, artemether (Fig. 11.1), is soluble in sesame or peanut oil, and can be administered by intramuscular injection. Artemisinin-based combination therapies (ACT) are recommended by the World Health Organization (WHO) to treat especially multi-drug–resistant forms of malaria as currently used medications have become increasingly ineffective [9].

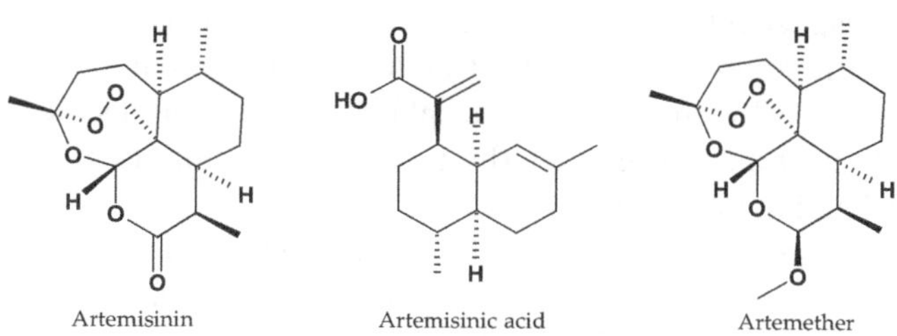

Figure 11.1 Structures of artemisinin and artemisinic acid, which were separated by SFC, and artemether used in "artemisinin-based combination therapies".

Petroleum ether, hexane, chloroform, dichloromethane, or toluene is commonly applied in the liquid extraction of artemisinin from *A. annua*. However, supercritical fluid extraction (SFE) is an interesting alternative to avoid using large amount of potentially hazardous solvents. In an optimization of experimental parameters, SFE was compared with two conventional liquid solvent extraction processes, one sonicated with an ultrasonic probe and the other vigorously mixed with a dynamic mixer [10]. The analysis of the SFE extract and those from two classical liquid solvent extraction processes gave not only comparable artemisinin concentration in chromatogram profiles but also similar relative standard deviations (RSDs) in repeatability studies.

The extraction with pure CO_2 showed a slow kinetic profile and some adsorption problems into capillaries from the restrictor to the collecting solvent. Based on the solvating power for artemisinin, methanol, ethanol, methanol–water (50/50) and toluene were tested as the modifier. Their kinetic curves were compared with various concentrations. A small addition of the modifiers was sufficient to achieve a rapid and quantitative extraction.

Just located in the glandular trichomes, artemisinin is weakly retained within the foliar tissues of *A. annua* and readily extracted by supercritical carbon dioxide. It is noteworthy that the small portion of a modifier may not be necessary to enhance the diffusion of the analyte through the plant material, but should play essential role to transport the artemisinin from the extraction cartridge to the collection vessels [11].

Under optimized conditions, both artemisinin and artemisinic acid were quantitatively extracted at a flow rate of 2 mL/min in less than 20 min with a supercritical fluid composed of carbon dioxide and 3% methanol at pressure and temperature fixed at 15 MPa and 50°C, respectively. In comparison with the standard liquid solvent extraction, supercritical fluid extraction of artemisinin from *A. annua* is more selective to obtain clean plant extracts and less tedious in operation. It also minimizes the risk of artemisinin degradation in the extraction procedure [10, 11].

As a thermolabile and UV transparent natural product, artemisinin cannot be easily analyzed by conventional GC and HPLC-UV but should be an excellent candidate for capillary SFC, which combines the high efficiency of capillary columns with the broad range of selective and sensitive GC detectors.

Three capillary columns were successfully used for the analysis of artemisinin with stationary phases of increasing polarity, including dimethylpolysiloxane DB-1, cycanopropylphenyldimethylpolysiloxane DB-1701 and polyethylene glycol DB-WAX. Density gradient elution was applied with pure CO_2 as mobile phase. A flame ionization detector (FID) was coupled with capillary SFC to obtain a good sensitivity and a large working dynamic range [12].

With validated methods, all three columns gave similar performance in the analysis of artemisinin with good resolution and in less than 25 min. However, the peak of artemisinic acid was asymmetric with DB-1 and DB-1701, which were apolar stationary phases. In the quantification of artemisinin and artemisinic acid in plant extracts, the polar DB-WAX column had higher efficiency for artemisinic acid and was also able to separate it from interfering compounds. Detailed capillary SFC conditions with DB-WAX column are summarized in Table 11.1, together with the quantitative results of artemisinin.

Faster separation and shorter running time are the advantages with packed-column supercritical fluid chromatography. This was confirmed in the comparison study to analyze artemisinin and artemisinic acid [13]. Three stationary phases with different polarities were tested, including polar bare silica, aminopropyl silica, and hydrophobic C18 silica. Evaporative light scattering detector (ELSD) was coupled with packed-column supercritical fluid chromatography, which is a universal mass detector for nonvolatile compounds. Methanol was used as the modifier to elute both compounds from the column.

The experimental data with the standard solution of artemisinin and artemisinic acid showed that they can be separated with very small amount of methanol, 1.6% for bare silica, 0.8% for C18, and 3% for aminopropyl silica column, respectively. Aminopropyl silica material was more selective for artemisinic acid. With an optimized methanol gradient elution on aminopropyl silica column, artemisinin and artemisinic acid were separated from interfering compounds in complex plant extracts in less than 8 min with capacity factors of 1.4 and 5.1, respectively.

The response of ELSD is related to the concentration of the analyte through an exponential relation. Therefore, the peak area of artemisinin in SFC-ELSD analysis should be a function of its concentrations on a logarithmic scale. Its linearity was verified

Table 11.1 Separation and quantification of artemisinin by capillary and packed–column supercritical fluid chromatography with three detectors

	SFC-FID	SFC-ELSD	SFC-APCI-MS
SFC conditions			
SFC type	Capillary SFC	Packed-column SFC	Packed-column SFC
Flow rate	10 mL/min at 0.25 g/mL	4 mL/min	2 mL/min
CO_2 pressure	—	170 bar	200 bar
Modifier	—	MeOH	MeOH (10%)
Elution	CO_2 density gradient, initial density 0.25 g/mL hold for 7 min, increased at 0.04 (g/mL)/min to 0.75 g/mL, with a final hold for 7 min	Isocratic/gradient, initial 1% for 3 min, increased to 10% in 12 sec and held for 5 min	Isocratic
Oven temperature	100°C	40°C	50°C
Stationary phase	DB-WAX	NH_2	CN
	(Polyethylene glycol phase, film thickness 0.1 μm)	(Aminopropyl, 5 μm)	(Cyanopropyl, 5 μm)
Column dimension	20 m × 0.1 mm id	12.5 cm × 4 mm id	25 cm × 4.5 mm id

	SFC-FID	SFC-ELSD	SFC-APCI-MS
Quantification			
Repeatability (RSD%)	5.7	7.9	0.40
	(n = 6)	(n = 6)	(n = 18)
Linearity range	0.1 to 2.0 mg/mL	0.1 to 1.0 mg/mL	1 ng to 250 ng
	(n = 7)	(n = 5)	(n = 5)
Coefficient of correlation	0.999	0.990	0.998
Slope	1.008	2.09	29390
Intercept	–0.045	3.53	61980
Detection limit	20 µg/mL	10 µg/mL	0.37 ng
Quantification limit	70 µg/mL	40 µg/mL	—

by correlation coefficients and the repeatability expressed by the relative standard deviations (RSD) in Table 11.1.

The data in Table 11.1 show that the results of packed-column SFC-ELSD are comparable with those in capillary SFC-FID. Therefore, there is no significant difference between both SFC methods in artemisinin analysis except the shorter separation time in packed-column SFC-ELSD.

Packed-column SFC in combination with atmospheric pressure chemical ionization mass spectrometry (SFC-APCI-MS) was also evaluated for the qualitative and quantitative analysis of artemisinin in *A. annua* [13]. As a soft ionization technique, APCI-MS provides protonated molecular ion and informative structural fragmentation for artemisinin. This is particularly helpful to unambiguously verify the artemisinin peak in the chromatograms with complex plant extracts.

The separation was achieved on a column, packed with 5 μm CN (cyanopropyl) stationary phase, using methanol (10%) as the modifier. The flow rate of 2 mL/min produced a pressure of 200 bar at the column inlet. The SFC was interfaced with APCI-MS for the analysis of artemisinin. To increase the sensitivity and selectivity, single ion monitoring (SIM) in positive ion mode with the most abundant ion at m/z 251 (100% relative intensity) was used to quantify artemisinin. Optimized experimental condition and quantification details were presented in Table 11.1.

In addition to retention time, which was the only identification criterion in SFC-FID and SFC-ELSD, mass spectrum and fragmentation pattern were also used in SFC-APCI-MS to ensure the reliable identification of the artemisinin peak in around 150 alcoholic extract samples [13]. The excellent repeatability with low RSD (0.40%), the outstanding linearity over a large concentration range (1–250 ng) and the significantly lower detection limit (370 pg) were additional advantages of SFC-APCI-MS.

11.2.2 Diterpenes

Diterpenes contain 20 carbons with four isoprene units and are very widely distributed in nature. They are all derived from the C_{20}-isoprenoid intermediate geranylgeranyl diphosphate, which is converted into derivatives of the various skeletal types by the action of diterpene syntheses (cyclases) [8].

Ginkgolides are diterpenetrilactones isolated from the leaves of *Ginkgo biloba* L., also known as the Maidenhair tree, which is unique species of tree with no close living relatives. *G. biloba* is widely cultivated and introduced since an early period in human history, and has various uses as a food and traditional medicine. Ginkgolides are chemically unique platelet activating factor (PAF) antagonists and responsible for the beneficial effects of ginkgo phytopharmaceuticals, such as blood flow improvement and memory enhancement.

Packed-column supercritical fluid chromatography coupled with evaporative light scattering detector (SFC-ELSD) was successfully applied for the analysis of main ginkgo constituents, including ginkgolides A, B, C, and J, and closely related sesquiterpenetrilactone bilobalide [14]. Their structures are presented in Fig. 11.2.

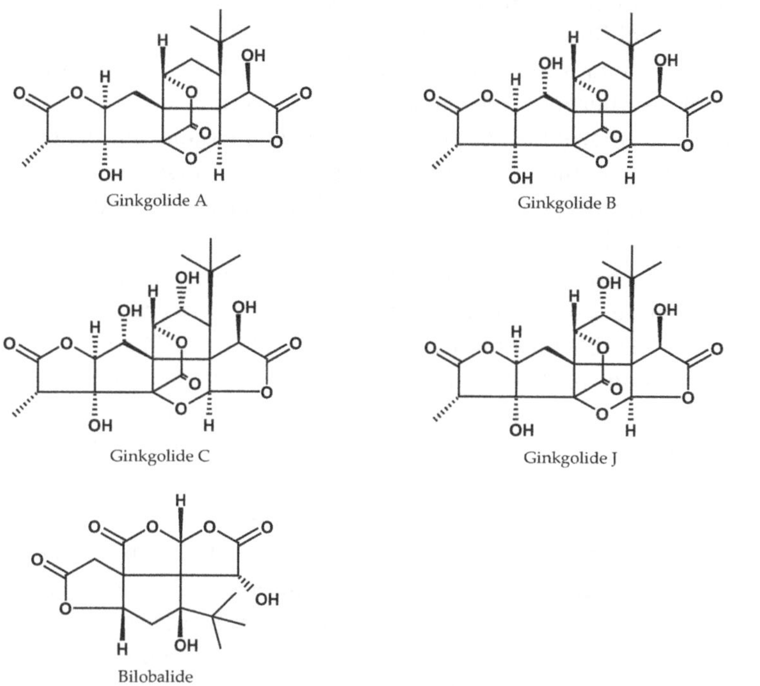

Figure 11.2 Structures of ginkgolide A, B, C, and J, and bilobalide, which were well separated by packed-column SFC-ELSD with a deactivated aminopropyl column.

The stationary phase selectivity plays an essential role in the separation of these ginkgo constituents. In the initial column screening, the columns packed with cyanopropyl-, octyl-, and octadecyl derived silica were tested, but no separation could be achieved at all. With two more polar poly(ethylene glycol) and aminopropyl columns, bilobalide could be separated from ginkgolide A/B and J/C pairs, however, with some tailing. In the efforts to reduce the column activity by adding some water into the mobile phase, separation improvement for the A/B and C/J pairs could be observed. However, no baseline separation for the pairs could be achieved.

Finally, instead of deactivation by adding water to mobile phase, a deactivated aminopropyl column (Deltabond) only with MeOH as modifier gave excellent baseline separation for all five components in nice peak shape. In the further evaluations, the optimum conditions for percentage modifier (12%), pressure (280 atm), temperature (40°C) and flow rate (3.5 or 4 mL/min) could be finalized. A good baseline separation could be achieved within 10 min under very simple isothermal, isobaric and isocratic conditions. The detection limits were around 20–40 ng for ginkgolides and bilobalide. This optimized method was also applied for the analysis of standard ginkgo extracts and a liquid ginkgo phytopharmaceutical drug.

As the plant metabolite isolated from the bark of the pacific yew tree *Taxus brevifolia*, paclitaxel is an important anti-cancer drug for the treatment of ovarian breast, lung and skin cancers. It is marketed as an injectable formulation under the trade name Taxol.

A rapid packed-column supercritical fluid chromatography method for the separation of paclitaxel and related compounds was developed [15]. Six columns, including Brownlee RP-18, Deltabound C_{18}, Octyl, Phenyl and Cyano, as well as LiChrospher Diol were evaluated and compared in the column screens.

In the elution method evaluation on the Diol column, retention time (min), capacity factor (k'), resolution (R_s), and asymmetry factor (T_f) were used for the comparison of pressure gradient and modifier gradient. The elution with modifier gradient in 4 min gave superior results than that with pressure gradient in 20 min.

Under final optimized conditions, the Diol column was used at isothermal oven temperature 30°C with isobaric pressure at 150 bar. Methanol was selected as modifier and run in a gradient mode: 8% between 0 and 3 min, ramped to 28% from 3 to 28 min, ramped to 35% from 28 to 33.7 min, and 35% from 33.7 to 37.7 min. An impressive baseline separation of paclitaxel from 15 of its impurities (e.g., 7-epitaxol, 10-deacetyltaxol and 10-deacetyl-7-epitaxol in Fig. 11.3) and degradation products was achieved with comparable results to those with HPLC, but only in half the run time. The SFC method was validated and could be used for impurity/purity profiling and can also easily be adopted for the stability-indicating purposes.

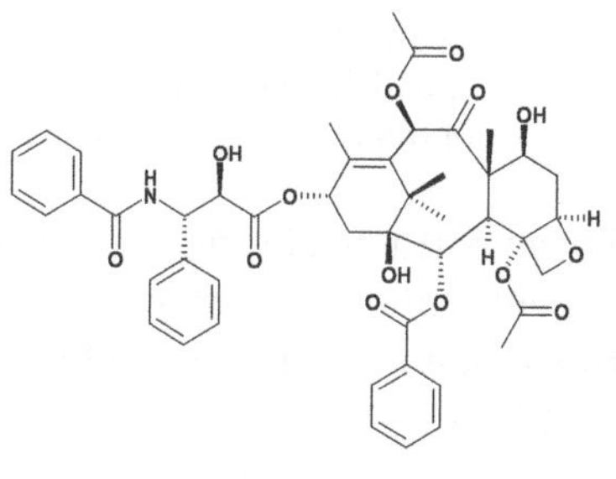

Paclitaxel

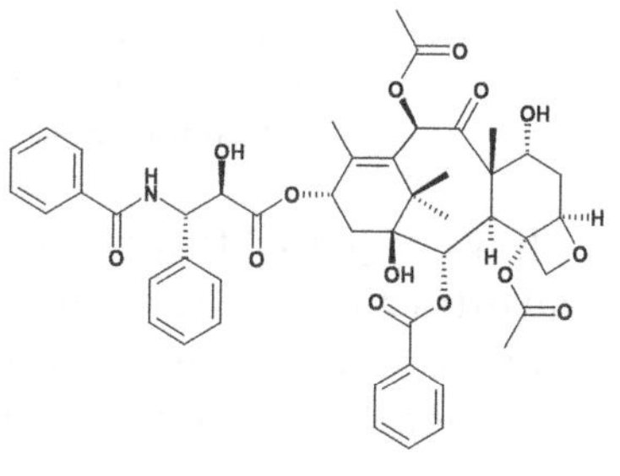

7-Epitaxol

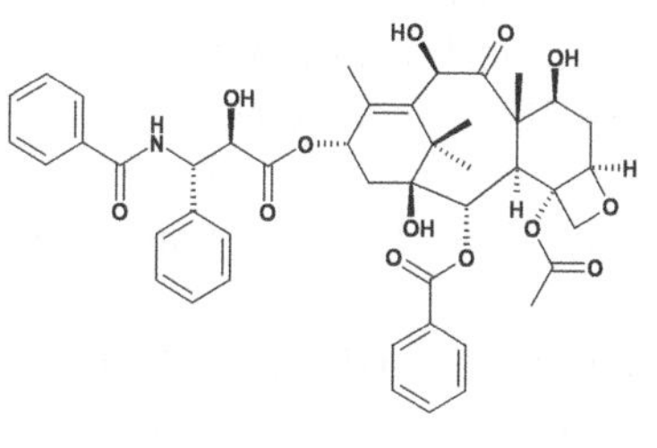

10-Deacetyltaxol

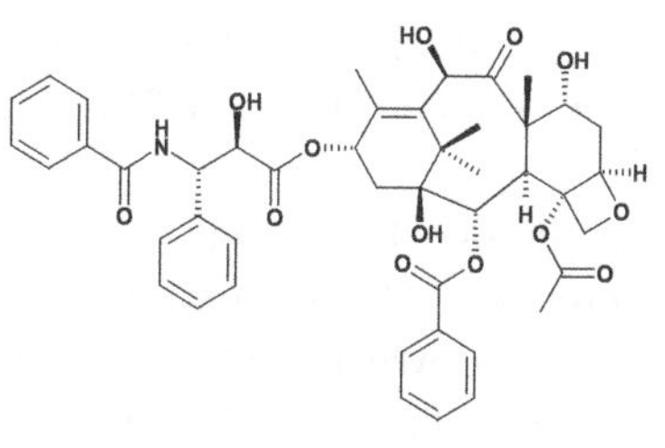

10-Deacetyl-7-epitaxol

Figure 11.3 Structures of paclitaxel, 7-epitaxol, 10-deacetyltaxol and 10-deacetyl-7-epitaxol, which were separated by packed-column SFC on a Diol column.

Owing to the slow-growing nature of the pacific yew tree *Taxus brevifolia* and the very low yield of paclitaxel from its bark, semi-synthetic preparations from more abundant and renewable sources have become a very attractive alternative, which is much more efficient than the total synthesis. Taxanes are promising precursors for the semi-synthesis of paclitaxel and its derivatives, such as 10-deacetylbaccatin III and taxicins. Their structures are presented in Fig. 11.4. They are either found as major components with much higher yield in the bark extract or available from renewable plant parts of yew trees.

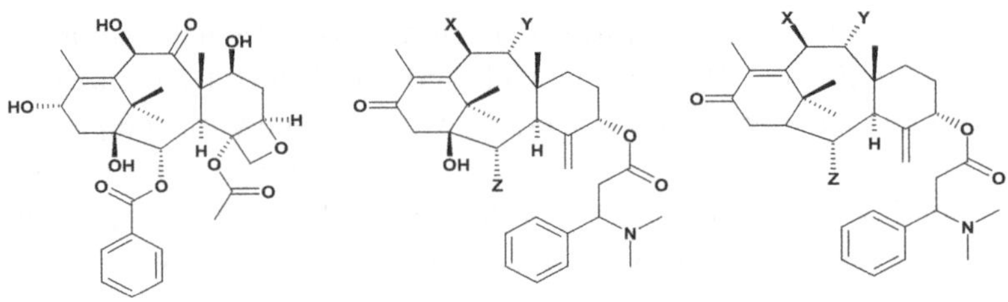

10-Deacetylbaccatin III　　Taxicin I　　Taxicin II

Figure 11.4 Structures of 10-deacetylbaccatin III, taxicin I and taxicin II (where X. Y and Z = OH or OCOMe), which were separated by packed-column SFC with a nitrile column.

It was reported that taxicins had been extracted by supercritical fluid extraction (SFE) from the dried needles of the English yew tree, *Taxus baccata,* and analyzed by both capillary and packed-column supercritical fluid chromatography (SFC) [16].

In the optimization of extraction conditions, 10% methanol was added into carbon dioxide. The oven temperature at 50°C made the carbon dioxide–methanol mixtures fully miscible but minimized the risk of decomposition. The pressure of 400 atm gave rise to a high fluid density to ensure high solubility. After assessing the cumulative weight of extracts in different times from 15 to 105 min, 90 min were determined for the extraction.

As shown in Fig. 11.4, three positions of taxicins, X, Y, and Z, can be substituted by either a hydroxyl or an acetylated group. To simplify subsequent analyses, the taxicin extracts were subjected to elimination and acetylation reactions, which eliminated dimethylamine and produced a *trans* cinnamate group, as well as reduced the isomer numbers by full acetylation at positions X, Y and Z.

In the analysis with capillary SFC, the more polar Carbowax column gave better baseline separation of taxicins I and II than the biphenyl capillary column. However, even under optimized conditions, the column was highly susceptible to overloading. Very diluted solution should be used in order to avoid unacceptable fronting of peaks. FID detector was couple with SFC for the quantitative analysis.

Better results were obtained with packed-column SFC. Although the ODS column gave only broad, overlapping, and poorly resolved peak in the chromatogram, the more polar nitrile column with methanol as modifier delivered excellent separation of taxicins I and II with greater efficiency, selectivity and resolution. The analysis time, which was 30 min with capillary column, could be reduced to 12 min. The accuracy of packed-column SFC was also better than that of capillary SFC.

11.2.3 Triterpenes

Triterpenes are formed biosynthetically from six isoprene units. The various polycyclic ring systems of triterpenes are the result of the different folding possibilities for its parent precursor squalene on the enzyme surface [8].

Azadirachtin A belongs to tetranortriterpenoid skeleton and is a highly oxidized triterpenoid with a decalin segment and a modified furan segment (Fig. 11.5). It was isolated from seed kernels of the neem tree, *Azadirachta indica*. As a natural biopesticide, azadirachtin A is effective as an anti-feedant and ecdysis inhibitor against a wide spectrum of insect species at ppm level. Packed-column SFC has been used for the separation of azadirachtin A, B and other triterpenoids identified from the same plant, including nimbin with anti-inflammatory, anti-tumor, immune-stimulating and bactericidal activities, and salannin with insect repellent and growth-disrupting properties [17].

In the study to optimize packed-column SFC, special efforts were made with the focus on the herbal analysis problems where the components have quite similar chemical structures and physicochemical properties. Improved resolution with extended column length was achieved by coupling two columns in series. Moreover, a cyanopropyl column followed by an amino column significantly decreased tailing factor values and improved symmetry of the peaks. The oven temperature, modifier concentration and CO_2 pressure were also evaluated based on the retention time, capacity factor, selectivity, tailing factor and number of theoretical plates. Optimization details are summarized in Table 11.2, together with the criteria used in the validation process.

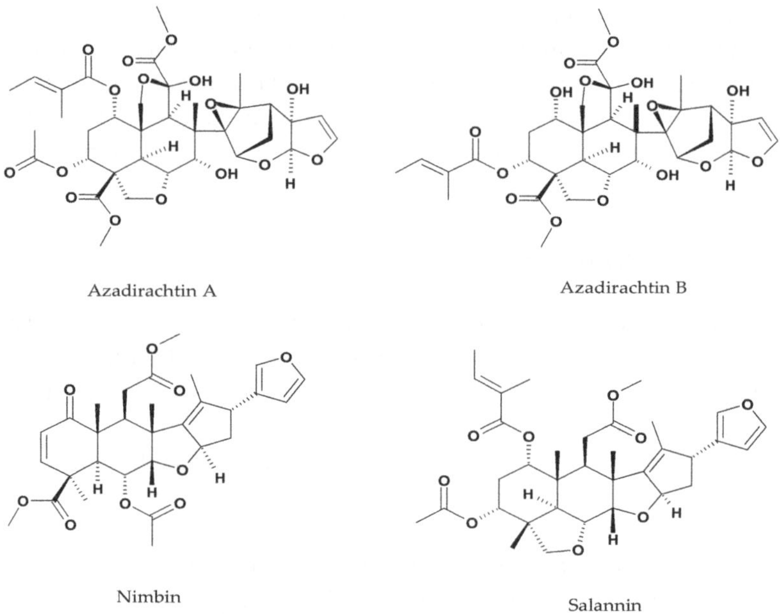

Figure 11.5 Structures of azadirachtin A, azadirachtin B, nimbin and salannin, which were separated by packed-column SFC with cyanopropyl and amino columns connected in series.

Table 11.2 Optimization and validation of packed-column supercritical fluid chromatography for the crude extract analysis of neem seed kernels of *Azadirachta indica*

	Evaluated condition	**Optimized condition**	**Optimization criteria**	**Validation criteria**
Columns	Cyanopropyl, amino	Cyanopropyl and amino connected in series	- Retention time (t_R) - Resolution (R_s)	- Calibration linearity - Precision in repeatability
Modifier concentration	MeOH 5.0–17.5%	MeOH 10%	- Capacity factor (k')	- Robustness by variation
Flow rate	2 mL/min	2 mL/min	- Selectivity (α)	- Limit of detection
Pressure	5–25 MPa	20 MPa	- Theoretical	- Limit of
Temperature	35–65°C	65°C	plate (N)	quantitation

The optimized conditions were successfully used to the quantitative analysis of crude extracts of neem seed kernels, as well as neem-based commercial formulation. The peaks in obtained chromatograms were sharp and symmetric with clear baseline separation. Peak purity checking tools with a diode array detector (DAD) coupled with SFC enabled the singularity check of peaks. The limit of detection (LOD) and the limit of quantitation (LOQ) were 2.5, 3.2, 2.0, 1.5 and 4.5, 5.0, 4.0 3.8 μg/mL, respectively, for azadirachtin A and B, salannin and nimbin.

11.2.4 Alkaloids

Alkaloids are basic natural products occurring mostly in plants, as well as in many microorganisms and marine organisms. Although including some related compounds with neutral and even weakly acidic properties, they contain mainly one or more basic nitrogen atoms, usually in heterocyclic rings [8]. With significant biological activities, alkaloids from plants have a long history for broad medical uses, e.g., morphine for analgesic, reserpine for anti-hypertensive, vinblastine for anti-tumor and quinidine for anti-arrhythmic.

Catharanthus roseus (Madagascar periwinkle) is a species native and endemic to Madagascar but has long been cultivated for herbal medicine and as an ornamental plant. Vinblastine, isolated from *C. roseus*, is an anti-microtubule drug used to treat certain kinds of cancer. As a microtubule disruptive drug, it binds tubulin, thereby inhibiting the assembly of microtubules.

C. roseus is one of the most thoroughly investigated plants, of which more than 70 alkaloids have been isolated. Therefore, the analysis of crude extracts from this plant is quite challenging for both separation and identification. Coupled with UV and mass spectrometry for getting much structure information of individual components, supercritical fluid chromatography was evaluated for the analysis of complex indole alkaloids from leaves of *C. roseus* [18].

An amino-bonded column was selected with the oven temperature at 65°C. Methanol was used as modifier with a gradient elution from 5% to 15%. Most of alkaloid peaks in the chromatogram were sharp and well shaped with numerous near-baseline separations. The pharmaceutically important bisindole group, including leurosine, 3′,4′-anhydrovinblastine and vinblastine (Fig. 11.6), was well separated from the majority of the monoindole alkaloids. The analogs with very similar structures and differing only slightly from each other in their function group, such as desacetoxyvindoline, vindorosine, vindoline and deacetylvindoline, were also well separated in the chromatogram.

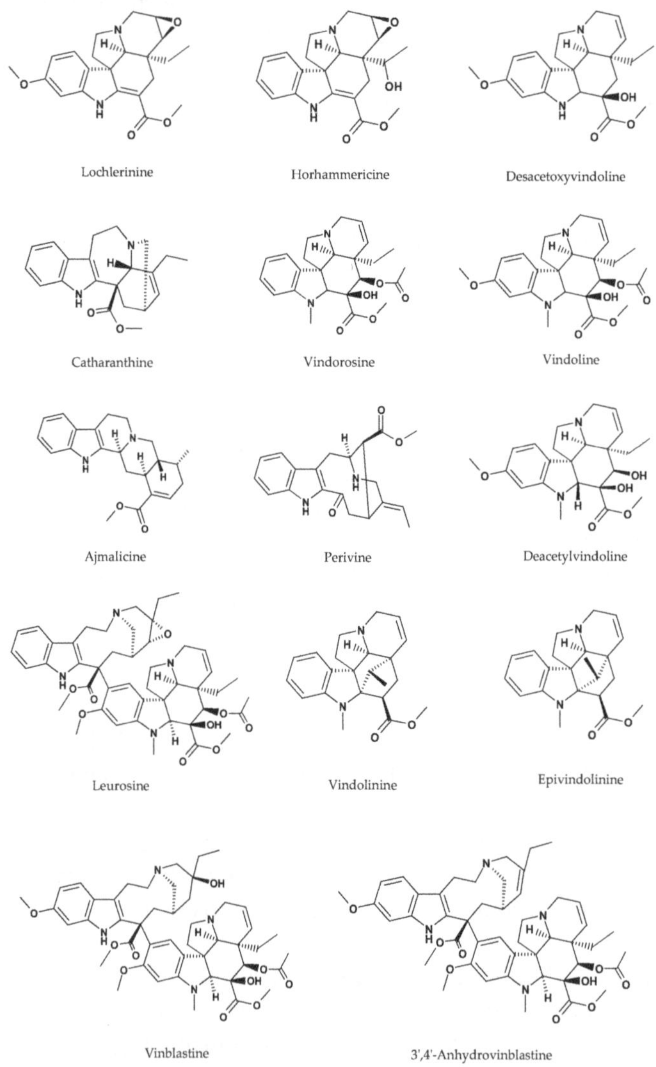

Figure 11.6 Structures of 14 indole alkaloids identified from leaves of *Catharanthus roseus* by packed-column SFC with an amino-bonded column.

In comparison with HPLC, the use of a large proportion of a highly volatile mobile phase in SFC is an attractive feature for its easy interfacing with a mass spectrometer. The MS spectra in high quality obtained with hyphenated SFC-MS provide significant structure information and can well serve as "fingerprints" in identifying known alkaloids in complex mixtures. Based on the reference MS spectra obtained from authentic samples, 14 alkaloids (Fig. 11.6) could be unambiguously identified, which represent more than 95% of the alkaloids of the mature leaves of *C. roseus*. The total number of detected compounds was around 60 in the extract analysis with SFC-MS .

11.2.5 Flavonoids

Occurringinallhigherplants,flavonoidsareimportantphenylpropane derivatives possessing the C_{15} basic skeleton of flavane. Herbal drugs containing flavonoids are used for protection against venous diseases and for improving coronary perfusion et al. [8].

As a special secondary metabolite group, polymethoxylated flavones are present in certain *Citrus* species. Their concentration in the fruit peel is higher than that in leaves and juices. Since the flavonoid determination is useful in chemotaxonomic studies on *Citrus* and to authenticate *Citrus* oils, fast separation of polymethoxylated flavons was evaluated with supercritical fluid chromatography [19].

A bare silica column with a high specific surface area was selected due to its higher selectivity than a partitioning stationary phase. Methanol was used as modifier in the range of 5–30% to evaluate the retention behavior of polymethoxylated flavons on the bare silica column at the constant temperature 40°C and under pressure 200 atm. The relative standard deviation (RSD) of peak capacity factor was less than 0.5% with four times injections. There was no selectivity improvement with methanol in the range of 5–20%.

By comparing the capacity factors of various polymethoxylated flavons, it was observed that their retention is determined not only by the number of methoxy groups in the molecule, but also by steric effects. The experimental data indicated that a methoxy group in position 3 decreases the retention, whereas a methoxy group in position 3′ increases the retention. In addition, two methoxy groups in both positions 3′ and 4′ contribute to an increase in the capacity factors.

An established method was applied to analysis of six polymethoxylated flavons in orange oil and tangerine oil, including tangeretin, heptamethoxyflavone, nobiletin, sinensetin, tetramethylscutellarein and hexamethoxyflavone (Fig. 11.7). The SFC procedure was faster than HPLC with good resolution and adequate accuracy for quantification.

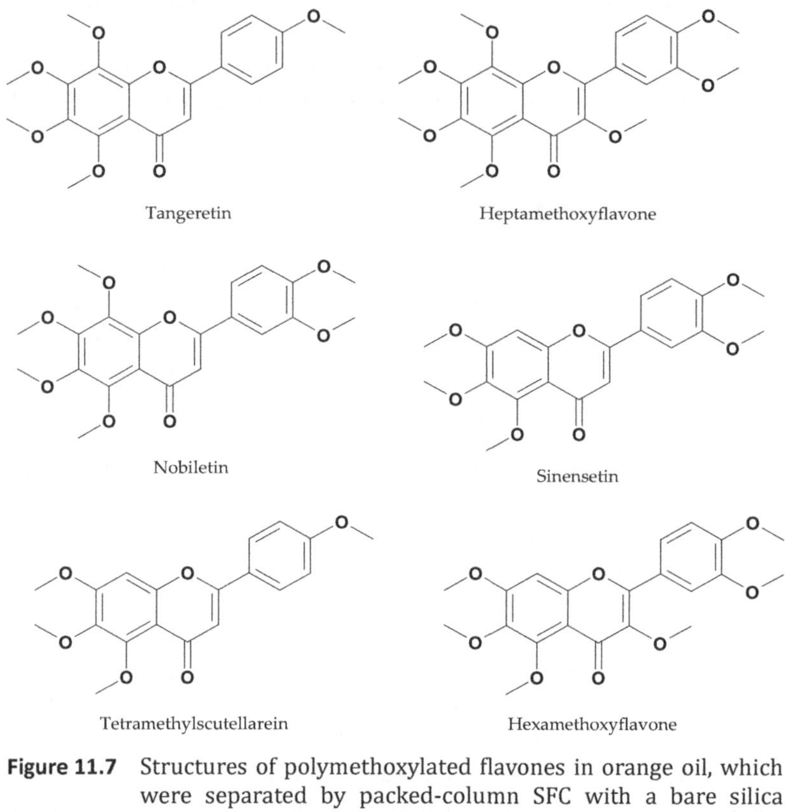

Figure 11.7 Structures of polymethoxylated flavones in orange oil, which were separated by packed-column SFC with a bare silica column.

11.2.6 Kava Lactones

Kava lactones are a class of lactone compounds found in the Kava shrub *Piper methysticum*, which grows throughout the South Pacific. The extracts of the Kava roots have been used to induce relaxation and reduce stress level. Kava lactones are alpha-pyrones with a methoxy group attached to the non-aromatic ring and have varying physiological effects *in vivo* depending on their molecular structures, e.g., the degree of saturation and benzene substitution. The feasibility of both supercritical fluid extraction and supercritical fluid chromatography was demonstrated with seven Kava lactones [20], which are presented in Fig. 11.8.

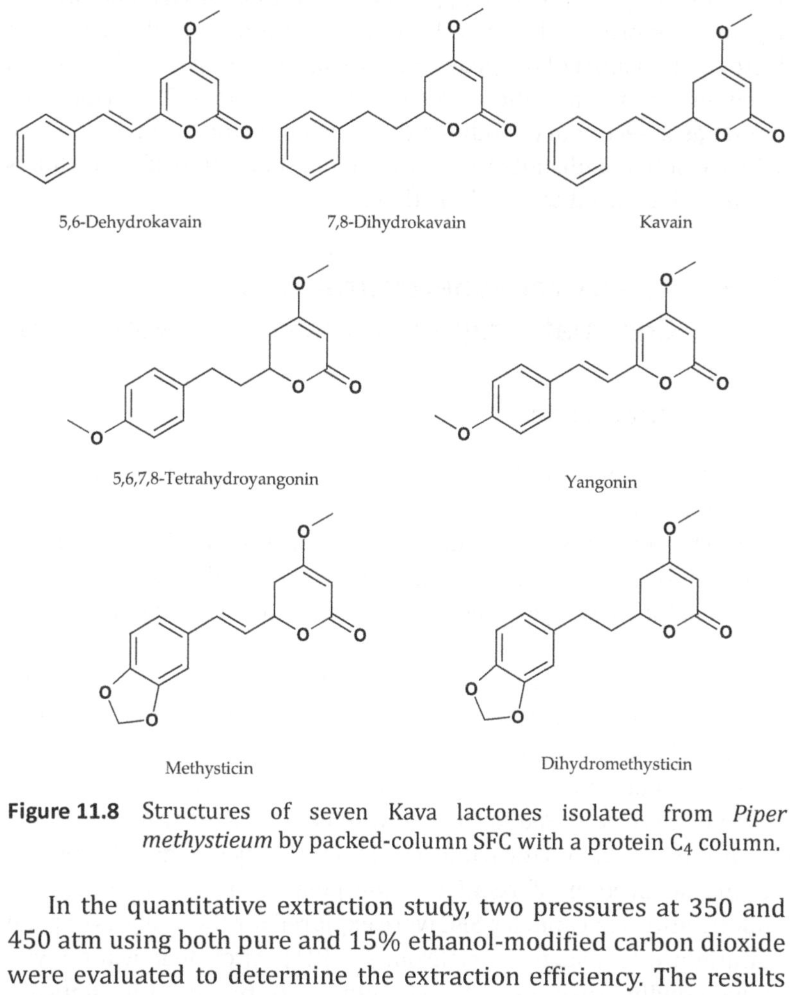

Figure 11.8 Structures of seven Kava lactones isolated from *Piper methystieum* by packed-column SFC with a protein C_4 column.

In the quantitative extraction study, two pressures at 350 and 450 atm using both pure and 15% ethanol-modified carbon dioxide were evaluated to determine the extraction efficiency. The results were compared with the recovery of solid–liquid extraction under sonication using dichloromethane–methanol (1:1). With an efficiency greater than 90% relative to conventional solvent extraction, most of Kava lactones can be extracted by pure carbon dioxide at 60°C with a flow rate of 2 mL/min for 60 min. The extraction efficiency did not increase significantly with 15% ethanol-modified carbon dioxide.

In the optimization of supercritical fluid chromatography to separate Kava lactones, five columns were screened with the same dimension and particle size, but different stationary phases, including NH_2, Cyano, Diol, C_4 protein and Diphenyl. NH_2 and protein C_4 columns provided almost baseline separation of all seven Kava lactones, and most of the peaks were resolved better with the protein C_4 column. In the modifier screening, the addition of 0.1% isopropyl amine as a secondary modifier to methanol prior to mixing in-line with carbon dioxide not only improved the separation of the lactones but also decreased the analysis time.

11.3 Analytical Supercritical Fluid Chromatography of Microbial Metabolites

11.3.1 Macrolides

Macrolides are natural compounds containing a lactone ring with more than 10 ring atoms and their spectrum of biological activities is very broad. Many macrolides contain one or more O-glycosidically bound rare sugars, e.g., deoxyhexoses or aminodeoxyhexoses. The macrolactone ring is formed on the polyketide biosynthetic pathway [8].

Erythromycin is a macrolide containing a 14-membered lactone ring and two sugars, L-cladinose, and D-desosamine (Fig. 11.9). It is produced from an actinomycete strain *Saccharopolyspora erythraea*, and has an antimicrobial spectrum similar to that of penicillin. This antibiotic is often used for people who have an allergy to penicillins.

Although supercritical fluid chromatography is generally devoted to the separation of relatively non-polar compounds, necessary modifications in capillary SFC were explored for the analysis of polar natural product such as erythromycin [21]. After deactivation with polymethylhydrosiloxane, the columns were prepared by statically coating fused silica with polysiloxane stationary phases and cross-linking the phases with azo-*tert*-butane. It was demonstrated that well-deactivated columns at 40°C and properly constructed flame ionization detector can be used to analyze erythromycin.

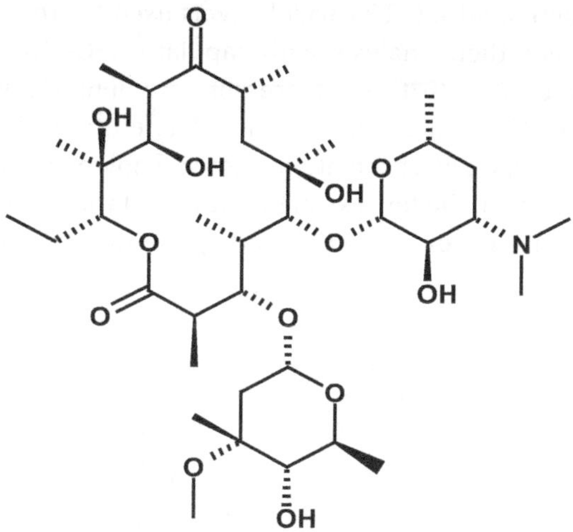

Erythromycin

Figure 11.9 Structure of erythromycin from an actinomycete strain *Saccharopolyspora erythraea*, which was analyzed by capillary SFC-FID.

FK 506, also known as tacrolimus or fujimycin, is a macrolide isolated from an actinomycete strain *Streptomyces tsukubaensis* (Fig. 11.10). It is an immunosuppressive drug that is mainly used after allogeneic organ transplant to reduce the activity of the patient's immune system and so lower the risk of organ rejection. FK 506 binds to the immunophilin FKBP12 (FK506 binding protein) and the heterodimeric complex interacts with and inhibits calcineurin thus inhibiting both T-lymphocyte signal transduction and IL-2 transcription.

Rapamycin, also known as sirolimus, is another macrolide discovered from an actinomycete strain *Streptomyces hygroscopicus* (Fig. 11.10). It is also an immunosuppressant drug used to prevent rejection in organ transplantation. Unlike FK 506, rapamycin is not a calcineurin inhibitor and does not inhibit the production of IL-2. It inhibits the response to interleukin-2 (IL-2), and thereby blocks activation of T- and B-cells.

For clinical monitoring of FK 506 and rapamycin in whole blood, feasibility study was performed with capillary supercritical fluid chromatography coupled with a FID detector [22]. A solid phase extraction with C-18 cartridge was used for the whole blood extraction. For their analysis with capillary SFC-FID, SB-biphenyl column was used at 70°C with a pressure gradient of carbon dioxide from 100 to 300 atm at 10 atm/min. The elution of FK 506 and rapamycin in their SFC chromatograms was shown with the standard samples. However, the low extraction rate and the low sensitivity of FID detection did not allow their adequate quantitation in whole blood.

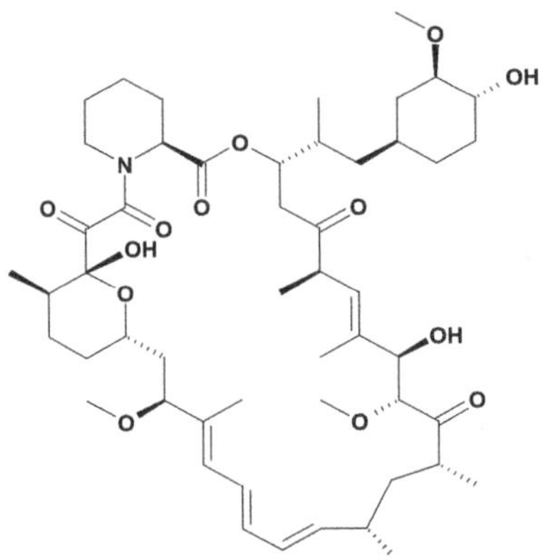

Rapamycin

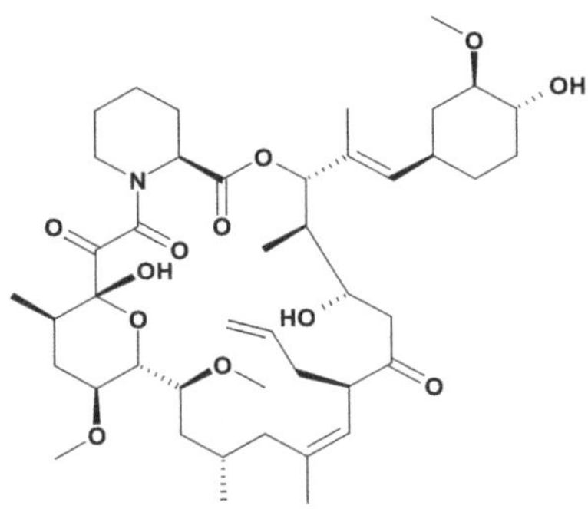

FK 506

Figure 11.10 Structures of two macrolides, FK 506 and rapamycin, which were analyzed by capillary SFC-FID.

11.3.2 Cyclic Peptides

Cyclic or linear peptide antibiotics contain often unusual, non-proteinogenic amino acids and exhibit broad biological activities against bacteria, fungi, viruses, as well as immunosuppressive and anti-tumor properties. Macrocyclic peptides with one or more ester bonds in the ring are known as peptide lactones or depsipeptides [8].

Cyclosporin A was initially found from the fungus *Tolypocladium inflatum*, which was isolated from a soil sample collected in Norway. It is a cyclic peptide of 11 amino acids and contains a single D-amino acid, which is rarely encountered in nature (Fig. 11.11). As an immunosuppressant drug used in organ transplant to reduce the activity of the immune system, cyclosporin A binds to the cyclophilin and the complex subsequently inhibits calcineurin, which is a responsible element for activating the transcription of interleukin 2.

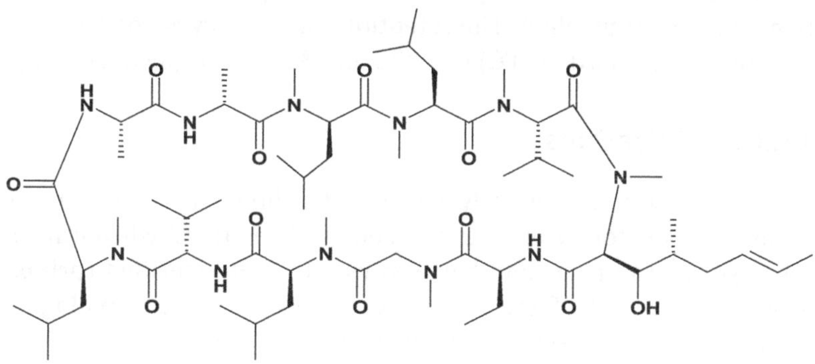

Cyclosporin A

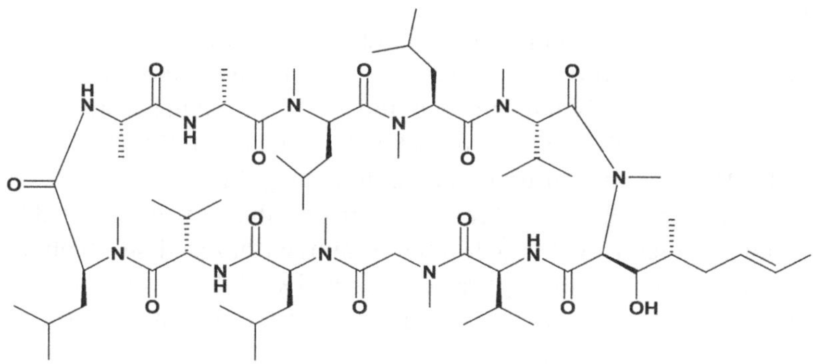

Cyclosporin D

Figure 11.11 Structure of cyclosporin A and D isolated from the fungus *Tolypocladium inflatum*, which were analyzed by capillary SFC-FID.

Supercritical fluid chromatography was investigated for its potential in the clinical monitoring of cyclosporin A, which is crucial to ensure its therapeutic range for different organ transplants and to minimize its undesirable side effects. In the feasibility study, cyclosporin A was spiked in whole blood samples and a solid phase extraction with C-18 cartridge was used in the sample preparation. Capillary SFC-FID was performed with a SB-biphenyl column at 70°C.

The pressure gradient of carbon dioxide from 175 to 250 atm at 2 atm/min. In the analysis of a whole blood extract of a bone marrow transplant patient, about 220 μg/L of cyclosporin A was observed, within the therapeutic range of 150 to 400 μg/L recommended for bone marrow transplant. The retention times for cyclosporin A and D (internal standard in SFC) were 35 and 33 min, respectively [22].

11.3.3 Polyethers

Polyether antibiotics usually consist of a linear chain of carbon atoms as skeleton, which is cyclized to form tetrahydrofuran or tetrahydropyran rings and has always a free or esterified carboxy group at one end of the chain. They show numerous biological activities against bacteria, fungi, viruses, protozoa, insects and worms. Due to their ability to form well crystallizing complexes with mono- or divalent cations, they are also designated as ionophores [8].

Anticoccidial ionophore semduramicin was isolated from *Actinomadura roseorufa.* Two closely related ionophores are antibiotics for birds and chickens, and are also used in livestock feed to aid fermentation in the stomachs of animals that digest cellulose [23, 24]. Their structures are presented in Fig. 11.12.

A baseline separation for these two ionophores was achieved by capillary supercritical fluid chromatography with a fused-silica column having a film of DB-5 (95% dimethyl–5%-diphenylpolysiloxane) at 100°C. The pressure of carbon dioxide first held at 225 atm for 5 min, then increased in linear gradient to 306 atm at 4 atm/min. A flame ionization detector was coupled with capillary SFC for the detection. Since both antibiotics are organic salts, the highlight of this study was the separation of ionic compounds by SFC using an unmodified carbon dioxide mobile phase.

11.3.4 Trichothecenes

Trichothecenes are a group of sesquiterpene mycotoxins with a common 12, 13-epoxy-trichothec-9-ene ring system. They are produced by various ascomycetes and fungi that attack agricultural products and grow on grain. Trichothecene mycotoxins are all contact poisons and possess a wide range of biological properties, such as phytotoxic, insecticidal, anti-fungal, and anti-viral activities [8].

An investigation of capillary column supercritical fluid chromatography in combination with mass spectrometry was made for the separation and detection of trichothecene mycotoxins [25]. The structures of investigated trichothecene mycotoxins are given in Fig. 11.13.

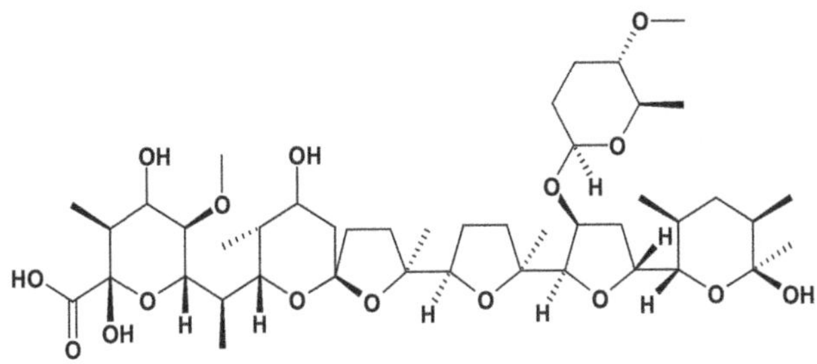

Semduramicin

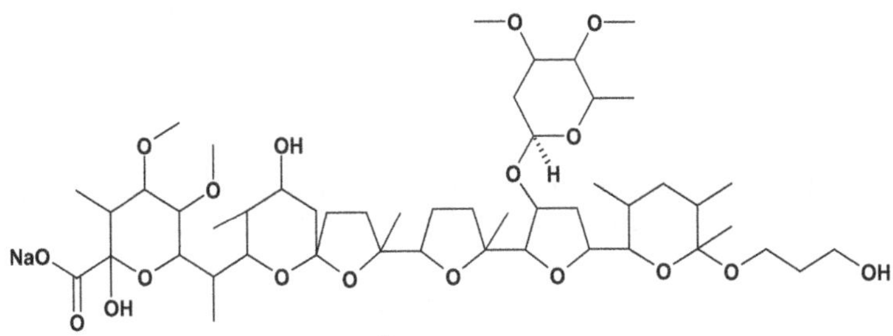

1

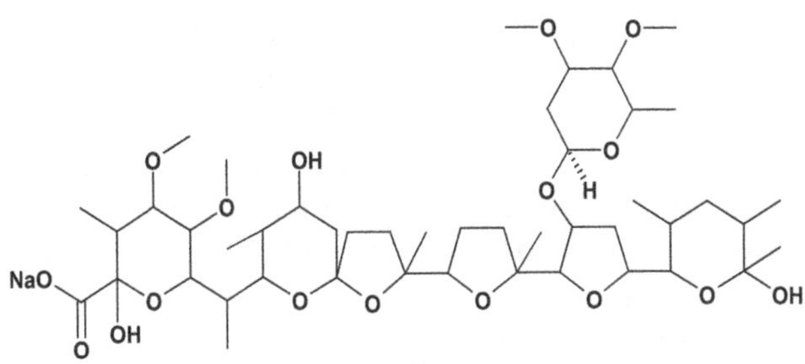

2

Figure 11.12 Structure of semduramicin isolated from *Actinomadura roseorufa* and two ionophores (1 and 2), which were separated by capillary SFC-FID.

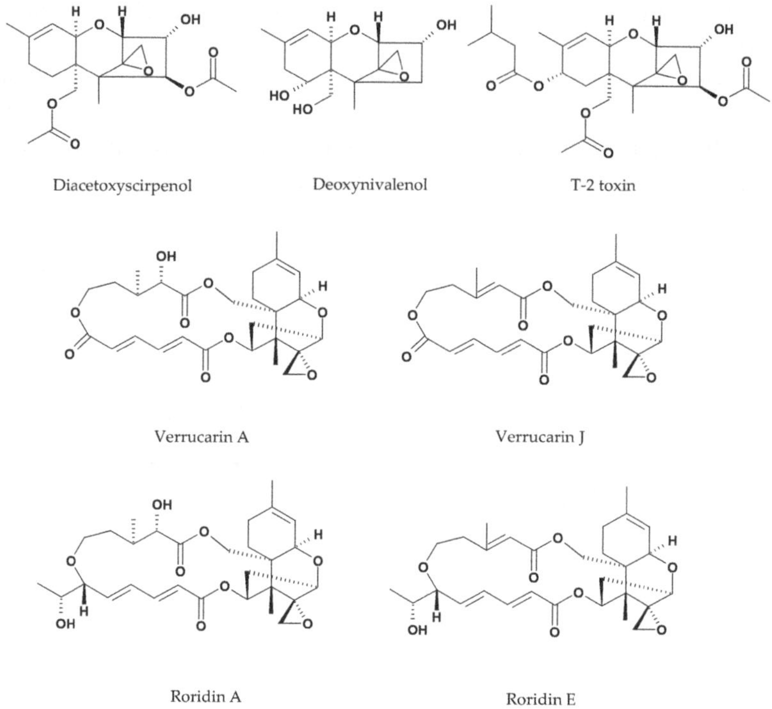

Figure 11.13 Structure of trichothecene mycotoxins, which were separated by capillary SFC-MS.

The solubility study indicated their sufficient solubility in supercritical carbon dioxide at 40°C. The fused silica capillary SFC columns were deactivated and coated with a thick film of a 5% phenyl polymethylphenylsiloxane phase (SE-54) rendered non-extractable by extensive cross-linking. Due to the low pressure drop across capillary SFC columns, effective fast capillary SFC separation was investigated by rapid pressure ramp (e.g., 100 atm/min), together with selected ion monitoring techniques in mass spectrometry for an efficient detection. The conditions in normal capillary SFC and fast capillary SFC are summarized in Table 11.3, together with the separation examples of trichothecene mycotoxins.

Various trichothecene mycotoxins were resolved within a few of min in fast capillary SFC with shorter and smaller columns and rapid pressure ramp. This exploratory work clearly indicated its potential for improved rapid separation with capillary SFC.

Table 11.3 Separation examples of trichothecene mycotoxins and the conditions in normal capillary SFC and fast capillary SFC

	Normal capillary SFC	Fast capillary SFC
Column	15 m × 50 μm id	1.75 m × 50 μm id 0.8 m × 25 μm id
Stationary phase	SE-54, 0.25 μm	SE-54, 0.25 μm SE-54, 0.2 μm
Pressure	70 to 350 atm	75 to 300 atm
Oven temperature	100°C	100°C
FID temperature	350°C	350°C
Separation examples	Diacetoxyscirpenol; deoxynivalenol; T-2 toxin (in 60 min). Diacetoxyscirpenol; verrucarin A; verrucarin J; roridin A ; roridin E (in 70 min)	Diacetoxyscirpenol; T-2 toxin; verrucarin J; roridin A (0.8 m column, in 1.5 min). Diacetoxyscirpenol; verrucarin J; verrucarin A; roridin E; roridin A (1.75 m column, in 4.5 min)

11.3.5 Chloramphenicol

Chloramphenicol, isolated and identified from an actinomycete strain *Streptomyces venezuelae*, contains nitro and dichloroacetyl groups, which are a most unusual combination for natural products [8]. As the first antibiotic manufactured synthetically on a large scale, it stops bacterial growth by inhibiting protein chain elongation and is effective against a wide variety of Gram-positive and Gram-negative bacteria. Its most serious adverse effect is bone marrow toxicities, including bone marrow suppression and aplastic anemia.

Due to its very broad spectrum of antibiotic activity, chloramphenicol was also used as veterinary drug in the disease treatment of pigs, cows and fish. However, its toxic side effects in humans led to restriction or banning of the drug for the treatment of livestock that forms part of the food chain. As a possible technique to monitor veterinary antibiotics, supercritical fluid chromatography was evaluated for the analysis of chloramphenicol, together with two synthetic antibiotics, levamisole and furazolidone, and one microbial metabolite lincomycin from the actinomyces *Streptomyces lincolnensis* [26]. The structures of these four veterinary antibiotics are presented in Fig. 11.14.

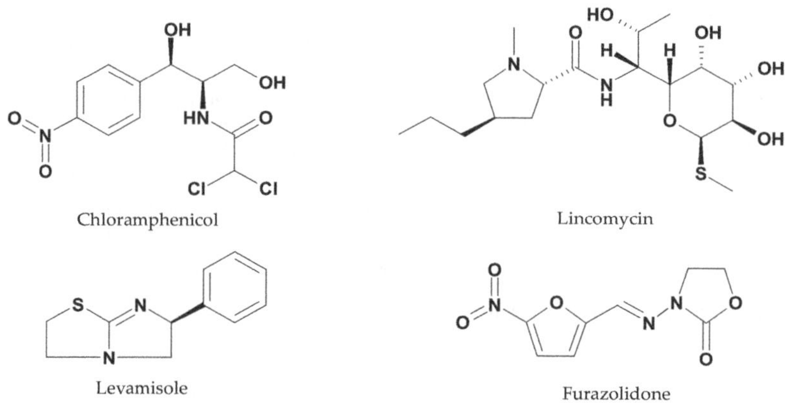

Figure 11.14 Structures of four veterinary antibiotics, chloramphenicol, lincomycin, levamisole and furazolidone, which were separated by packed-column SFC-MS with an amino-bonded column.

In the method optimization with packed-column SFC-MS, the more acidic stationary phases such as Cyano, ODS2 and silica were of very little use, because lincomycin and levamisole remained adsorbed to these stationary phases even eluted with high percentages of modifier while furazolidone gave a badly tailing peak. The amino-bonded column could resolve all four veterinary antibiotics with baseline separation.

The effects of modifiers on the retention times of four antibiotics were individually investigated with 20% of methanol, 2-methoxy-ethanol, propylene carbonate, 2-propanol and dimethylformamide. Peak shape and tailing extent were also used as evaluation criteria. With 15% methanol as modifier, all four veterinary antibiotics were well separated with flow rate 4 mL/min at oven temperature 75°C and under column pressure 351 bar.

11.4 Preparative Supercritical Fluid Chromatography of Natural Products

11.4.1 Preparative Supercritical Fluid Chromatography of Plant Metabolites

In comparison with analytical supercritical fluid chromatography, the separation and purification of natural products in preparative scale have been much less investigated. There are only a few reports in the literature on the secondary metabolites' separation from crude plant extracts by preparative SFC.

In a recent study, packed-column SFC was used for the preparative purification of anti-malarial artemisinin from the crude extract of *Artemisia annua* [27]. The structure of artemisinin has already been presented in Fig. 11.1. In the optimization study, various modifiers in different concentrations, CO_2 velocities, pressures and temperatures have been evaluated with a C_{18} column. Experimental evaluation details, evaluation criteria and optimized conditions are summarized in Table 11.4. Under optimized condition, 143 mg of artemisinin with a purity of 74.83% was isolated directly from the crude plant extract, which contains 7.29% of artemisinin.

Table 11.4 Optimization of preparative supercritical fluid chromatography for the artemisinin purification from the extract of *A. annua*

	Evaluated range	Evaluated values	Evaluation criterion	Optimized condition
Modifier				
MeOH	0–10%	5	Resolution (R_s)	0
EtOH	0–10%	5	Resolution (R_s)	0
CO_2 velocity	6–30 g/min	7	Resolution (R_s)	22 g/min
Pressure	10–16 MPa	7	Resolution (R_s) Capacity factor (k')	11 MPa
Temperature	313–338 K	6	Resolution (R_s) Capacity factor (k')	313 K

Kava shrub *Piper methysticum* can induce relaxation and reduce stress level. In an investigation on preparative separation of Kava extract by SFC, two protein C_4 columns must be connected in series since a single one did not have enough efficiency to separate all seven Kava lactones, which are presented in Fig. 11.8 [20]. A gradient of methanol-modified carbon dioxide with pressure at 125 atm was used with 2% MeOH in CO_2 containing 0.1% isopropyl amine for 3 min, and then increased to 10% MeOH at rate of 0.4%/min. The oven temperature and flow rate were 80°C and 4 mL/min, respectively. Sufficient separation for most of the Kava lactones was achieved in this preliminary investigation.

11.4.2 Preparative Supercritical Fluid Chromatography of Microbial Metabolites

Few publications on preparative supercritical fluid chromatography of microbial metabolites could be found in literature. In the second annual international conference on packed-column SFC (SFC 2008), we presented the development of supercritical fluid chromatography for microbial metabolite separation and its integration into our high throughput purification platform with flash LC chromatography on normal phase and preparative HPLC-UV-MS on reversed phase [28].

In our development, we evaluated sample loading capacity and recovery rate, and also optimized the column dimensions, stationary phase, flow rate, temperature, backpressure, modifier and other parameters specifically for the separation of microbial secondary metabolites. An automated column screening system was implemented with various stationary phases, including Cyano, Diol, Pyridine, Amino and Pyridine-amide.

Much shorter running time, faster scale-up and easiness in modifier removal meaningfully increased the purification throughput. Improved sample quality facilitated structure elucidation of elaborated microbial metabolites, particularly for the minor or unstable ones. Stacked injections significantly reduced the time needed in preparative separation.

In our established process, supercritical fluid chromatography was mainly used in the final purification step for those samples, which cannot be well separated by HPLC-UV-MS. In our technology evaluation, development and validation, 124 secondary metabolites of actinomycetes and fungi were purified by supercritical fluid chromatography. Eighty-four structures were elucidated, including 16 from actinomycetes and 68 from fungi. A total of 65 identified novel metabolites, representing 77.38% of all identified structures, have not been described so far in literature.

11.5 Conclusions and Prospects

The high chemical diversity and broad biological activities of natural products have been well demonstrated by the secondary metabolites from both plants and microorganisms, which were selected as representative examples in this chapter. The successful separation of these highly diversified compounds composed by different structure scaffolds has also clearly showed that supercritical fluid chromatography is a versatile and powerful technique for natural product separation and purification.

Two supercritical fluid chromatography techniques, including capillary SFC and packed-column SFC, have been applied in the natural product analysis. Different optimization strategies, methodologies and processes for both SFC techniques have been clearly demonstrated with selected representative examples. Established methods and optimized parameters should be very useful for the analysis of each major metabolite types. Concise technique features in each reported case studies, as well as some technical failures in the early feasibility studies, may also support the method development for new applications.

The SFC mobile phase in supercritical state exhibits lower viscosity and higher diffusivity than the solvents in liquid chromatography, thus allows higher flow rates at lower column pressure. Therefore, supercritical fluid chromatography has many unique advantages and attractive features, such as short running time with fast separations, high separation efficiency with rapid column re-equilibration, minimal consumption of organic solvents with environmental benefit, significant reduction in evaporation time of collected fractions, cost efficiency with recent acetonitrile shortage and increased peak capacity with high orthogonality.

SFC has been well integrated in chiral analysis, particularly in the pharmaceutical industry process. In the meantime, it is also becoming an emerging technique for the achiral separation and purification in new drug discovery [29]. In recent years, technology development with the emphasis on packed-column SFC has been making rapid and significant progress, such as ultra high-performance SFC with sub-2 μm stationary phase [30], analytical and preparative SFC columns with different selectivity [31], parallel SFC with multiple channels [32], two-dimensional SFC [33], UV and MS triggered semi-preparative and preparative systems [34, 35]. The integration of these innovative SFC technologies will certainly increase both throughput and performance of natural products separation and purification. As a complimentary method to other techniques, supercritical fluid chromatography will make its unique contributions to the high throughput purification and play an important role to further increase the chemical diversity of natural products for new drug discovery.

Acknowledgments

Special acknowledgment goes to Frank Petersen, Philipp Krastel, Eric Francotte and Georg Diehl, as well as other involved colleagues and coworkers for stimulating discussions and constructive contributions. Weizheng Wang is particularly thanked for her excellent editorial assistance.

References

1. Cragg, G. M., Grothaus, P. G., and Newman, D. J. (2009) Impact of Natural Products on Developing New Anti-Cancer Agents, *Chem. Rev.*, **109**, pp. 3012–3043.
2. Mishra, B. B., and Tiwari, V. K. (2011) Natural products, an evolving role in future drug discovery, *J. Eur. Med. Chem.*, **46**, pp. 4769–4807.
3. Carter, G. T. (2011) Natural products and Pharma 2011: strategic changes spur new opportunities, *Nat. Prod. Rep.*, **28**, pp. 1783–1789.
4. Schmitt, E. K., Moore, C. M., Krastel P., and Petersen F. (2011) Natural products as catalysts for innovation: a pharmaceutical industry perspective, *Curr. Opin. Chem. Biol.*, **15**, pp. 1–8.
5. Hostettmann, K., Hostettmann, M., and Marston, A. (1997) *Preparative Chromatography Techniques—Applications in Natural Product Isolation,* 2nd Ed. (Springer Verlag, Berlin Heidelberg).
6. Sarker, S. D., Latif, Z., and Gray, A. I. (2006) *Natural Products Isolation,* 2nd Ed. (Humana Press, New Jersey).
7. Smith, R. M. (1999) Supercritical fluids in separation science – the dreams, the reality and the future, *J. Chromatogr. A*, **856**, pp. 83–115.
8. Steglich, W., Fugmann, B., and Lang-Fugmann, S. (2000) *Römp Encyclopedia Natural Products,* translated by Dunmur, R. (Georg Thieme Verlag, Stuttgart).
9. Kuhn, T., and Wang, Y. (2007) *Progress in Drug Research* **65** (eds. Petersen, F., and Amstutz, R.), Chapter "Artemisinin—an innovative cornerstone for anti-malaria therapy" (Birckhäuser Verlag, Basel), pp. 383–422.
10. Kohler, M., Haerdi, W., Christen, P., and Veuthey, J. (1997) Extraction of artemisinin and artemisinic acid from *Artemisia annua* L. using supercritical carbon dioxide, *J. Chromatogr. A*, **785**, pp. 353–360.
11. Kohler, M., Haerdi, W., Christen, P., and Veuthey, J. (1997) Supercritical fluid extraction and chromatography of artemisinin and artemisinic

acid. An improved method for the analysis of *Artemisia annua* samples, *Phytochem. Anal.*, **8**, pp. 223–227.

12. Kohler, M., Haerdi, W., Christen, P., and Veuthey, J. (1997) Determination of artemisinin and artemisinic acid by capillary and packed supercritical fluid chromatography, *J. High Resolut. Chromatogr.*, **20**, pp. 62–66.

13. Dost, K., and Davidson, G. (2003) Analysis of artemisinin by a packed-column supercritical fluid chromatography-atmospheric pressure chemical ionisation mass spectrometry technique, *Analyst*, **128**, pp. 1037–1042.

14. Strode III, J. T. B., Taylor, L. T., and van Beek, T. A. (1996) Supercritical fluid chromatography of ginkgolides A, B, C and J and bilobalide, *J. Chromatogr. A*, **738**, pp. 115–122.

15. Jagota, N. K., Nair, J. B., Frazer, R., Klee, M., and Wang, M. Z. (1996) Supercritical fluid chromatography of paclitaxel, *J. Chromatogr. A*, **721**, pp. 315–322.

16. Heaton, D. M., Bartie, K. D., Rayner, C. M., and Clifford, A. A. (1993) Application of supercritical fluid extraction and supercritical fluid chromatography to the production of taxanes as anti-cancer drugs, *J. High Resolut. Chromatogr.*, **16**, pp. 666–670.

17. Agrawal, H., Kaul, N., Paradkar, A., and Mahadik, K. R. (2005) Standardization of crude extract of neem seed kernels (Azadirachta indica A. Juss) and commercial need based formulations using HPTLC and extended length packed column SFC method, *Chromatographia*, **62**, pp. 183–195.

18. Balsevich, J., Hogge, L. R., Berry, A. J., Games, D. E., and Mylchreest, I. C. (1988) Analysis of indole alkaloids from leaves of *Catharanthus roseus* by means of supercritical fluid chromatography/mass spectrometry, *J. Nat. Prod.*, **51**, pp. 1173–1177.

19. Morin, P., Gallois, A., Richard, H., and Gaydou, E. (1991) Fast separation of polymethoxylated flavones by carbon dioxide supercritical fluid chromatography, *J. Chromatogr.*, **586**, pp. 171–176.

20. Ashraf-Khorassani, M., Taylor, L. T., and Martin, M. (1999) Supercritical fluid extraction of kava lactones from kava root and their separation via supercritical chromatography, *Chromatographia*, **50**, pp. 287–292.

21. Richter, B. E. (1985) Modified flame ionization detector for the analysis of large molecular weight polar compounds by capillary supercritical fluid chromatography, *J. High Resolut. Chromatogr.*, **8**, pp. 297–300.

22. Wong, S. H. Y., Ghodgaonkar, B., Fong, P., Campbell, B., Burdick, J. F., and Doctor, F. (1994) Supercritical fluid chromatography for therapeutic

drug monitoring of immunosuppressants: selectivity for cyclosporine, FK 506 (tacrolimus), and rapamycin. *J. Liquid. Chromatogr.*, **17**, pp. 2093–2109.

23. Dirlam, J. P., Bordner, J., Chang, S. P., Crizzuti, A., Nelson, T. H., Tynan, E. J., and Whipple, E. B. (1992) The isolation and structure elucidation of CP-120,509, a new polyether antibiotic related to semduramicin and produced by mutants of *Actinomadura roseorufa*, *J. Antibiotics*, **45**, pp. 1544–1548.
24. White, C. M., Gere, D. R., Boyer, D., and Pacholec, F. (1988) Analysis of pharmaceuticals and other solutes of biochemical importance by supercritical fluid chromatography, *J. High Resolut. Chromatogr.*, **11**, pp. 94–98.
25. Smith, R. D., Udseth, H. R., and Wright, B. W. (1985) Rapid and high resolution capillary supercritical fluid chromatography (SFC) and SFC/MS of trichothecene mycotoxins, *J. Chromatogr. Sci.*, **23**, pp. 192–199.
26. Perkins, J. R., Games, D. E., Startin J. R., and Gilbert, J. (1991) Analysis of veterinary drugs using supercritical fluid chromatography and supercritical fluid chromatography–mass spectrometry, *J. Chromatogr.*, **540**, pp. 257–720.
27. Pang, F., Lü, H., Liu, Y., and Zhang, M. (2010) The study on purification of artemisinin by supercritical fluid chromatography, *J. Chem. Eng. Chin. Univ.*, **24**, pp. 569–573.
28. Wang, Y., Stutz, G., Roggo, S., Diehl, G., and Francotte, E. (2008) Development of supercritical fluid chromatography for high throughput purification in natural product drug discovery, lecture in the 2nd annual international conference on packed column SFC (SFC 2008, 1–2 October 2008, Zurich, Switzerland), http://www.greenchemistrygroup.org/pdf/2008/DO%20Novartis-Wang% 20 Natural%20Products.pdf.
29. White, C., and Burnet, J (2005) Integration of supercritical fluid chromatography into drug discovery as a routine support tool, II. Investigation and evaluation of supercritical fluid chromatography for achiral batch purification. *J. Chromatogr. A*, **1074**, pp. 175–185.
30. Berger, T. A. (2010) Demonstration of high speeds with low pressure drops using 1.8 μm particles in SFC, *Chromatographia*, **72**, pp. 597–602.
31. Berger, T., and Berger, B. (2010) A review of column development for supercritical fluid chromatography, *LC-GC North Am.*, **28**, pp. 344–357.

32. Zeng, L. (2007) Parallel supercritical fluid chromatography/mass spectrometry system for high-throughput enantioselective optimization and separation, *J. Chromatogr. A*, **1169**, pp. 193–204.
33. Zeng, L. (2011) Two-dimensional supercritical fluid chromatography/mass spectrometry for the enantiomeric analysis and purification of pharmaceutical samples, *J. Chromatogr. A*, **1218**, pp. 3080–3088.
34. Ebinger, K., Weller, H. N., Kiplinger, J., and Lefebvre, P. (2011) Evaluation of a new preparative supercritical fluid chromatography system for compound library purification: the TharSFC SFC-MS prep-100 system, *JALA*, **16**, pp. 241–249.
35. McClain, R. T., Dudkina, A., Barrow, J., Hartman, G., and Welch, C. J. (2009) Evaluation and implementation of a commercially available mass-guided SFC purification platform in a high throughput purification laboratory in drug discovery, *J. Liquid. Chromatogr.*, **32**, pp. 483–499.

Chapter 12

Polarimetric Detection in Supercritical Fluid Chromatography*

Gary W. Yanik
PDR-Separations LLC, 3 Old Meadow Way, Palm Beach Gardens, FL 33418, USA
gwyanik@pdr-separations.com

12.1 Introduction

Polarimetric detection offers numerous benefits to supercritical fluid chromatography (SFC) applications. Laser polarimeters operating at 670 nm only respond to optical activity and ignore absorbance—all peaks are enantiomers or optically active and gradients do not affect the baseline. Detection is based on polarization (phase) changes, not absorbance (amplitude) changes. Polarimeters respond to enantiomers with a unique positive or negative peak with amplitude proportional to concentration and specific rotation—regardless of elution order or separation method. In analytical applications, polarimeters show peaks with optical activity, even when analytes lack a chromophore. In method development applications,

*All separations, purifications, chromatograms, observations, etc., were done in the author's laboratories.

Supercritical Fluid Chromatography: Advances and Applications in Pharmaceutical Analysis
Edited by Gregory K. Webster

ISBN 978-981-4463-00-3 (Hardcover), 978-981-4463-01-0 (eBook)
www.panstanford.com

polarimeters show elution order of enantiomers and confirm that peaks are enantiomers. In preparatory applications, polarimeters control peak collection based on peak polarity rather than just slope and do not suffer from the overload and non-linear problems of absorbance-based detectors (if too much light is absorbed, there is nothing left to detect).

12.2 Theory of Operation

Measured optical activity for a particular analyte depends on a variety of parameters, including wavelength. At wavelengths between 200 and 400 nm, optical activity values typically change quickly with wavelength, and zero crossings (sign inversions) are not unusual. Variations in value and zero crossings are far less common above 500 nm. Traditional polarimeters used the sodium D line emission at 590 nm and many modern laser polarimeters use laser diode emitters at 670 nm. Measurements at these wavelengths (590 and 670 nm) have proven to be extremely stable and reproducible with no absorbance-related interferences while shorter wavelengths prove problematic in some cases. Measured optical activity is also affected by sample matrix (including the diluent) and to a lesser extent by temperature.

Light can be described as possessing properties similar to that of a traveling wave (see Fig. 12.1a). Wave motion results from vibrations (oscillations) that produce crests and troughs. The distance between successive crests or troughs, called wavelength, corresponds to light's color. The magnitude of vibration (see Fig. 12.1b) corresponds to its intensity or amplitude. Most light sources emit waves of light vibrating in many different planes (see Fig. 12.1c).

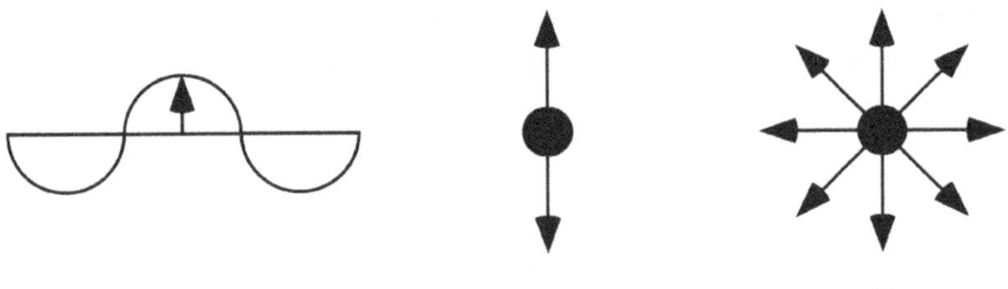

(a) (b) (c)

Figure 12.1 Wave motion of light.

If passed through a polarizing filter, the emerging light will be limited to waves vibrating in a single plane (see Fig. 12.2). The resulting light is called plane-polarized because the light is polarized in a single optical plane.

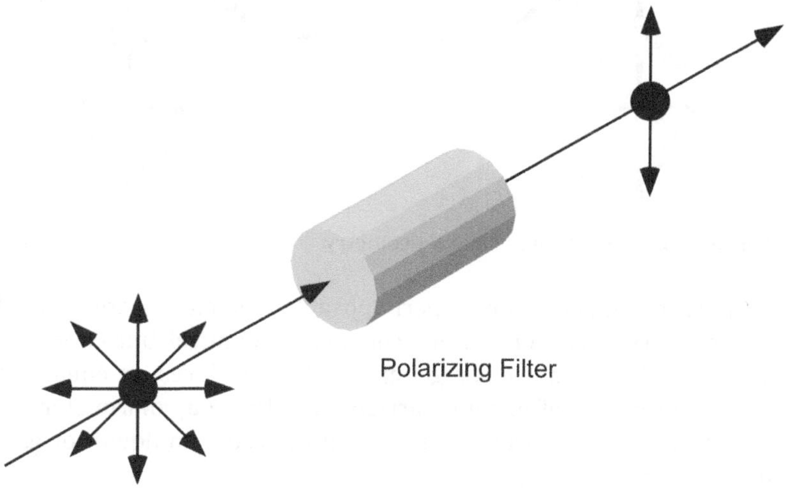

Figure 12.2 Producing plane polarized light.

So what is optical activity? To answer that question, we shift gears from physics to chemistry.

Some organic compounds have a molecular geometry that affects their transmission of plane polarized light. These compounds contain a carbon atom to which four different functional groups are attached (see the phenylalanine example in Fig. 12.3). Because a singly bonded carbon atom has a tetrahedral geometry, the functional groups can be attached in either of two configurations with one molecule being a mirror image of the other (similar to your right and left hands). Otherwise the molecules are structurally identical. This geometry rotates (changes) the angle of the plane of polarized light, is called optical activity, and is unique. The "rotation" of the plane of polarization is actually an optical phase phenomenon that locally affects the light beam inside the flow cell. Optical activity is defined as the ability of a compound to rotate plane polarized light in one direction or another (positive versus negative rotation).

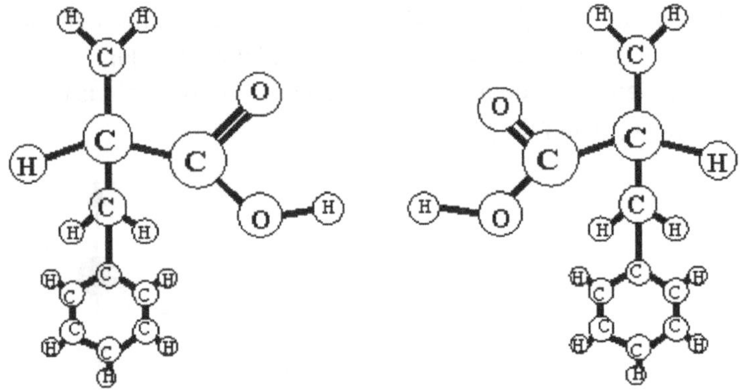

Figure 12.3 Phenylalanine chiral geometry.

Mirror image, non-superimposable isomers are called enantiomers. Enantiomers are chemically identical but contain a difference in 3D molecular geometry that produces an equal but opposite rotation of plane polarized light. Using a polarimeter, we can measure the optical rotation (α) produced by a liquid- or gas-phase sample (see Fig. 12.4).

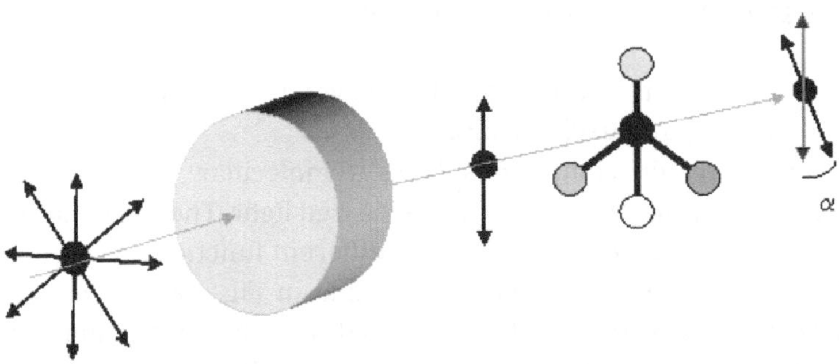

Figure 12.4 Rotation of plane-polarized light by analyte.

The optical system of a typical laser polarimeter is illustrated in Fig. 12.5 and consists of a laser diode, polarizing prism, Faraday rotator, flow cell, analyzing prism, and photodiode.

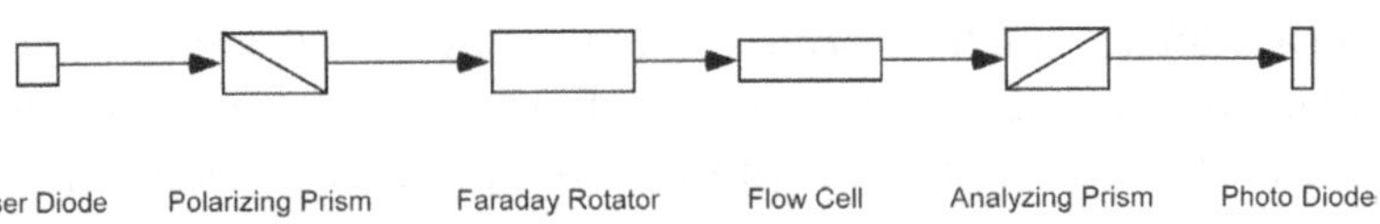

Figure 12.5 Laser polarimeter block diagram.

Polarized light from a laser diode is further filtered with a polarizing prism, and this highly polarized light passes through a Faraday rotator and then a flow cell. The Faraday rotator imparts a sinusoidal rotation to the plane of polarization. A sample exhibiting net optical activity flowing through the flow cell will offset the average sinusoidal rotation relative to zero rotation. Proper alignment of the analyzing prism and special electronic circuits allow the magnitude and sign of the net optical activity to be extracted from the photodiode signal. Since Faraday oscillation is typically above 500 Hz, integration of results is not required and real time values are available continuously. Faraday oscillation is part of a noise rejection scheme.

12.3 Comparison of ALP, UV, and CD

UV (ultra violet absorption) and CD (circular dichroism) are both absorbance-based detection schemes and share some characteristics. They require a chromophore in the analyte and excitation with a wavelength in the chromophore's absorbance band. Their linear dynamic range is naturally limited (more so for CD) because as absorbance increases the photon flux at the photodetector approaches zero and noise effects prevail. In addition to a chromophore, CD requires a chiral center coupled to that chromophore for detection of optical activity. Both UV and CD are often affected by solvent (absorbance) gradients and CD suffers from other absorbance related interferences.

Polarimeters do not require chromophores or wavelength tuning to match an absorbance band because they are not absorbance-based detectors. Polarimeters detect optical activity by continuously measuring the angle of rotation of a highly polarized laser beam—a phase measurement not an amplitude measurement.

To compare sensitivity is complicated because of variations in analyte characteristics and chromatographic conditions as well as variations in instruments. For most pharmaceutical compounds, UV is more sensitive than CD or a polarimeter. Exceptions include compounds with no chromophore (sugars, antibiotics, etc.) where only polarimeters are sensitive. If CD is first used with an appropriate analyte stopped in the flow cell, a stopped flow wavelength scan is performed, and the wavelength of maximum signal is selected;

then subsequent analysis of that particular analyte by CD will be competitive with polarimeters. In automated analytical screening and preparatory purification applications, polarimeters almost always perform better than CD because no stop flow scans or adjustments are required and the linear dynamic range is much larger. CD is not very useful for preparatory chromatography because of saturation (too much absorbance, not enough photon flux arriving at photodetector).

12.4 Analytes

This section will briefly discuss analyte classes that are successfully being detected by laser polarimeters.

12.4.1 Small Molecule Pharmaceutical Candidates

The majority of small molecule pharmaceutical candidates are chiral compounds containing one or more chiral centers. Most have specific rotation values between 10° and 40°. Laser polarimeters are very beneficial for development, scale up, and production of these optically active compounds.

See the chromatogram in Fig. 12.6 and structure in Fig. 12.7 of the antihypertensive Labetalol, a diastereomer. The two enantiomeric pairs are separated into four separate components. UV is blue and ALP is red. With UV alone, it is not possible to determine which peaks are from which pair of enantiomers because each +/– pair has the same absorbance. The laser polarimeter clearly shows that the first and third peaks are from one enantiomeric pair, and the second and fourth peaks comprise the other pair. There is one positive and one negative peak pair with the same area and another positive and negative peak pair with a different area in the laser polarimeter plot. This is the result of different specific rotation values for each +/– pair (polarimeter)

12.4.2 Antibiotics and Sugars: Compounds without Chromophores

Most antibiotics and sugars have no useful chromophore but exhibit specific rotation values above 100°. This means that absorbance-based detectors (e.g., UV and CD) are not useful and polarimeters enjoy increased sensitivity (higher specific rotation) as compared to the sensitivity of polarimetric detectors for small molecule pharmaceutical candidates.

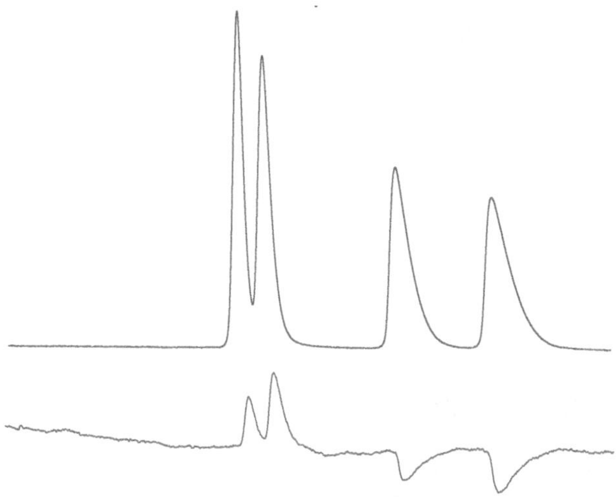

Figure 12.6 Labetalol, a diastereomer, UV and laser polarimeter.

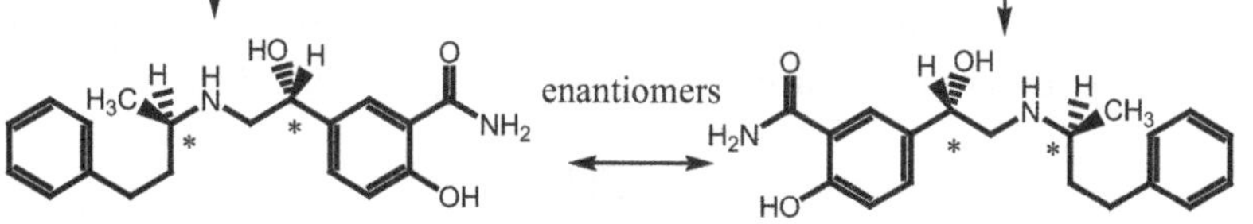

Figure 12.7 Labetalol structures.

Antibiotics like gentamicin and erythromycin (see Fig. 12.8) occur with multiple structurally similar analogs. Mass differences between the analogs are usually less than 2% and UV detection is usually not possible above 220 nm. There are large differences in specific rotation for these analogs even though structural differences are small. Thus, polarimeters can not only easily detect but often fingerprint antibiotics because the ratio of analogs is usually process specific.

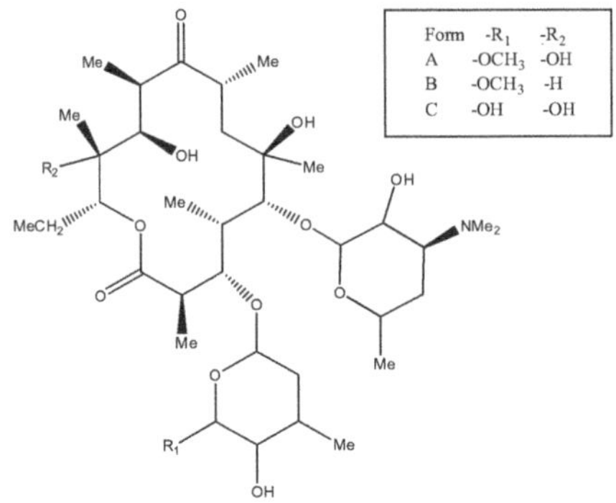

Figure 12.8 Structure of erythromycins.

12.4.3 Amino Acids

Polarimeters give consistent measurements (sign and amplitude of rotation) of optically active compounds and are very useful in differentiating and quantitating amino acids without the use of external standards.

12.4.4 Natural Products

Nature is chiral, so natural products exhibit optical activity. Many compounds only appear naturally in one enantiomeric form. Some compounds such as limonene, an oily compound found in citrus fruit, caraway, dill, and bergamot, occur in both *d* and *l* forms. The specific rotation of limonene is approximately 101° (see Fig. 12.9).

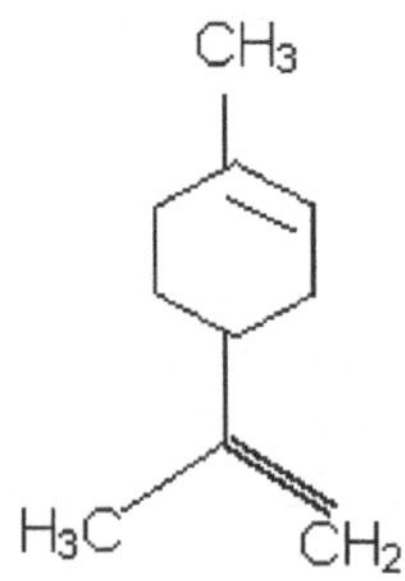

Figure 12.9 Structure of limonene.

12.4.5 Foods, Flavors, and Fragrances

Both taste and smell are chiral effects, so foods, flavors, and fragrances are successfully analyzed by polarimeters.

12.4.6 Fertilizers and Pesticides

Many new and developing fertilizers and pesticides are enantiomer enriched for the same reasons that most small molecule pharmaceuticals are—improved specificity and efficacy with reduced/enhanced toxicity. Laser polarimeters are very useful in analyzing and controlling the processing of these compounds.

A laser polarimeter was used to control peak collection for purification of cypermethrin enantiomers using SFC (see Fig. 12.10). Cypermethrin has three chiral centers and eight enantiomers. The ability of a laser polarimeter to identify enantiomers by peak polarity and area was a critical aid to interpretation of chromatograms during method development and peak collection during preparatory scale separations. The last UV peak in Fig. 12.10 is an impurity, not a cypermethrin enantiomer. This is easy to determine with a laser polarimeter plot.

12.5 Applications

Laser polarimeters are routinely used in SFC analysis, method development, and preparatory purification applications.

12.5.1 SFC Method Development

Competitive and economic factors are driving pharmaceutical companies toward increased efficiency and cost reduction. New pharmaceutical candidates produced by medicinal chemists need to be analyzed and purified quickly in order to decide which compounds are worth further development—often creating a throughput bottleneck. While SFC provides a linear improvement compared with HPLC, good method development planning can make further improvements.

To develop a good method automatically, one should intelligently screen a well-chosen set of columns and solvents with an automated sequence. Table 12.1 lists useful columns and solvents.

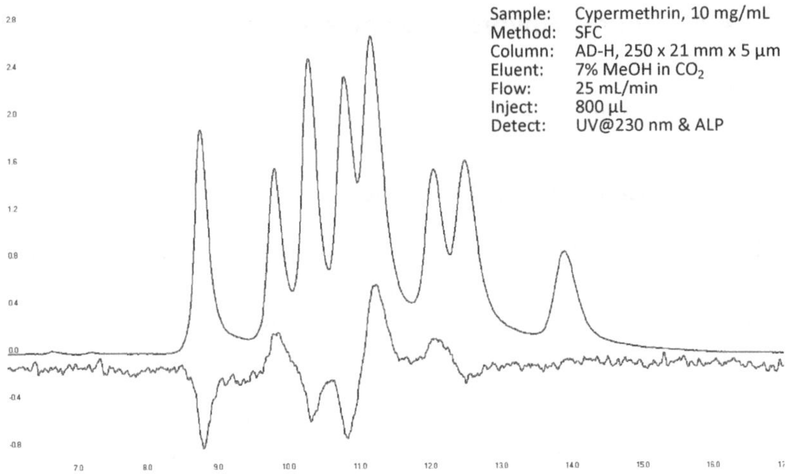

Figure 12.10 Cypermethrin.

Table 12.1 Useful columns and solvents

Column	Manufacturer		SFC solvents
OD-H	Daicel	1	CO_2
AD-H	Daicel	2	Isopropanol
AS-H	Daicel	3	Ethanol
OJ-H	Daicel	4	Methanol
IA	Daicel	5	Acetonitrile
IB	Daicel	6	Methyl tert butyl ether
IC	Daicel	7	Ethyl acetate
ID	Daicel	8	Tetrahydrofuran
IE	Daicel	9	Methylene chloride

An automated method screening system should have columns and solvents arranged as shown in Fig. 12.11. This enables the analyst to build master sequences to automate method screening.

Gradients should be used in screening methods because they save time and sample by exposing a single injection to a varying eluent composition, and wash the column as solvent strength rises. Methods should be arranged in a sequence from strong to weak relative solvent strength to minimize interference from retained peaks and all columns should be washed at the end of the sequence to remove retained peaks.

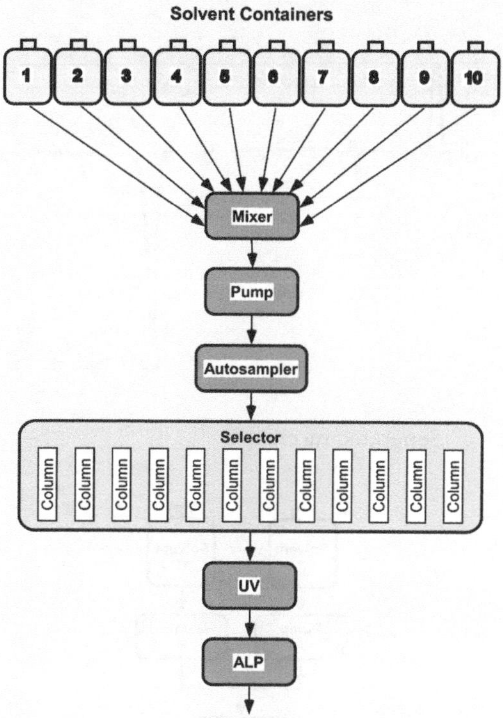

Figure 12.11 Method screening flow diagram.

Polarimeters enable the identification of the elution order of enantiomers. For preparatory applications, it is preferred if the desired enantiomer elutes first; it will be more concentrated and of higher purity. In the racemic separation of Fig. 12.12, the two small square peaks eluting before the void volume spike are internal calibrates showing –100 and +1000 microdegrees of rotation. The first eluting enantiomer is negative and the second enantiomer is positive. This elution order can be reversed using a different column and the same eluent.

12.5.2 SFC Preparatory Purification: Peak Collection

Typical SFC preparatory purification systems have two pumps (CO_2 and modifier), sample injector, column, detectors, and a collection valve as shown in Fig. 12.13.

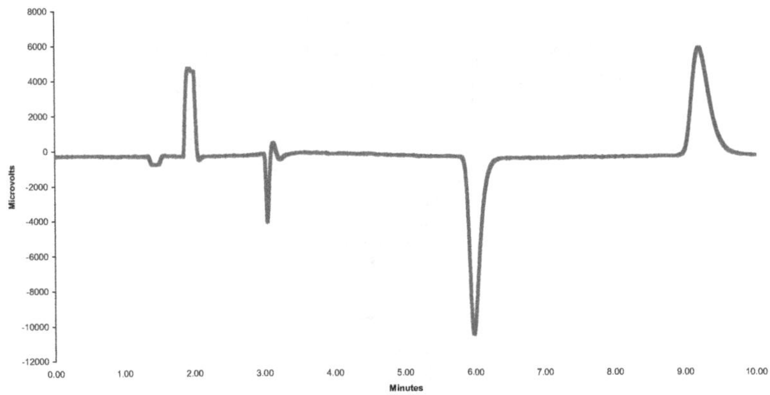

Figure 12.12 Separated racemate.

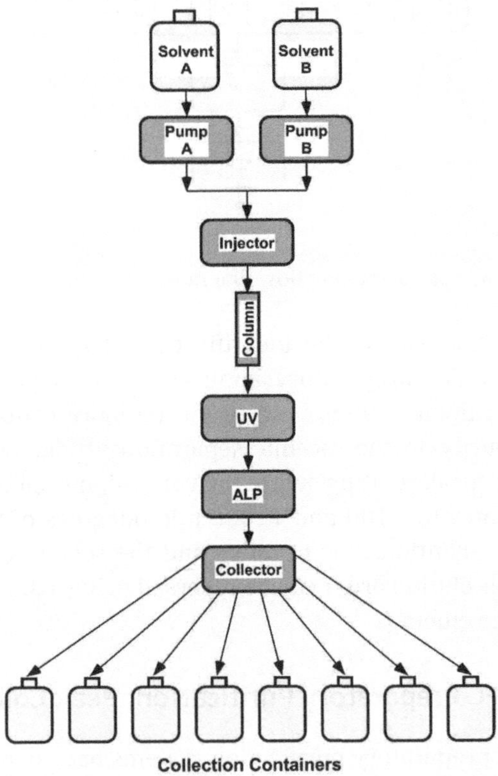

Figure 12.13 Preparatory purification system.

The collection valve diverts eluent flow from waste to collection vessels and back to waste under command of software. Collection decisions can be based on a combination of parameters, including time, detector amplitude and sign, detector derivative and sign, and enantiomeric excess. Enantiomeric excess (ee) is calculated and plotted as a "signal" by mathematically combining UV and laser polarimeter data and is especially useful when peaks overlap. Collection based on laser polarimeter derivative amplitude and sign are simple to use and very robust—always putting the correct signed (+/−) peak in the correct collection vessel, regardless of elution order and without a need for time windows. In the author's labs, we routinely make multi-day continuous runs, thousands of injections, with no one on-site outside normal working hours—sometimes processing up to a kilo of sample via SFC on a 2 cm × 25 cm column in a week using the system photographed in Fig. 12.14.

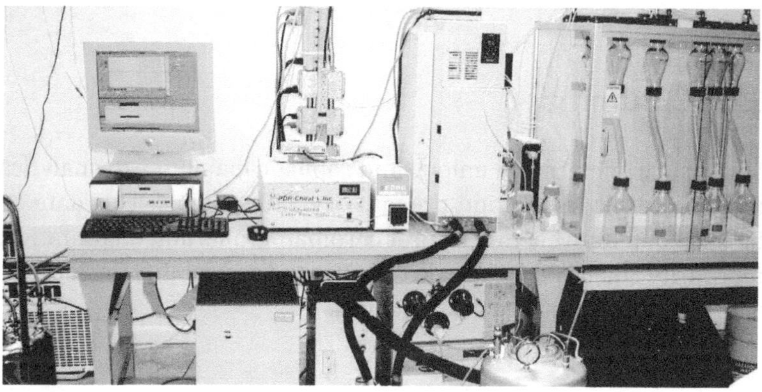

Figure 12.14 AutoPrep SW and ALP on Berger Multigram II.

The chromatogram in Fig. 12.15 shows stacked-injection preparatory scale collection triggered by laser polarimeter derivative amplitude and sign. This robust collection mode is dynamically adaptive to variations in loading and retention time. The more dilute injections on the right are collected properly, even though loading was reduced to 1/3 of the expected amount. In this experiment, more sample solvent was added to rinse the sample solution bottle, so the syringe was injecting the same volume of a less concentrated solution. This mode of collection is not dependent on programmed time windows and so does not make mistakes if peak retention times vary.

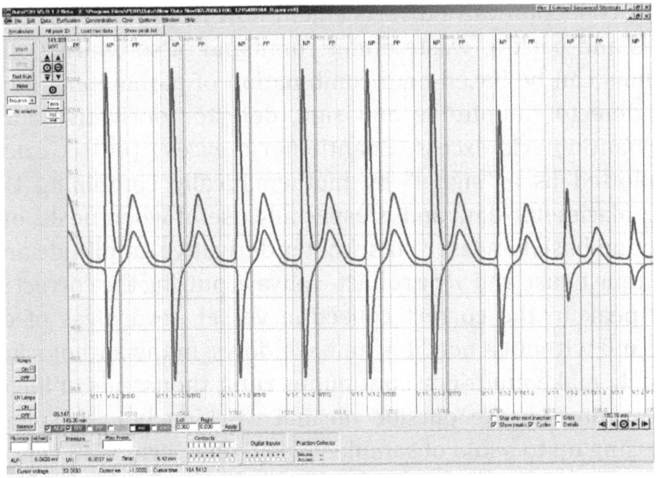

Figure 12.15 Peak collection by laser polarimeter derivative amplitude and sign.

12.6 Summary

Laser polarimeters add unique and robust capabilities to analytical, method development, and preparatory SFC systems. This data can be used qualitatively to interpret chromatograms and quantitatively for purity analysis and peak collection. Applicability includes all compounds exhibiting optical activity.

Chapter 13

Supercritical Fluid Chromatography with Ultra-Performance Particles

Ziqiang Wang
Purification Business Operations, Waters Corp., 159 Lukens Dr., New Castle, DE, USA
Ziqiang_Wang@waters.com

13.1 Introduction

13.1.1 Supercritical Fluid Chromatography

As readers may have learned from other chapters of this book, supercritical fluid chromatography (SFC), as its name implies, utilizes the supercritical fluids instead of conventional solvents that are used as the common carrier/mobile phases in liquid chromatography (LC). The unique advantages of supercritical fluids of low viscosity, higher solute diffusivity and liquid-like density make it a highly desired tool to complement traditional chromatography (HPLC/GC), which is widely used in various application fields.

Supercritical CO_2, the major preferred form used in modern SFC practices, from its inertness, green aspect, mild conditions,

Supercritical Fluid Chromatography: Advances and Applications in Pharmaceutical Analysis
Edited by Gregory K. Webster

ISBN 978-981-4463-00-3 (Hardcover), 978-981-4463-01-0 (eBook)
www.panstanford.com

and economical factors, resembles similar physical and chemical properties of hexane in terms of chromatographic performance. For this reason, SFC has been evaluated to be compared and developed as a complement and replacement tool for normal-phase (NP) chromatography.

For example, the first success of SFC application in chromatography comes from the use of SFC as the tool of chiral chromatography in which the normal phase–based chromatography with polymeric polysaccharide and brush-type Pirkle-type stationary phases are used as a main and prevailing format. Chiral SFC has shown significant improvements in all aspects, including efficiency, throughput, applicability and productivity. In today's global pharmaceutical industry, for which the major application of chiral chromatography is for drug discovery and development, chiral SFC has already been the first method of choice for such application.

While SFC has achieved great success with chiral separations, it also can possess huge potential for the non-chiral separation as the complementary tool to the modern reverse-phase liquid chromatography (RPLC). As the powerful modern HPLC/UPLC has been developed, RPLC has its limitation on the applicability of molecules it can handle with regard to the polarity range, aqueous solubility, extent of acidity, etc. Literature has shown that the combination of both HPLC and SFC for non-chiral separations would cover more than 96.3% of processed molecules, compared with just 80% of either techniques used alone [1], as shown in Fig. 13.1.

	SFC/MS	LC/MS
% Total hits	75.0	79.4
% Total hits + strong signal	87.0	89.9
% hit with one, but not other	3.7	8.1

Figure 13.1 Success rate for SFC and LC for drug molecules separation. From [1].

It has been reported [2] that for achiral development, SFC has been heavily relied upon for the analysis and purification of a diverse set of compounds from the medicinal chemistry library. Supporting high-speed SFC quality control of the purified compounds is made possible at high-flow-rate SFC. The flexibility of SFC systems has been extended for use with more non-polar compounds and enhancements in signal-to-noise. Further SFC applications include difficult-to-separate hydrophobic peptides.

Moreover, for preparative scale purifications, SFC is the clear winner in terms of simplicity of process, increased productivity, significant cost reduction, and green environmental impacts [3]. The implementation of such strategy has shown improvement in productivity and success rate in handling high-throughput screening and purifications [4].

13.2 Current Status of SFC Performance

While the "liquid-like" density and lower viscosity make SFC compatible with sample format and safety limitation on stationary phases, the real reason underlying SFC's improvement in chromatography comes from the higher diffusivity of supercritical CO_2.

Liquid-based chiral chromatography, for example, in its old days, in spite of its high specificity, suffered from very low efficiency of the inherent slow kinetics from high molecular polymeric stationary phase, in addition to smaller chiral three-dimensional distinguishing forces compared with hydrophobicity/adsorption-based non-chiral chromatography (HPLC). While any regular non-chiral HPLC column typically exhibits about 20,000 plate numbers from almost every vendor around the world today, the same-size chiral column can only show plate number around 2000 from LC, a pathetic 10% performance compared with modern HPLC.

The use of supercritical fluids in chiral chromatography improved the solute mass transfer process to one to two orders of magnitude faster than that of pure liquid solvents. This comes from the higher mass diffusivity with supercritical fluids [5]. This change significantly lowers the C-term contribution in the Van de Emter equation to total plate height. This factor alone, as indicated by the literature, has improved the column efficiency to at least three times higher than LC.

In fact, to date, the major achievements brought by the adaptation of SFC into chiral chromatography can be fairly attributed to just this improvement in mobile phase kinetics alone, as the development status of all other factors in SFC such as stationary phases, hardware design (pumping, detection, etc.), and other environmental settings is basically at par with the regular liquid chromatography development, or even just directly adapted from LC with minimal changes. As depicted in Fig. 13.2 [6], the improvement in mass diffusion (*Y*-axis, the change from C-term), combined with faster flow due to lower viscosity of supercritical fluids (*X*-axis), has given enough justification to most practitioners to switch to SFC and be satisfied with its state of the art.

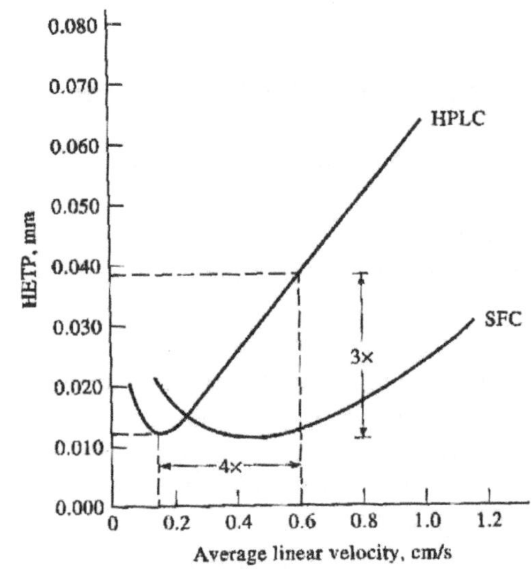

Figure 13.2 Performance comparison between HPLC and SFC with particles of the same size.

It is therefore safe to say that with as much extent as it can be, the current status of SFC development on packed column field has just been dealt with in less than 50% of the efforts at the big picture level.

The chromatographic process, in a nutshell as the way Tswett originally developed it, is to initiate mass transfer within binary or multiple phases to gain degrees of separation. The mobile phase is just one of the many factors in this complex process. Other factors, such as stationary phase and environmental settings (temperature, detections, etc.) should also be explored for improvement. Any such improvement in efficiency and productivity would be welcomed and in great demand in today's world, while shrinking resources and pursuit for incessant productivity breakthrough justify the needs for innovations from all spectrums.

The past decade has seen the surge in the development of the commercialization of ultra-performance particles in LC. With smaller size particles (less than 3 μm in diameter), the chromatographic performance has been brought to a much higher level, namely, higher efficiency with shorter run time, with the help of the modern instrumentation state of the art. Both the fundamentals [7] and applications [8] have been well demonstrated, and great commercial success (UPLC) has also been achieved, implying that it has been widely accepted as the direction for performance and productivity improvement for the industries.

The major characteristic of ultra-performance particles that brings improvement in liquid chromatography comes from the reductions in both eddy diffusion and mass transfer contributions to plate height, originated from the stationary phase at the cost of ultra-high pressure pumping and instrumentation development in LC. This has been achieved successfully with diligent and collective efforts by practitioners in the field.

While SFC is a niche technique in the past 20 years or so, and witnessed cycles of development interests, it is always a legitimate thought process that to explore the potential benefit of novel development from general chromatography fields, and this includes particle development. However, as it turns out, due to the size of the SFC market and scales of resources of the major players of SFC, this has not been brought to any level of significance until recently. Lack of new particles in desired sizes and appropriate phase chemistries are the two main causes, in addition to the insufficient knowledge of dealing with such particles at the ultra-performance level, in both instrumentation and technical support.

Along with the maturation of UPLC development and market evolution, in recent years it has become more applicable and practically possible to apply ultra-performance particles with supercritical fluids to fill the void of the missing pieces of development in SFC.

With the faster mass diffusivity of supercritical fluids, it can be reasonably expected that the adaptation of ultra-performance particles in SFC shall have complementary advantages that combine from both mobile and stationary phase enhancements.

13.3 Characteristics of Ultra-Performance Particles for SFC Considerations

Several factors need to be discussed for such development, as they may offer different requirements for the technique and the fundamental mechanism.

First, with lower viscosity of supercritical CO_2, the increased pressure drop across the column will not be as steep as in UPLC. It is just around 10–30% higher than the pressure range in HPLC at the highest point in most cases. This is actually good news, as the developed UPLC pumping/hardware is to be easily adapted for SFC with minimal modifications.

Second, while the improvement by even lower C-term/mass transfer contributes to the plate height with smaller particles, on the top of the lowering impact by higher diffusivity of supercritical CO_2, there is also a legitimate concern that this improvement could be compromised, diminished, or even make it worse with increased supercritical mobile phase density accompanying the increased pressure drop. Unlike in UPLC/HPLC, where the solvent strength of the mobile phase remains relatively flat along the column, in SFC, it varies and increases with the pressure build-up going higher. With increased mobile phase strength, the peaks elute earlier and the resolution may decrease.

To sort out this phenomenon, the argument needs to be checked not only in qualitative terms but also in quantitative terms.

Figure 13.3 shows the CO_2 density diagram in practical operational pressure ranges [9]. It shows the range of change is about less than two times the difference between low and high points (from 0.6 g/mL to close to 1.0 g/mL). In supercritical liquids, the molecular solubility is directly proportional to the liquid density in the applicable working range, and this can be translated arithmatically to the retention changes on the column. In other words, theoretically, the probability of the peak retention of loss (hence loss of efficiency) due to CO_2 density increase are limited to several folds in most cases. At the same time, the mass transfer difference is at least one to two orders of magnitude higher with supercritical CO_2 in terms of diffusivity. With mathematical relationship in governing chromatography theorem, this favors the overall impact, in most cases, a high probability toward improvement.

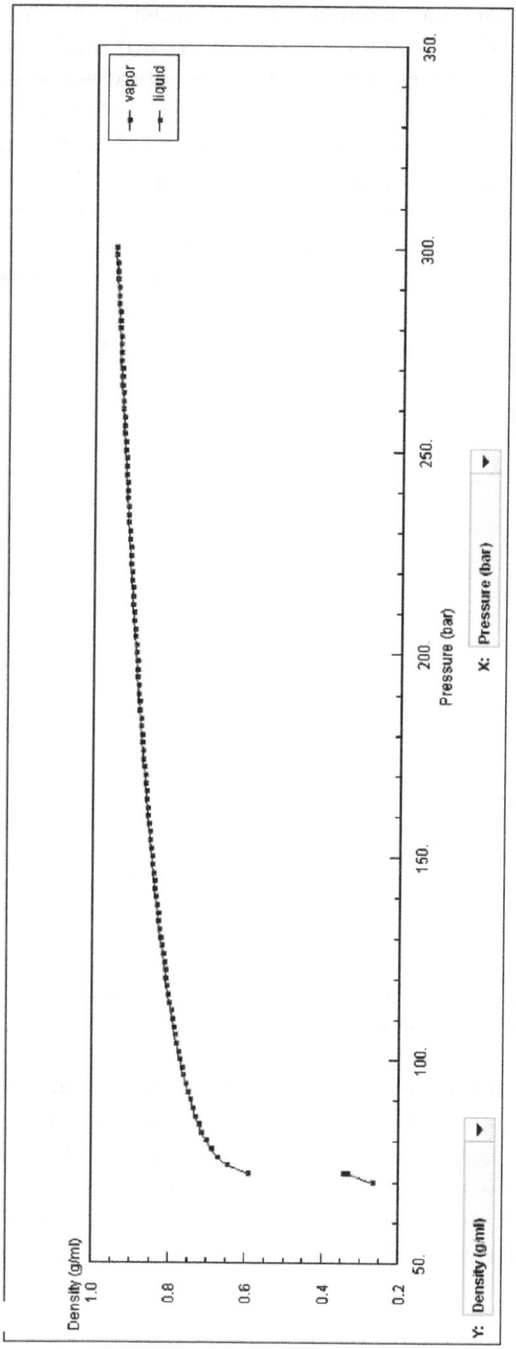

Figure 13.3 CO_2 density within typical working pressure range of SFC at 30°C.

This rationale has been confirmed experimentally by recent research and development of SFC with ultra-high-performance particles. Figure 13.4 shows the HETP plot of experimental data on an SFC prototype with various sizes of ultra-performance particles [10]. With decreased particle size, the net efficiency of the column has actually increased accordingly, but not diminished or going the other way. Also, this data shows that the original details of argument are both true for the relatioship of two conflicting factors; it shows that the minimum HETP is not as low as the same HETP in UPLC case under similar conditions [11], meaning the loss of retention by density increase does have some extent of impact.

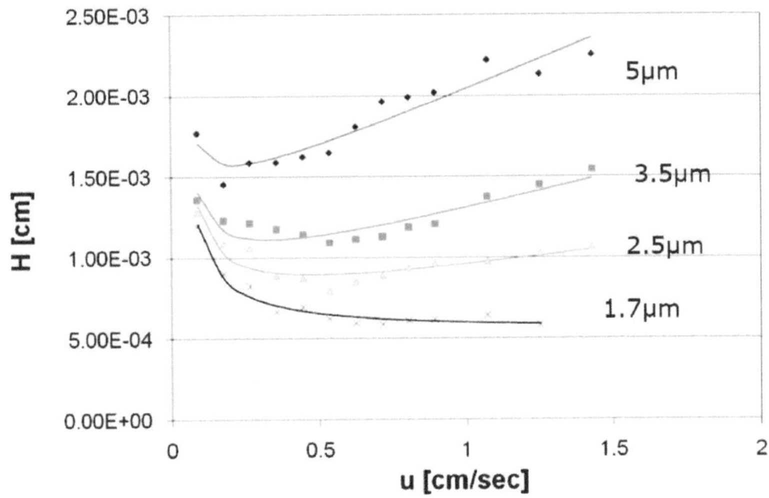

Figure 13.4 HETP experimental data of SFC with various sizes of ultra-performance particles.

Overall, through both thereotical rationing and experimental verification, it has been demonstrated that the SFC with ultra-performance particles improves the efficiency significantly, and this will take SFC to a higher level of performance with more and wider applicability. The bottom line is that it will become a better tool with higher productivity and economical impact to complement UPLC/HPLC.

13.4 Published studies on SFC with Ultra-Performance Particles

In the past decade, researchers and developers have been diligent in pursuing expected advantages of supercritical fluids with ultra-performance particles on both existing/customized and prototyped SFC instrumentation.

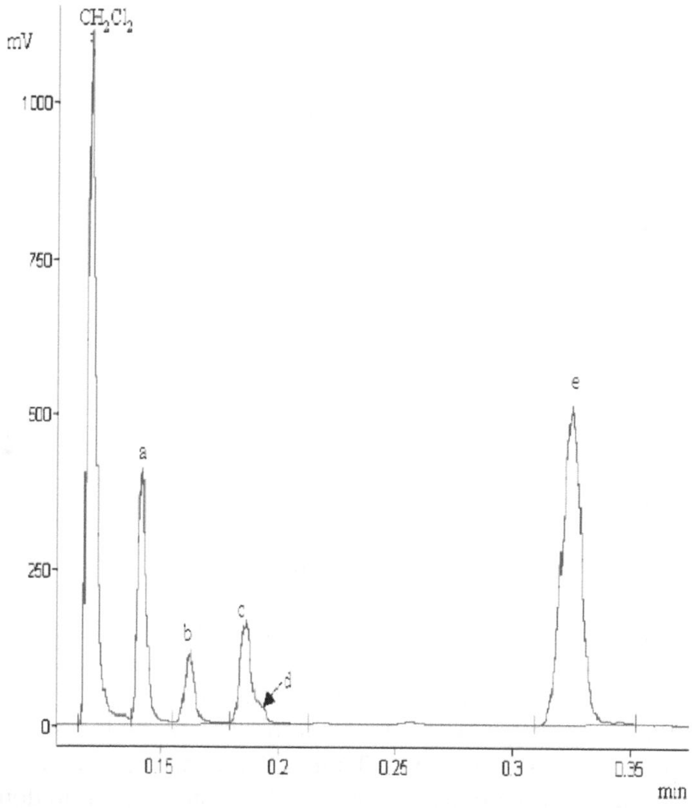

Figure 13.5 Fast SFC separation with 1.8 μm particles.

Figure 13.5 shows the improvement in separation speed with 1.8 μm particles [12]. Naphthalene (a), dodecylbenzene (b), octahydroanthancene (c), anthracene (d), and Benzanthracene (e) were separated in 19 sec; a similar efficiency was observed with UPLC. The flow rate is 5 mL/min of 100% CO_2 on Nucleodur Gravity

3 mm × 50 mm, 1.8 μm column. Backpressure is at 150 bar, and detection at UV 205 nm. The generated pressure drop across the column is about 230 bar. The calculated peak capacity is 22.

Figure 13.6 shows another separation on the same ultra-performance column [12], the sample is Polywax 655 with $n = 16$ to $n = 84$. Flow is 100% CO_2 at 2 mL/min, $T = 100°C$; pressure gradient from 80 to 300 bar at 99 bar/min; FID detection.

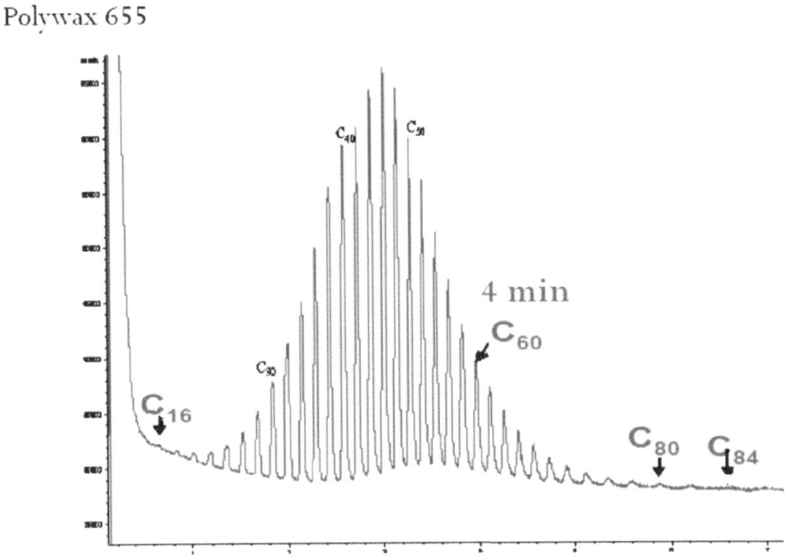

Figure 13.6 Polymer separation with SFC.

Figure 13.7 shows the fast separation of polymers on a C4 column [12]; the run time is less than 2.5 min. Polymer range of $N = 20–80$ was successfully separated. Flow rate is 2.5 mL/min CO_2, $T = 100°C$, and the pressure gradient is from 80 to 370 bar (99 bar/min). The column is Sepax C4 4.6 mm × 50 mm, 1.9 μm particle; FID detection. This is two to three times faster with the same separation done by GC.

In another study, the comparison of some types of 2 and 3 μm particles with 5 μm LC particles was carried out in different conditions [13]. Figure 13.8 shows that the use of small particle diameter would offer very high performance, similar to it does in HPLC. At the optimum conditions, the theoretical plate number reached is equal to 240.00 by meter of column, with 2.2 μm particles, and 92,000 with 5 μm particles. At optimum H values, which do not obviously correspond to maxima for N/t_0, the reduced equivalent plate height (h) is closed for the two particle sizes, h = 2.8 with 2.2 μm particles and h = 2.5 with 5 μm particles. However, these studies also show that efficiency variations in SFC on packed columns are not identical to the one observed in HPLC in terms of temperature. Higher temperature does not always offer higher efficiency as in UPLC. Another observation is that the pressure build-up is significant lower with these particles than with UPLC, which is a granted benefit for better mechanical stability and longer column lifetime, from the lower viscosity of supercritical fluids.

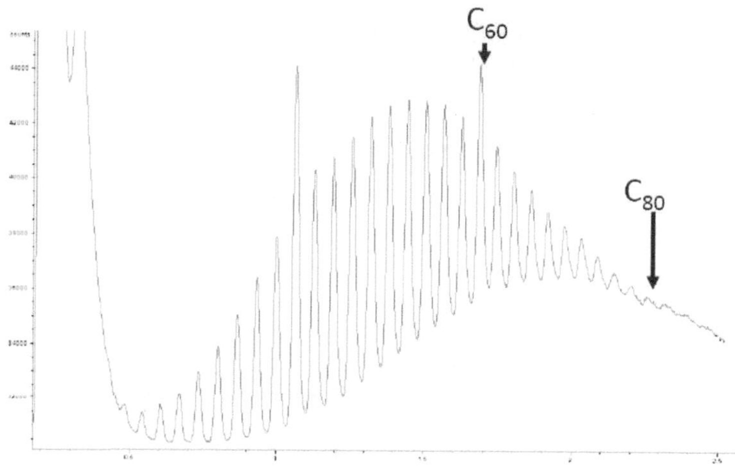

Figure 13.7 Polymer separation with C4 column.

Figure 13.9 [14] shows the application of 1.8 μm HILIC particles in SFC. The column is EPIC HILIC Silica 2.1 × 100 mm, 1.8 μm. The same phase chemistry HILIC column with 5 μm size particle is used as comparison. The resolution between test compounds theobromine and acetaminophen was relatively consistent across flow rates for the 5 μm column, whereas the 1.8 μm column reached a maximum Rs value between 0.7 and 0.8 mL/min. Tailing factors indicated that both particle sizes afforded equally good peak symmetry for the given standard components. The Van de Emter curves exhibit a classical, flatter shape for small particle columns at increased flow rates and with greater efficiencies. In a similar trend to previously discussed

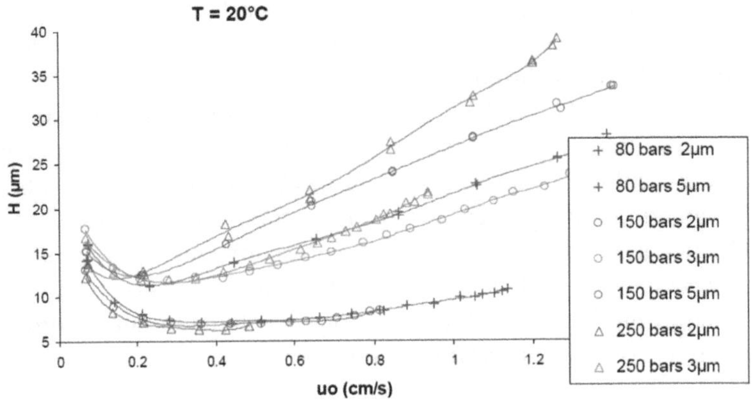

Figure 13.8 Investigations on 2 and 3 µm particles compared to regular 5 µm particles. Isocratic 5% methanol.

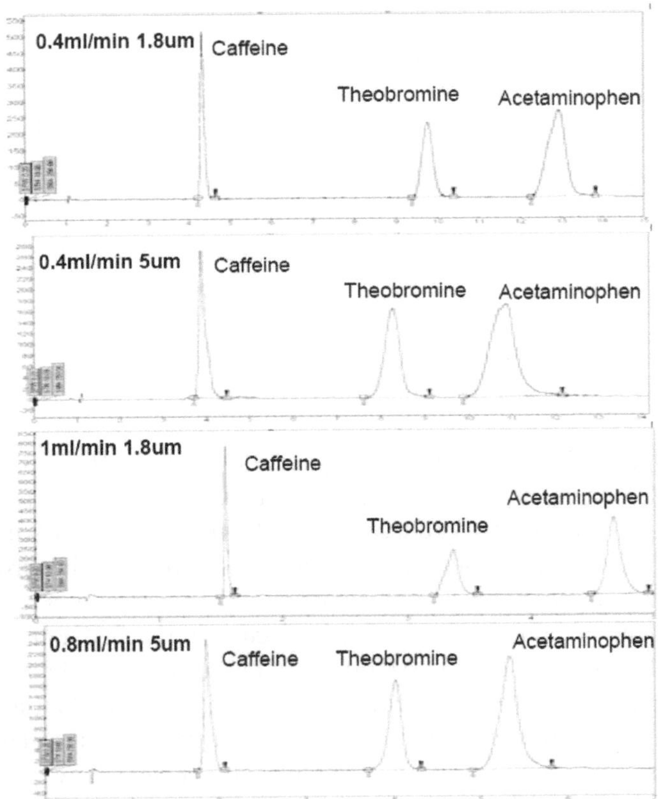

Figure 13.9 1.8 µm HILIC with SFC.

experiments, the results for column efficiency clearly demonstrate the number of theoretical plates (*N*) of the 1.8 μm column to be much greater than the 5 μm column. The maximum *N* for the 5 μm column for two of the three selected standard components was exhibited between flow rates of 0.8–0.9 mL/min, which would equate with ~4 mL/min on a standard 4.6 mm i.d. column typically used for analyses. Representative chromatograms at 0.4 mL/min for both columns demonstrate the chromatography at typical analytical flows for a 4.6 mm-diameter column. The 1.0 mL/min, 1.8 μm chromatogram and the 0.8 mL/min, 5 μm chromatogram represent flow rates at maximum *N* for each column. The flow rates for the <2 μm column did not exceed 1.25 mL/min due to a pressure drop of ~160 bar across the column. One must keep in mind that a 1.25 mL/min flow on a 2.1 mm, 1.8 μm column would be comparable to the flow rate of 6 mL/min when scaling up to 4.6 mm i.d. column, thereby, allowing the potential for greater speed without forsaking resolution and efficiency. The investigations have demonstrated that smaller-particle columns offer higher performance as in HPLC.

Figure 13.10 presents application of multiple detections (UV and light scattering (ELSD)) with sub 2 μm particle columns used in SFC for separations of oligomeric materials [15]. The study confirms that rapid, high-efficiency separations appear possible with these columns; ELSD as a versatile detection tool can be successfully integrated in SFC for high-efficiency separation industrial applications; no severe loss of selectivity is observed at test conditions; no significant degradation of column performance due to possible frictional heating or extra column volume was observed and also frequent column plugging was not observed with the 1.8 um particle columns used in this study. Overall, high-resolution chromatograms at reduced analysis time appear possible, which also allows reduced mobile phase usage.

As with the development of ultra-performance particles for LC, the fused-core porous shell particle has also attracted many interests as an alternative tool. This is of no exception in SFC. Figure 13.11 [16] has demonstrated their applicability for separation of pharmaceutical compounds. Also Figs. 13.12 and 13.13 show separations on 1.8 μm particles for compounds with various physical properties, such as polarity, acidity, and therapeutic functional groups [17]. These data

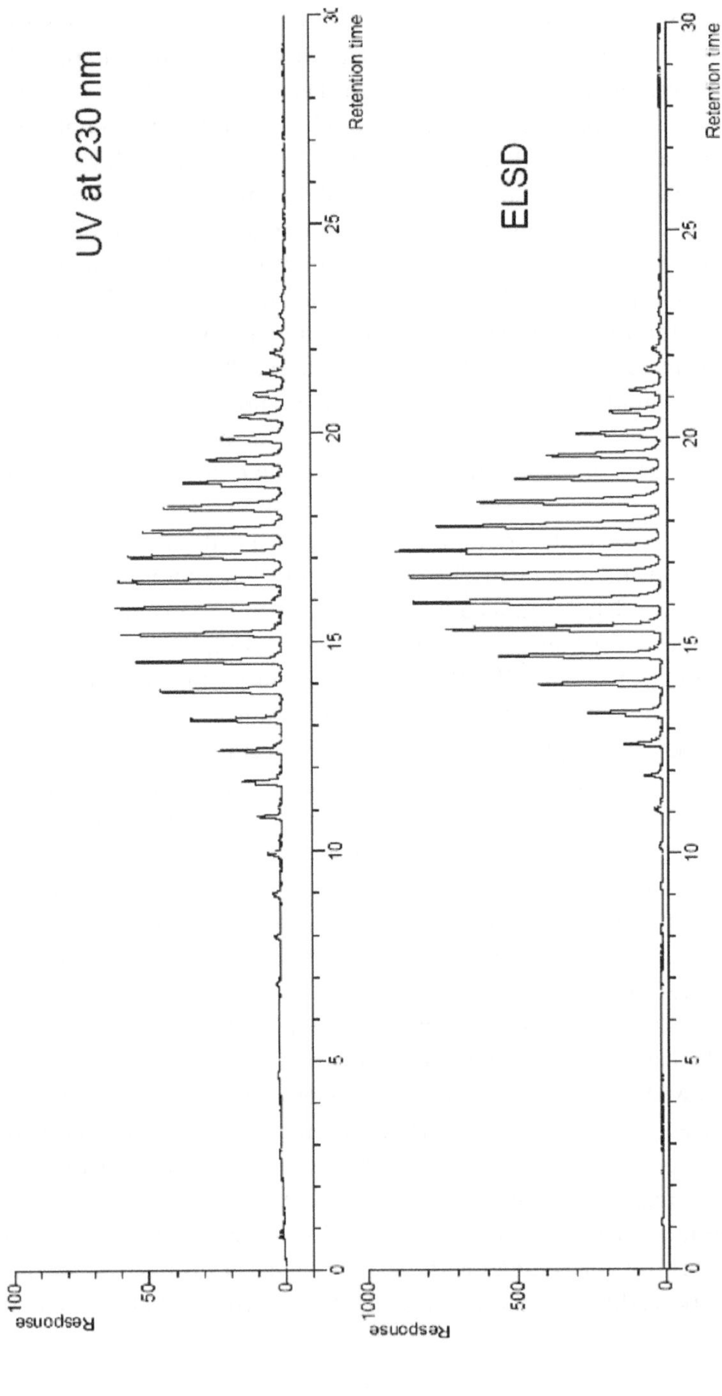

Figure 13.10 Ultra-performance SFC with ELSD.

show improves efficiency with these applications, especially at shortened run times.

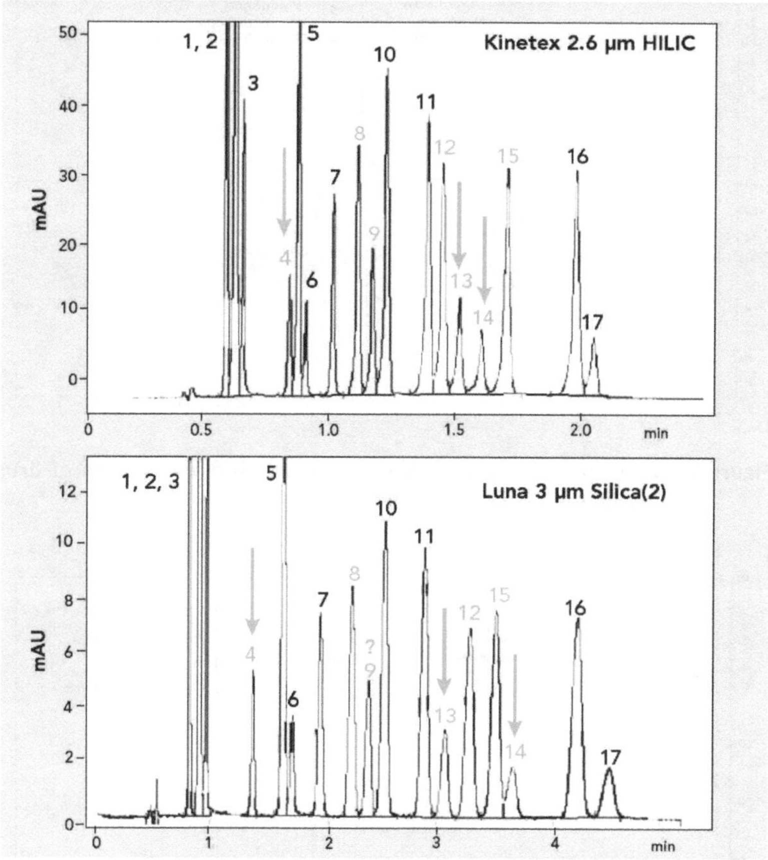

Figure 13.11 Porous shell particles on separation of 17 drug-like compounds and comparison to 3 μm porous particles: (1) flurbiprofen, (2) naproxen, (3) ketoprofen, (4) warfarin, (5) theophylin, (6) caffeine, (7) thymine, (8) uracil, (9) cortisone, (10) prednisone, (11) aceamidophenol, (12) hydrocortisone, (13) sulfamethizole, (14) prednisilone, (15) sulfadimethoxine, (16) sulfaquinolaline, (17) sulfamethizide.

The newest report on performance is with 1.7 μm hybrid particle made by BEH technology and run on UPSFC platform that is under developing. Figures 13.14–13.18 demonstrate the excellent performance of several applications with this configuration.

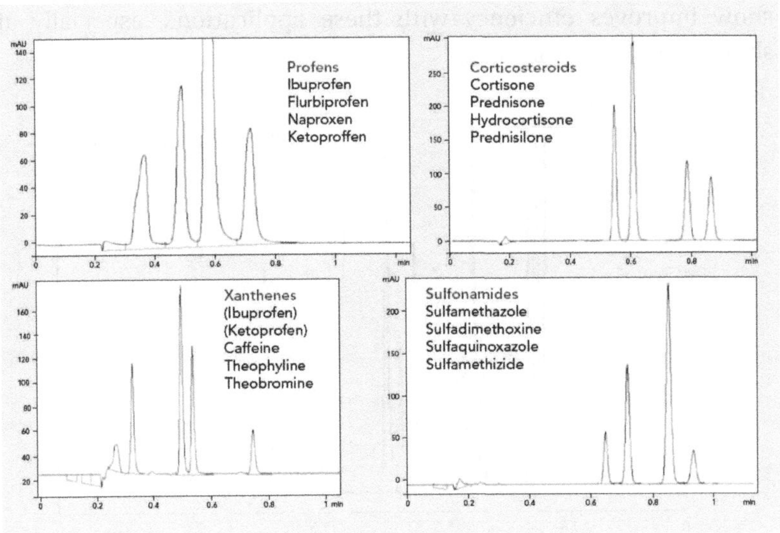

Figure 13.12 Separation on 1.8 μm particles for different groups of drug compounds. Graph from ref. [17].

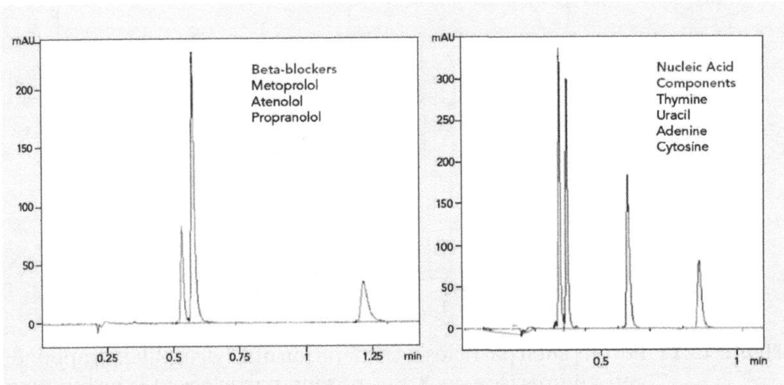

Figure 13.13 Separation on 1.8 μm particles for different groups of drug compounds. Graph from ref. [17].

Figures 13.14 and 13.15 show the chromatogram and performance measure of a complex mixture of 18 compounds with gradient elution [18]. The column has shown excellent selectivity and higher efficiency. Also shown are the excellent reproducibility and peak symmetry, which denotes the superior quality of the particle chemistry.

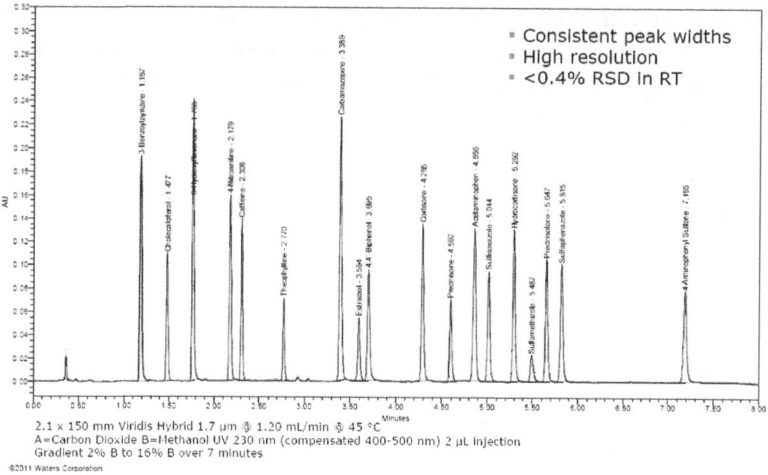

Figure 13.14 Gradient separation with 1.7 μm particles on UPSFC .

	3-Benzoylpyridine	Cholecalciferol	6-Hydroxyflavanone	4-Nitroaniline	Caffeine	Theophylline
	1.187	1.477	1.766	2.179	2.308	2.77
	1.184	1.476	1.765	2.178	2.307	2.769
	1.192	1.483	1.772	2.185	2.313	2.774
	1.191	1.483	1.771	2.184	2.312	2.773
	1.194	1.486	1.773	2.186	2.313	2.774
	1.197	1.484	1.772	2.183	2.312	2.773
Mean	**1.191**	**1.482**	**1.77**	**2.183**	**2.311**	**2.772**
Std. Dev.	**0.005**	**0.004**	**0.003**	**0.003**	**0.003**	**0.002**
% RSD	**0.38**	**0.269**	**0.195**	**0.158**	**0.121**	**0.075**

	Carbamazepine	Estradiol	4,4'-Biphenol	Cortisone	Prednisone	Acetaminophen
	3.389	3.594	3.695	4.285	4.592	4.856
	3.389	3.595	3.696	4.286	4.594	4.858
	3.394	3.601	3.701	4.292	4.6	4.862
	3.393	3.599	3.7	4.29	4.597	4.86
	3.393	3.6	3.701	4.29	4.598	4.861
	3.391	3.597	3.698	4.287	4.595	4.859
Mean	**3.391**	**3.598**	**3.699**	**4.288**	**4.596**	**4.859**
Std. Dev.	**0.002**	**0.003**	**0.003**	**0.003**	**0.003**	**0.002**
% RSD	**0.063**	**0.081**	**0.072**	**0.064**	**0.059**	**0.044**

	Sulfisoxazole	Hydrocortisone	Sulfamethizole	Prednisolone	Sulfaphenazole	4-Aminophenyl Sulfone
	5.014	5.292	5.487	5.647	5.815	7.185
	5.016	5.294	5.489	5.649	5.817	7.186
	5.022	5.3	5.496	5.656	5.824	7.194
	5.019	5.298	5.493	5.653	5.821	7.191
	5.02	5.298	5.493	5.653	5.821	7.192
	5.017	5.295	5.49	5.65	5.817	7.187
Mean	**5.018**	**5.296**	**5.492**	**5.651**	**5.819**	**7.189**
Std. Dev.	**0.003**	**0.003**	**0.003**	**0.003**	**0.003**	**0.004**
% RSD	**0.062**	**0.057**	**0.059**	**0.052**	**0.06**	**0.055**

Figure 13.15 RSD measurements on 18-compound mix.

One advantage the ultra-performance particle brings with elevated efficiency is that it is now amenable for SFC technique to get into impurity identification study. Figure 13.16 shows the separation and measurement of impurities from estradiol [19] with comparison with the LC method. It can be seen that with the faster SFC method, the S/N is higher, the detection limit and accuracy of measurement can be improved with more impurities identified, which are crucial for this type of application.

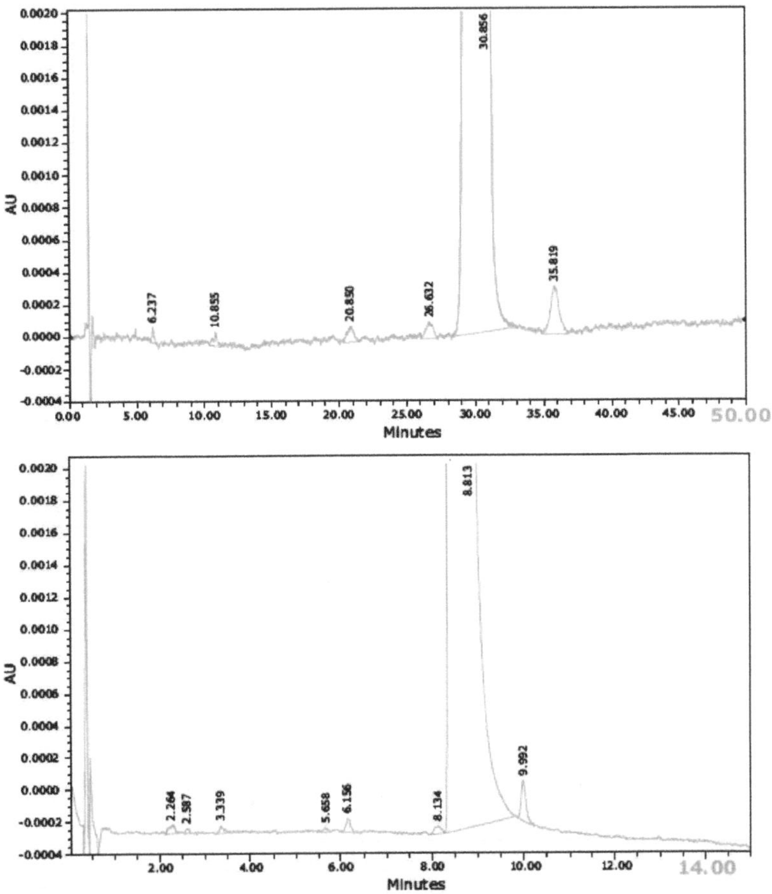

Figure 13.16 Impurity identification of estradiol: top trace: LC method; bottom trace: SFC method.

Figure 13.17 shows the converted SFC method for dapsone from USP HPLC normal-phase method [20]. The new method generates equal or better quality of data and is seven times faster.

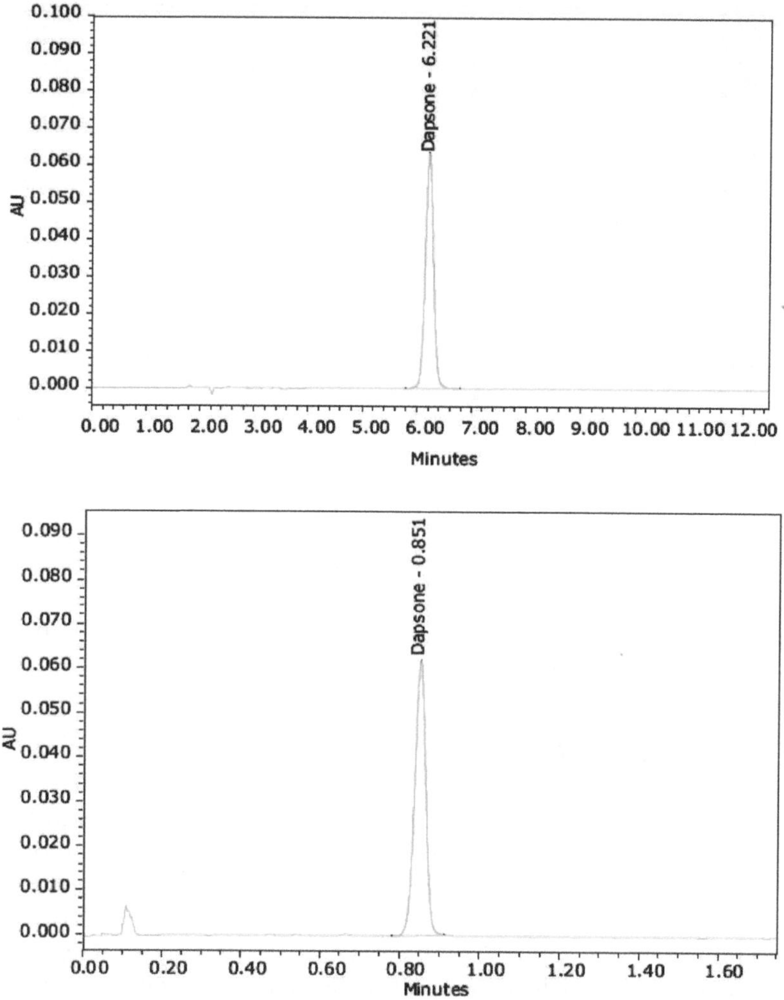

Figure 13.17 SFC method for dapsone and comparison to compendial USP HPLC method: top: HPLC method; bottom: UPSFC method.

Figure 13.18 shows another converted SFC method for compendia USP determination of tolbutamide [21]. It is 10 times faster with better data quality compared with the HPLC method, as is obvious from the peak performance.

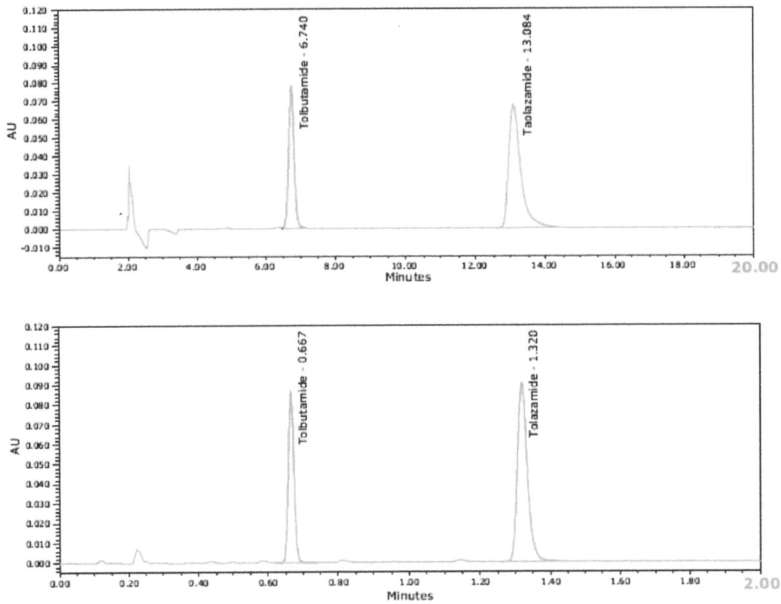

Figure 13.18 SFC method for tolbutamide: top: HPLC method; bottom: UPSFC method.

13.5 Summary

The investigations on applying ultra-performance particles in SFC have so far given positive results and helped to maintain interests from both research and commercial developments. The published results clearly demonstrate the increased efficiency, shortened runtime, and exceptional peak performance. These critical features support the rationale of the applicability of this thought process that such combination would promote the SFC technique and benefit chromatography practitioners in terms of higher performance, faster throughput, and higher productivity, just as UPLC has shown for the past decade after its commercialization.

13.6 Future Directions of SFC Development

There is no doubt that the path of SFC to become a mainstream tool for the chromatography lab has not been smooth, as proven by its history since its inception. Furthermore, the applicability of ultra-performance particles poses more requirements for its successful deployment. Certain factors need to be considered and carefully integrated to ensure that all benefits it promises according to the investigations reported so far can be harvested. For example, following factors are a must for the SFC technique to be developed toward this direction:

- low extra column dead volumes
- higher pressure systems
- effective mobile phase mixing
- need efficiently packed columns
- a wider variety of column stationary phases
- need column reliability in SFC conditions

In addition to hardware development, it will be helpful to have application development for more and in-depth knowledge within various application fields, such as quality control and impurity profiling, two areas that can really benefit from this technique. More resources and tools need to be deployed to help practitioners and developers to have a thorough and higher level understanding.

One reason for the bumpy road of SFC in the past 30 years or so is that it was seen as just a minor variation of certain conventional chromatographic techniques, be it GC or HPLC. The mindset of most developers is that SFC is no exception for following the general theorem of LC or GC. And with this thought, most skill sets for its development had been coming from just one part of the conventional fields (GC initially or LC recently), with lack of in-depth knowledge from other fields in chromatography and other disciplines, such as engineering and physics/kinetics.

The consequence of this non-comprehensive thought process is that SFC has never been treated on its own foot. The instrumentation is mostly adapted from the existing hardware from GC or LC, with minimum or even no customization in many cases. The developed methodology has always been framed in the same fashion as that of HPLC (or GC). This results in limited success in its development and understanding of its mechanism and process.

It is author's opinion that one core presumption in conventional chromatography theory is the segregated and static framework. For example, the plate height concept assumes that the column is divided into sequential and equal length plates that act independently on their own to the mobile solutes. The plate height is identical across the whole column, and on each plate the solute interacts with the plate according to same physical–chemical relationships. This presumption works to the most extent for LC and GC because the objects studied stay in relative single state (either pure gases or liquids). The mass transfer within the same state is governed by simple laws. The interaction on the interface of different states of a material is limited in range and easy to be simulated with straightforward formulas, such as simple arithmetic ones.

In SFC, however, this same framework may deviate much further from what truly happens. The supercritical phase is nothing like the normal phase in that the mass transfer is susceptible to much higher degree variation from its highly "tunable" physical and chemical properties. Accordingly, the interactions among the different states of materials bear more dimensional changes; the fundamental study on the supercritical phase has indicated that there is no simple arithmetic relationship capable of describing its behavior, and multi-dimensional modeling involving phase changes is necessary to give any good estimate.

After all, chromatography today is still a semi-empirical scientific exploration. With more complex processes from supercritical phases, it is believed that the best approach for its breakthrough development is to apply sound, high-degree mathematic algorithm/modeling to unlock its multi-dimensional mechanism. This would be impossible to achieve if the experimental tool is still developed based on simpler processes as in conventional chromatography. In other words, the approach in the past, i.e., mainly adaptation from the traditional hardware for the SFC study, will not offer the light at the end of the tunnel. A new *holistic* design that can better express the multi-variant nature of supercritical fluids, and hence offers more realistic implementation and higher degree control of the process, will be the only direction and foundation for its eventual success. For this reason, the projected UPSFC development, combined with ultra-performance particle technology that offers highly sophisticated insight into the process, will become a genuine tool for leading its chromatography application success.

References

1. Pinkston, J. D. et al. (2006) Comparison of LC/MS and SFC/MS for Screening of a Large and Diverse Library of Pharmaceutically Relevant Compounds, *Anal. Chem.*, **78**, pp. 7467–7472.

2. Bolaños, B. et al. (2004) SFC/MS in Drug Discovery at Pfizer, La Jolla, *Int. J. Mass Spectrom.*, **238**, pp. 85–97.

3. Wu, D. R., Leith, L., Balasubramanian, B., Palcic, T., and Wang-Iverson, D. (2006) The Impact of Chiral Supercritical Fluid Chromatography in Drug Discovery: From Analytical to Multigram Scale, *Amer. Lab.*, June 1, 2004, pp 914.

4. Ventura, M. et al., (2004) High-Throughput Preparative Process Utilizing Three Complementary Chromatographic Purification Technologies, *J. Chromatogr. A*, **1036**, pp. 7–13.

5. Wang, Z. (2011) *Chiral Separation Methods,* eds. Ahuja, S. Chapter 10 Separation of Chiral Compounds by SFC, (John Wiley & Sons, New Jersey) pp. 299–330.

6. Gere, D. R., Board, R., and McManigill, D. (1982) Supercritical Fluid Chromatography with Small Particle Diameter Packed Columns, *Anal. Chem.,* **54**, pp. 736–740.

7. Thompson, J. W., Mellors, J.S., Eschelbach, J. W., and Jorgenson, J. W. (2006), Recent Advances in Ultrahigh-Pressure Liquid Chromatography, *LCGC,* April 2006, pp. 16–20.

8. Swartz, M. E. (2005) UPLC™: An Introduction and Review. *J. Liq. Chromatogr. Relat. Technol.*, **28**, pp. 1253–1263.

9. Lemmon, E. W., McLinden M. O., and Friend, D. G. (2011) Thermophysical Properties of Fluid Systems in *NIST Chemistry WebBook*, NIST Standard Reference Database Number 69, Eds. Linstrom, P. J., and Mallard, W. G. National Institute of Standards and Technology, Gaithersburg MD, 20899, http://webbook.nist.gov, (retrieved December 31, 2011).

10. Waters Corp. (2011) Presentation Fundamentals and Advancements in Analytical SFC, *Pittcon* 2011, Atlanta, USA.

11. Waters Corp. (2005), Ultra Performance LC: Separation Science Redefined, http://www.waters.com/webassets/cms/library/docs/720001368en.pdf.

12. Thiébaut, D., Sarrazin, C., Dupouy, P., Sassiat, P., and Vial, J. (2008) Ultra High Performance/High Pressure Supercritical Fluid Chromatography On sub-2 micron Particles Packed Columns. *2nd International Conference on Packed Column SFC*, Zurich, Switzerland.

13. Fougère, L., and Lesellier, E. (2008) Study of Efficiency on Packed Column SFC with Different Particle Diameters. *2nd International Conference on Packed Column SFC*, Zurich, Switzerland.

14. Robeson, R., and Wentz, E. (2009) Evaluation of sub 2 μm Particle Packed Silica Columns in SFC. *3rd International Conference on Packed Column SFC*, Philadelphia, USA.

15. Campbell, R. M. (2011) Supercritical Fluid Chromatographic (SFC) Separations of Oligomeric Materials Using Columns Packed with sub 2-um Particles. *5th International Conference on Packed Column SFC*, New York, USA.
16. Berger, T. (2010) A Critical Review of the SFC Performance of a Kinetex™ Porous Shell 2.6 μm HILIC Column (4.6 × 150 mm). *4th International Conference on Packed Column SFC*, Copenhagen, Denmark.
17. Berger, T. (2010) The Use of Sub 2 μm Particle in SFC. *4th International Conference on Packed Column SFC*, Copenhagen, Denmark.
18. Waters Corp. (2011) Presentation, *Introducing Waters ACQUITY UPSFC™ System*, *Pittcon* 2011, Atlanta, USA.
19. Waters Corp. (2011), Chromatographic Purity of Estradiol using ACQUITY UPSFC System, http://www.waters.com/webassets/cms/library/docs/720003888en.pdf.
20. Waters Corp. (2011), Chromatographic Assay of Dapsone Tablets using ACQUITY UPSFC System, http://www.waters.com/webassets/cms/library/docs/720003893en.pdf.
21. Waters Corp. (2011), Chromatographic Assay of Tolbutamide using ACQUITY UPSFC System, http://www.waters.com/webassets/cms/library/docs/720003903en.pdf.

Chapter 14

Pilot and Production-Scale Supercritical Fluid Chromatography

Geoffrey B. Cox
PIC Solution Inc, PO Box 191, Media, PA 19065, USA
gcox@pic-solution.com

14.1 Introduction

Column chromatography has been associated with the purification and isolation of chemical species since its inception. The original work by Tswett [1] was for the isolation of plant pigments and it was only with the introduction of the instrumental methods of chromatography (first with gas chromatography [2] and later with HPLC [3]) that it became a mainstream analytical technique. Even then, these techniques were rapidly transformed into preparative processes, although the use of preparative gas chromatography has not expanded into the mainstream of preparative separation techniques in part due to the practical limitation that the compounds should have boiling points below 250°C and should be temperature stable.

Supercritical Fluid Chromatography: Advances and Applications in Pharmaceutical Analysis
Edited by Gregory K. Webster

ISBN 978-981-4463-00-3 (Hardcover), 978-981-4463-01-0 (eBook)
www.panstanford.com

Most large-scale chromatography is carried out at low pressures as the separations are predominantly of biomolecules that typically use large-particle ion exchange or affinity packing materials with low flow rates, although some applications at the production scale for the separation of small molecules exist. The largest scale small-molecule production liquid chromatographic process is probably the purification of p-xylene from C8 hydrocarbons by simulated moving bed (SMB) chromatography. Other large-scale SMB processes, for the production of high-fructose corn syrup [4] and for the isolation of the S enantiomer of etiracetam at the production scale [5], have been reported, although the scales are remarkably different, the HFCS being produced in vast quantities (annual consumption is 40 lb per capita!) while the annual need for S-etiracetam–and other pharmaceutical products—is much more modest and is of the order of a maximum of several hundred tons.

In supercritical fluid chromatography (SFC), the first large-scale preparative instrument was introduced by Prochrom in 1990 at the PREP symposium meeting in Ghent. This 10 cm unit was designed to be a pilot unit for the development of large-scale separation processes using SFC. Although it was anticipated at that time that large-scale SFC would be of great interest in many industries, it has turned out that relatively few large-scale SFC units have been installed, despite the potentially favorable production costs relative to HPLC purification.

14.2 Large-Scale SFC: The Potential

Over the past few years, the major application areas for preparative-scale SFC have been in pharmaceutical development, where the technique has been used for the isolation of small quantities—up to perhaps a few kilograms, but mostly in the 1–100 g range—of both chiral and achiral intermediates. These applications are discussed elsewhere in this book and are outside the scope of this chapter.

There are few reports of large-scale SFC separations in the literature, partly because there are currently few applications that have been scaled up to pilot or production levels and partly because for those that have been so scaled there are issues of confidentiality, as the products are mostly in the pharmaceutical industry, there are many and strong reasons for the details of the separations—or even of their existence—to be kept secret. However, as chromatography has the property of scaling in a linear fashion from the analytical to the production scale, one can make some calculations for preparative data published at the laboratory scale and extrapolate with a high degree of accuracy to what can be achieved at a larger scale.

One report of a SFC separation carried out at the laboratory scale is that of flurbiprofen. SFC conditions give a larger selectivity than can be achieved under HPLC conditions and so it can be expected that SFC would be the preferred method of purification of the flurbiprofen enantiomers. Of course, other methods are available to obtain the enantiomers at the large scale in purity adequate for their application; it is, however, instructive to calculate from the data available what would be the consequences of a decision to use SFC as a scale-up method. The published data describes a baseline separation between the enantiomers using a CHIRALPAK® AD-H™ column with methanol as modifier. The sample size on a column 250 × 20 mm was 350 mg of racemate and the cycle time (the time elapsed between the front of the first peak and the tail of the second peak) was 150 sec. Given that the column contained 47 g of chiral stationary phase (CSP), the productivity of the separation can be calculated as 2.1 kg (enantiomer) per kg (CSP) per day. This is a consequence not only of the high selectivity attained in the SFC process but also of the low mobile-phase viscosity. The mobile phase was 12% methanol in CO_2, running at 85 g/min with an operating pressure of 105 bar.

If one supposes that an annual production of 19 metric tons of the flurbiprofen enantiomer were required, then—again supposing that SFC were to be the preferred production method—one would need a column 250 mm long and 500 mm in diameter containing 63 kg of CSP. The flow rate would be approximately 53 L/min. At this scale, we can assume a high recovery of the organic solvent and at least a 95% recovery of CO_2. Supposing the column is replaced each year and using a 7-year amortization of the equipment cost ($5MM) and of the site costs, one can estimate that the cost of purification at this level would be of the order of $150/kg of enantiomer; this is not an unusual cost for an efficient chromatographic process.

Other workers have reported that this level of productivity is not unusual in SFC. Shaimi reported a productivity of 4 kg (enantiomer)/kg/day for guaifenesin working at sub-ambient temperature [6] while Stringham, Krueger and Marshall [7] using very short columns found a productivity of 1.2 kg/kg/day in an enantioselective separation and an astonishing 44 kg/kg/day for an achiral separation of a mixture of acetaminophen and antipyrine using a 2-ethylpyridine column. Thus, high levels of productivity are not unusual in SFC, while such values are normally attained only in SMB separations in the liquid phase. This implies that SFC separations should be cost-competitive with LC or SMB separations.

The costs of SFC separations relative to HPLC processes have been discussed by Van der Vorst et al. [8]. Relative to HPLC there are obvious cost savings since only 15–30% of the mobile phase is an organic solvent, whereas the rest is CO_2 gas, which is inexpensive and is easily removed from the product fractions before the evaporation of the mobile phase at the end of the separation. There are, however, additional costs in SFC relative to HPLC. The operating pressure is much higher in SFC, requiring more energy to raise the liquid CO_2/organic modifier mixture to the inlet pressure of the column. The mobile phase has to be cooled to around –5°C so that it can be pumped, following which it has to be heated to the operating temperature, often 35–40°C. The separation process results in a decrease in pressure as the mobile phase passes through the column and the back-pressure regulator. Provided a high enough back pressure is used, the density change as the solvent passes through the column is not excessive, so the adiabatic cooling is not great. On passing through the back pressure regulator, however, the supercritical/liquid CO_2 (depending on conditions) reverts to the gas phase with significant cooling, thus requiring a heating step to prevent frosting and problems with product collection. While this is not generally carried out at the lab scale, use of a suitable heat exchanger between the inlet CO_2 and the mobile phase issuing from the BPR could reduce the energy requirements of the phase changes and might result in important cost savings at the industrial scale. The conclusions of Van der Vorst were interesting as they calculated that the exergy costs arising from the production of liquid CO_2 and the use of electricity for heating and cooling became very significant in the operation of large-scale SFC to the point where HPLC seems to be favorable from an exergy viewpoint relative to SFC. However, it appears that much of the unfavorable aspects of the production of the CO_2 arise from the cumulative degree of perfection for the production process; it is not clear if the low CDP for CO_2 production is due to the fact that much CO_2 is produced as a by-product of other processes, in which case the analysis would need to take the other factors in these processes into account. In addition, the calculations did not take into account the recycling of the CO_2 in the chromatographic system, which reduces the energy requirements for the operation and also decreases the CO_2 consumption by 90%, which would have a great effect on the cumulative exergy extracted from the natural environment (CEENE), which is one of the major contributors to the unfavorable exergy requirements of SFC to HPLC in their calculations. It is clear from the analysis that at the process and the plant level, SFC is exergetically favorable compared with HPLC even though the two systems that were compared used different sources of heating and cooling (electricity for SFC and hot oil and steam for HPLC), with the addition of efficient CO_2 recycling it would also be expected to be favorable globally. One aspect not taken into account in the analysis is the faster overall speed of SFC; this would allow the SFC unit to be smaller than that used for HPLC to achieve a given production rate or would allow a much faster production of the desired quantity. However, no calculations of the exergy requirements for construction of the site or the equipment were included.

Of course, most comparisons are made for equivalent cases. Where there is a marked difference in the effectiveness of one process over the other, the balance between them would clearly be affected and in some cases either might be favored over the other. Although the exergetic life cycle analysis is important to assess the environmental impact of the separation process, the process economics are also of great importance to any production process and the end decision for the effective use of SFC or HPLC should always be made by a cost analysis of the specific process; in an ideal world the costs of the components of the operation should be related to their environmental impact.

14.3 Scale-Up Issues

The scale-up of SFC systems to production-scale units can raise several issues not met in production-scale HPLC systems. It is not at all intended that the issues noted below form a set of reasons not to employ the technique in production, but solutions are needed in order to progress to large-scale, efficient production SFC systems.

14.3.1 Column Size

Chromatography in general is intrinsically scalable from analytical to preparative and beyond. SFC is no exception. There are, however, some issues of hardware that can impact the scale of operation or, more particularly, how they may impact operation at very large scale.

Increasing the scale of a chromatographic process is generally carried out by maintaining the particle diameter of the packing material and the column length and increasing the column diameter. For HPLC processes, the consequences of this are often ameliorated by limiting the operating pressure of the processes (and therefore the wall thickness of the columns) by using larger particles. As an example, the production-scale SMB units that are used for enantioselective separations are filled with 20 micron particles and operate at a maximum of 50 bar pressure. In the case of SFC columns, the operating pressure is much higher due to the necessity to maintain the mobile phase at or near supercriticality. As has been noted, operation at low back pressures brings the process into what has been called the "Bermuda Triangle" of SFC [9], where small changes in conditions result in large changes in solvent density, sample solubility, etc. Thus, these authors recommend that SFC columns should be operated with outlet pressures of 140–150 bar (although it should be noted that at the laboratory scale, most processes work well at pressures between 100 and 120 bar). This high pressure requirement results in the need for columns with thick walls. This probably places an upper limit on the diameter of the columns for large-scale SFC, especially since the requirement for axial compression means that the columns must generally be bored from a solid block of steel in order to achieve the required uniformity and circularity of the bore and smoothness of the walls. Uniform ingots of steel of large dimensions are difficult to obtain; a practical limit of diameter for the SFC columns would probably be around 50–60 cm. Such columns would also be rather heavy; the wall thickness calculated for a 45 cm i.d. column capable of operating at 250 bar should be at a minimum of 5 cm and would probably be larger to allow an adequate safety margin; thus, the column tube alone would weigh around 200 kg before adding the outlet flange and compression piston.

14.3.2 Particle Size

There is, however, a further limit to the maximum operating pressure. Experiments have shown that silica-based media begin to fracture in wide-diameter dynamic axial compression columns at applied piston pressures of 50–100 bar, depending on the shape and internal structure of the silica particles [10]. This limits the pressure drop across the column to approximately 60 bar (for most spherical particles), which in turn places some restrictions on the particle diameter and the mobile-phase flow rate and composition. It is generally accepted that a given preparative separation requires a certain number of theoretical plates to obtain the desired purity and recovery. Starting from this number and the maximum and minimum practical column lengths (if they are too short the distribution of sample across the column is difficult, if they are too long the equipment becomes unwieldy and the quantity of packing material required becomes large), one can obtain some idea of the particle size and flow rates that can be used. Table 14.1 shows the results for two separations requiring 2500 and 5000 plates, respectively. In terms of productivity (kg product/kg packing material/day) the smallest particles are always the most appropriate, as long as the column length remains reasonable. This is in contrast to the actual production rate, which is directly proportional to the flow rates shown in the table. This indicates that the production rate will increase with increasing particle size (and therefore column length). These results should be tempered, however, by the cost of the packing materials. Silica-based particles show a strong relation of price per gram versus particle diameter, reflecting the difficulty in making small particles and also in handling them once made. If the packing material is pure silica or a simple bonded-phase material (such as a C18 phase) then the packing costs diminish greatly as particle size increases and packing materials of 10–15 microns often turn out to be optimal in terms of the additional cost per kilogram of product arising from the packing—for very large particles, the cost of so many kilograms of packing and the cost of the long columns required eventually overwhelm the reduction in cost due to the greater production rate, while the cost of even 1–2 kg of very small particle packing material becomes prohibitive.

Table 14.1 Calculated particle size, flow rates and relative productivity for SFC and HPLC

		2500 plates			5000 plates		
Column length (cm)	Qty packing	Particle size (µm)	Relative flow rate	Productivity	Particle size (µm)	Relative flow rate	Productivity
			SFC				
10	1	3.66	11.56	8.028	2.89	7.2	5.000
15	1.5	4.57	12.01	5.560	3.64	7.6	3.519
25	2.5	6.03	12.53	3.481	4.84	8.07	2.242
30	3	6.65	12.7	2.940	5.35	8.22	1.903
50	5	8.73	13.13	1.824	7.07	8.62	1.197
75	7.5	10.81	13.43	1.244	8.8	8.9	0.824
100	10	12.58	13.63	0.947	10.27	9.08	0.631
200	20	18.05	14.04	0.488	14.83	9.47	0.329
400	40	25.8	14.37	0.249	21.32	9.79	0.170

		2500 plates			5000 plates		
Column length (cm)	Qty packing	Particle size (μm)	Relative flow rate	Productivity	Particle size (μm)	Relative flow rate	Productivity
			HPLC				
10	1	3.66	2.31	1.604	2.89	1.44	1.000
15	1.5	4.57	2.4	1.111	3.64	1.52	0.704
25	2.5	6.03	2.51	0.697	4.84	1.61	0.447
30	3	6.65	2.54	0.588	5.35	1.64	0.380
50	5	8.73	2.63	0.365	7.07	1.72	0.239
75	7.5	10.81	2.69	0.249	8.8	1.78	0.165
100	10	12.58	2.73	0.190	10.27	1.82	0.126
200	20	18.05	2.81	0.098	14.83	1.89	0.066
400	40	25.83	2.87	0.050	21.32	1.96	0.034

Note: Assumes 60 bar pressure drop across the columns, diffusion coefficient for HPLC 2×10^{-10} m^2sec^{-1}, diffusion coefficient for SFC 1×10^{-9} m^2sec^{-1}.

Productivity calculated relative to the HPLC process with 5000 plates.

When the separation to be carried out is of a chiral molecule, the cost of the chiral selector is much greater than that of the base particle, rendering the cost of an enantioselective phase less dependent on the particle size of the silica. In this case, the minimum cost contribution of the packing material is generally found at the smallest particle size that is available in the quantity required for the separation. In this regard, SFC, which is around three to five times faster than HPLC and thus allows a column three to five times smaller than that needed for an HPLC process, can turn out to be more cost effective.

It has long been known that wide-diameter columns are difficult to pack efficiently without the formation of voids and channels either immediately on packing or after a certain time during which the packed bed consolidates. The solution for both HPLC and SFC has been to use axial compression systems [11] for columns with internal diameters much above 50 mm. Such columns are constructed with a piston that can move within the column that is used to compress the packed bed during both the packing process and the operation. This maintains the bed in a highly efficient configuration as long as the pressure exerted by the piston exceeds the back pressure within the column. At the small scale, it is possible to pack the column under axial compression conditions and subsequently block the piston to maintain the bed integrity while removing the compression pressure. This is more problematic with larger columns and becomes an issue with preparative SFC columns in that the pressure at the outlet of the column is maintained at elevated values. Thus, if the column outlet is maintained at 140 bar, the inlet pressure will be running at approximately 200 bar. During operation of the system, therefore, the dynamic axial compression pressure would have to be maintained a little in excess of the inlet pressure in order to maintain the bed compression. When the flow is stopped in the column and the pressure is released, unless the axial compression pressure is released at the same time, an excessive pressure would be applied to the packed bed with consequent damage to the packing material. Systems to accomplish this were developed at Prochrom and are believed to be available from Novasep, though there are no literature references to these.

14.3.3 Speed

Because of the low mobile-phase viscosity, SFC is intrinsically fast, often running four to five times faster than an equivalent HPLC separation. This can result in fast cycle times as a 10 min. HPLC separation can collapse to a 2 min. separation in SFC. This places some constraints on the operation of the hardware for large-scale SFC. Valves are required to operate quickly and the injection systems must also operate within the time scale of the separation. If an elution gradient is to be used, this, too, must function with appropriate speed and precision. At the laboratory scale, these constraints are not a problem as small valves have low inertia and can switch within a couple of milliseconds. This may not, however, be true for large, high-pressure valves as are needed for the injection devices and collection systems in production systems. This could become more of an issue should large-scale SFC-SMB systems be contemplated as switch times in modern SMB units using liquid mobile phases can often be less than 1 min, which means switch times of 10–15 sec for an equivalent SFC-SMB system.

14.3.4 Sample Introduction

In HPLC systems, samples are often introduced to the system in solution in the mobile phase used for the separation. This results in the minimum perturbation of the equilibrium in the column and thus the minimum effect on the shapes of the peaks in the chromatogram. In the majority of the small-scale SFC separations, however, it is more normal to dissolve the sample in the polar modifier used for the separation, thus injecting the sample in a strong solvent, which will distort the peak shape. This is because of the difficulty in dissolving the sample in the supercritical mobile phase and maintaining that sample in a supercritical state and finally in injecting a reproducible and controlled volume. At the larger scale, however, one can envision a supercritical extraction being used to prepare the sample, which is then directly transferred to the SFC unit. The control of the pressure of the injection is critical in this case as the supercritical fluid is compressible, and changes in the pressure will result in changes in the quantity of sample introduced, although this might be used to enhance sample solubility. However, for poorly soluble samples, this supercritical fluid extraction can be useful, allowing a preliminary cleanup of the sample prior to the chromatographic step; only material soluble in the mobile phase is introduced to the column, reducing fouling and extending column life.

14.4 Large-Scale Equipment

14.4.1 Batch Systems

Although the first preparative-scale SFC system was a pilot unit of 100 mm i.d. [12], there have been relatively few large systems reported. Most preparative-scale SFC units constructed have been at the laboratory scale, with columns ranging from 10 to 76 mm i.d. and the largest with flow rates up to 1 L/min. Unfortunately, the literature does not report larger systems, mainly because these are installed in manufacturing contexts, where confidentiality is paramount. There have been reports of systems with columns of 150 mm (apparently for the purification of fish oil esters) and of 350 mm (no application announced). Johnson Matthey has announced the availability of "the largest production-scale SFC in the USA"—a 200 mm i.d. system from Novasep—in their custom separations unit [13]. A unit with columns 350 mm has been described by the Malaysian Palm Oil Board [14] for the purification of phytonutrients (sterols, tocols, and carotenoids) from palm oil. Equally, there have been reports of SFC systems used for the purification of fish oils (eicosapentaenoic acid [EPA] and docosahexaenoic acid [DHA] esters) in Spain [15]. Patents have been issued for SFC purification of these oils to KD-Pharma [16].

Large-scale batch SFC systems mirror their smaller-scale laboratory counterparts. Those constructed for the separations of fish oils and palm oil generally have no provision for co-solvent addition (the low-polarity compounds separate well with pure CO_2 as mobile phase), which reduces the complexity of the hardware, although solutions for the handling of the viscous oils, which are the products, are required once they are trapped in the product cyclones. The hardware can be further simplified by the use of particles large enough that they can be packed into fixed-bed columns, thus removing the need for dynamic axial compression hardware, although this may result in rather long columns for the more challenging separations. Where the SFC system is to be used for multiple separations or those demanding high-performance separations (especially of enantiomeric species) both co-solvent and axial compression column hardware seem to be essentials.

14.4.2 Alternate Pumping Recycle in SFC

The production rate and the performance of SFC have been investigated by the use of recycle techniques analogous to the steady-state recycle (SSR) process used in HPLC [17]. As with all recycle techniques, this operation is mainly carried out for binary mixtures. In order to accommodate the phase changes needed to allow the mobile phase to be pumped accurately, the recycle system has to be configured with two columns instead of the single column used in HPLC recycle. The operation is shown in Fig. 14.1. The sample is injected into the first column (A) and the front of the early eluting peak is collected (B). The mixed fraction is directed into the second column (C) and the tail of the later eluting peak is collected (D) while switching the flow to the inlet of the second column further to separate the mixed band. At this point, it is also possible to inject some additional sample into the mixed band as it passes into the column to replace the material removed during the fraction collection. This would then be equivalent to the SSR systems used in HPLC. When the front of the first eluting component reaches the bottom of the second column (E) it is again collected. At this point, the first column is not connected (though the flow could be directed through it if desired). The mixed zone is passed to the first column and the tail of the second peak is collected (F), again, if desired, injecting more sample into the mixed zone as it passes into the top of the column. The flow is then switched to the inlet of the first column and the cycle is repeated. When used in the SSR mode, this system should improve the production rate of the SFC system significantly while avoiding some of the difficulties experienced with SFC-SMB. It is, however, difficult to find any publication on this technique although equipment for this process along with significant modification to the injection/collection scheme was described by Novasep [18] some time ago.

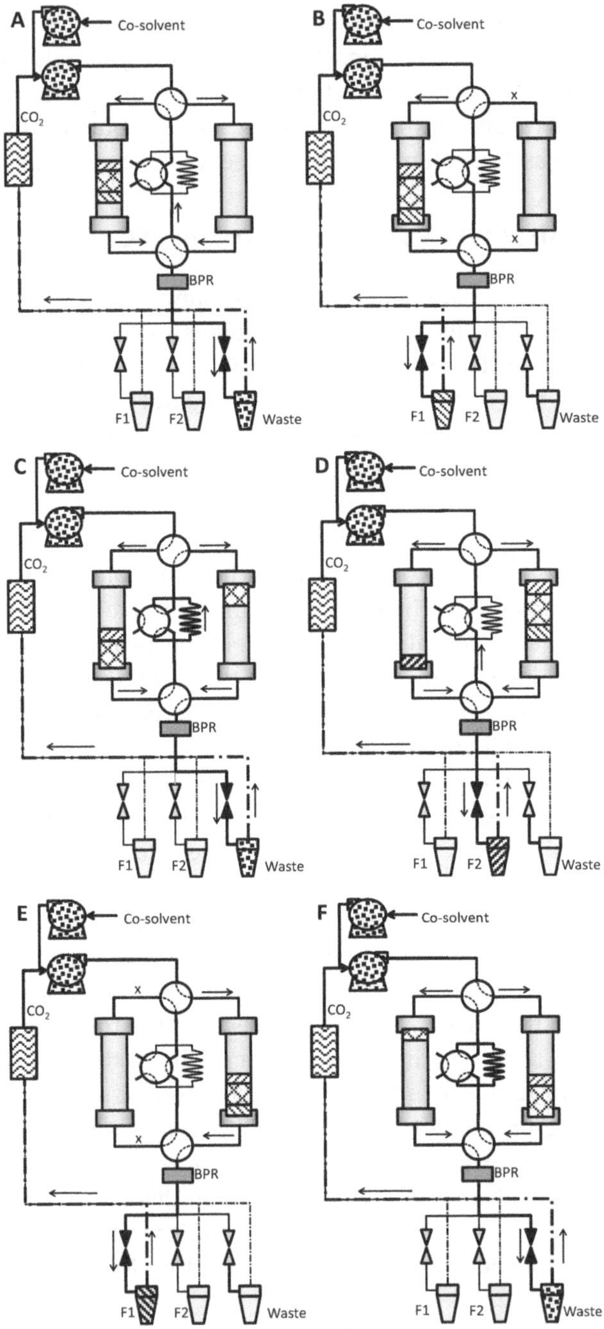

Figure 14.1 Alternate pumping recycle/steady-state recycle for SFC.

14.4.3 SMB-SFC

Simulated moving bed chromatography has been employed in the petroleum and food industries for many years. Introduced by UOP as an efficient and low-cost process for the purification of p-xylene from C8 hydrocarbon mixtures, it also became a method of choice for production of high-fructose corn syrup. Over the past 20 years it has been adopted as a high-performance technique for the resolution of enantiomers (mainly of drug molecules) from racemates. Its main features are the use of multiple columns with switching valves, which mimic a countercurrent chromatography process. This combines high production rates with low solvent consumption, both desirable attributes for a chromatographic process, which is normally associated with low productivity and high dilution of products.

The possibility of the combination of SFC and SMB techniques has been investigated by two groups, one at Novasep and ENSIC at Nancy [19] and the other at the Technical University of Hamburg [20].

The basic SMB principle is shown in Fig. 14.2. It works for binary separations; we will designate the faster eluting component as component A, the slower (i.e., more retained) component as component B. The simplest system starts with four columns each connected to a series of valves that allow the columns to be interconnected and to be connected to solvent and sample inputs and two product outputs. A proportion of the solvent flow emerges from the output ports, and the remainder continues through the column set. There is a continuous flow of solvent through the column set from the top of the left-most column (#1) to the bottom of the right-most column (#4) and then back to the top of column 1. The sample feed is pumped continuously into the top of column 3 in Fig. 14.2A. As the mixed feed zone travels down the column, a band of pure component A forms at the front of the zone, followed by the feed mixture (Fig. 14.2B). This behavior is well known as frontal elution. Eventually the pure zone reaches the bottom of column 3, and part of this emerges from the output port at that point. The remainder continues into the top of column 4 (Fig. 14.2C). Eventually, the mixed zone reaches the bottom of column 3 and would also emerge from the outlet port, contaminating the pure product previously collected (Fig. 14.2D). To avoid this, all the columns are moved one position to the left (see Fig. 14.2E). Now the feed enters the top of column 4, into the mixed zone that is emerging from column 3. The frontal elution system continues in column 4, the pure zone of component A running in front of the mixed zone (Fig. 14.2F, column 4). Meanwhile, the mixed zone remaining in column 3 is eluted with clean solvent from column 2. This has the effect of leaving a zone of the slower-moving pure component B running at the tail of the mixed zone in column 3 (still Fig. 14.2F). Eventually, the pure zone of component A reaches the bottom of column #4, eluting from the outlet port and the pure zone of component B reaches the bottom of column 3. If the flow rates in the columns are adjusted correctly, the mixed zone in column #4 reaches the output port at the same moment as the pure zone of component B reaches the bottom of column 3 (Fig. 14.2G). In order to avoid contaminating the pure component A, the columns are once more moved one position to the left (Fig. 14.2H). This has the effect of allowing the pure zone of component B to elute from the other output port, currently at the outlet of column 3. The cycle now continues with the columns moving from right to left and the main solvent flow moving through them in the opposite direction. Component A emerges cyclically from the output port downstream of the feed inlet while component B emerges from the upstream outlet port. The flow through each column is adjusted by the relative flows of the eluent, feed and the two outlet streams to ensure that the products emerge from the appropriate column at the appropriate time. In practice, a four-column SMB system does not produce pure components and additional columns are needed to reach the 99% purity that is often demanded of chromatographic separations, especially of chiral pharmaceutical intermediates and APIs. With a conventional liquid chromatographic SMB process, six to eight columns are usually employed; an alternative, closely related process (the "Varicol" process [21]) that moves the columns asynchronously allows the use of five columns to reach similar purity levels.

SFC-SMB processes follow the same overall scheme. There are some obvious technical challenges such as reproducibly controlling the flow rates and proportioning the flow between the outlets and the succeeding column. In HPLC, with non-compressible mobile phases, this is done with pumps or with proportioning valves. The compressible supercritical fluid poses some challenges for the engineer; pumping can only successfully be done after cooling and condensing the fluid to a liquid state. This means that the recycle system, often placed between two of the columns in a liquid SMB system has to be placed after an additional outlet valve with a condensation step so that the solvent can be pumped.

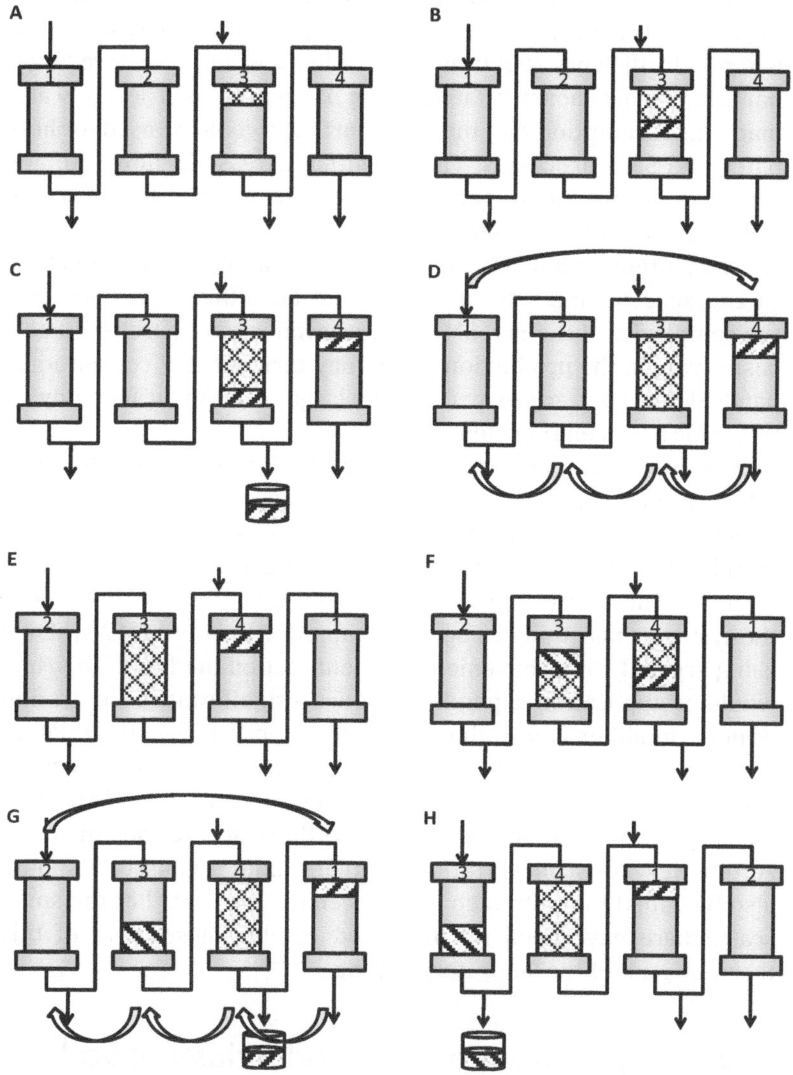

Figure 14.2 Schematic of the SMB process.

There has been one comparison of batch SFC and SFC-SMB [22]. The conclusion of this work was that the batch technique was more cost effective than SFC-SMB, largely due to the investment costs of the large, high-pressure columns. However, the study was limited by choice of an upper flow rate limit, corresponding to an analytical scale flow rate of around 6 mL/min, not extraordinarily high in terms of preparative-scale SFC, especially in the case of SMB, where the efficiency requirement is rather low and therefore larger particles can be chosen relative to the batch process. The study also made the assumption that the same particles would be used for both batch and SMB processes. The potential increase in flow rate and therefore the production rate of the SMB process could be sufficient to bring SMB-SFC into cost competition with the large-scale batch process, although any advantage would be application dependent. Of course, economic advantage may not be sufficient to overcome the complexity of the SFC-SMB system even where there could be a cost saving in the production and it may turn out that conventional batch SFC is a better prospect for the large scale. So far no study on this topic has been reported.

One interesting possibility in SMB-SFC is the potential to affect retention in the various zones of the SMB system by adjusting the internal pressure in the columns [23]. As the pressure is increased, the mobile-phase density increases, which in turn increases its solvent strength. This allows the variation of retention factors throughout the system, a task that has been attempted with liquid SMB systems using arguably less convenient gradient elution methods. This, too, would be application dependent as once a significant concentration of polar modifier is introduced into the mobile phase, the relation between solvent strength and density is less important and the main contribution to retention is the modifier concentration. Thus, this technique would be applicable only to those separations in which zero or very low concentration of polar modifier is used. It should also be added that although it has been demonstrated at the small scale, there have been no published results on scale-up of this technique out of the laboratory setting.

14.5 Applications at Pilot and Industrial Scale

Over the past few years, the major application areas for preparative-scale SFC have been in pharmaceutical development, where it has been used in the isolation of small quantities of both chiral and achiral intermediates. These applications are discussed elsewhere in this book and are outside the scope of this chapter. Anecdotally, several separations have been carried out at the multi-kilogram level for pharmaceutical intermediates, but no detailed reports are available, for the usual reasons of confidentiality.

14.5.1 Fish Oils

Probably the major application area for industrial-scale SFC has been the purification of fish oil components—the ethyl esters of EPA and DHA. The consumption of these ω-3 unsaturated fatty acids has been promoted for many indications, primarily for cardiovascular health and reduction in blood triglyceride levels, although brain and visual development are also cited. Numerous articles have appeared covering the effects of the ω-3 unsaturated fatty acids on psoriasis, bowel disease, mental illness, and prevention of cancer [24]. There are, however, conflicting reports on many of these indications [25].

HPLC separations of the esters of EPA and DHA were developed using both reversed-phase and normal-phase conditions, the latter with silver-impregnated resins. Other work with HPLC using ion exchange resins loaded with silver ions [26] reported high-purity (>97%) for methyl esters of EPA and DHA from different resins, albeit at the small scale.

The first reports of the use of SFC in fish oil purification were in the late 1980s, the focus switching to SFC because the mobile phase for the separation is pure CO_2, which is easily removed, leaves no residues and is low-cost. Separations have been reported using silver nitrate coated silica gel as the stationary phase [27], C18 reversed-phase silica packing material [28] and alumina [29]. High-purity samples of the esters are reported to be obtained at the large scale (1.2 kg crude per injection) using a pressure or a temperature gradient with an aminopropyl bonded-phase column [16]. This process was carried out using a column 185 × 24 cm packed with 20 to 40 micron particles.

A recent report detailing the comparison of several stationary phases developed for SFC in the purification of EPA, arachidonic acid (AA) and DHA showed the isolation of EPA in 90–95% purity and DHA in 74–83% purity using GreenSep PFP, Nitro and aminophenyl columns with pure CO_2 as eluent and at a temperature of 60°C and a back pressure of 170 bar [30]. This work suggests that there is probably some advantage to be gained by further study of the more recently developed phases for SFC in this application.

14.5.2 Cyclosporin

A report on the development of a pilot-scale isolation of cyclosporin A was made in 1998 [31]. The process was developed to eliminate the large quantities of organic solvents used in the liquid chromatography process—around 1 ton of hexane, toluene, and methanol per kg of product. Cyclosporin is produced by fermentation and is extracted in 22% purity. The process uses two steps with silica as the stationary phase and isopropanol in CO_2 as the mobile phase. Cyclosporin A was isolated in 98% purity from the process, which used a column 200 mm in diameter and 280 mm in length packed with 10 micron Kromasil silica. The purification cost was calculated to be around $560 per kg of product, reported to be an order of magnitude less than the conventional HPLC process. One problem area identified was a slow modification of the retention parameters with the number of injections due to fouling of the silica. This was minimized by the use of low-cost, large-particle silica in the first, preliminary purification step, allowing the use of higher value silica for the final cleanup step.

14.5.3 Palm Oil Components

A study on the purification of phytonutrients from palm oil was reported in 2008 [14]. Palm oil is used in food production and as a biodiesel fuel. Although it mainly comprises fatty acid esters of glycerol, it contains small quantities of "phytonutrients" such as tocols, carotenes, and phytosterols, which are calculated to be worth approximately $970 in a ton of biodiesel produced from palm oil. As such, it was considered worth investigating the recovery of the phytonutrients and purification of the individual materials. The separations were carried out using a pilot-scale SFC system with a dynamic axial compression column 350 mm i.d. and co-solvent and CO_2 recycling although other details were not included in the paper.

14.6 Future

The immediate future of production-scale SFC seems to be tied to the production of purified fish oils. The daily dose can be as high as 3 g per day [32], which implies a potential annual consumption well into hundreds of metric tons if even only a small fraction of the world's population decides to consume this product. If the products become approved as pharmaceutical ingredients instead of diet supplements, then a much higher purity will be required, which will probably require chromatographic processing, much as is currently used for the high-purity EPA and DHA products on the market.

Given the costs involved in chromatographic processing, it is likely that production-scale SFC will be largely carried out within the pharmaceutical industry. This industry is also probably the most receptive to the use of chromatographic processing—though there seem to be many barriers to its implementation outside the world of bioprocessing. Some of these relate to the perceived operating costs and complexity of chromatographic systems, which are not always well understood in the industry, while others appear to relate to the availability of capital investment for a new plant. While there is no intrinsic problem that would prevent the implementation of SFC at the production scale, the technique awaits the "killer" application for which SFC is significantly less expensive than any other process.

Another area where SFC could be attractive would be in the processing of smaller quantities of high value products; the SFC systems are versatile, can be turned around rapidly and typically produce materials at a speed three or four times that of HPLC. This, together with the low consumption of organic solvents, could lead to SFC becoming an economically viable proposition for manufacturing organizations that need flexibility and fast turnaround between small production runs.

References

1. Tswett, M. (1905). On a new category of adsorption phenomena and on its application to biochemical analysis, *Trudy Varhavskago Obshchestva Estestvoispytatelei, Otdelenie Biologii,* **14**, pp. 20–39.
2. James, A. T., and Martin, A. J. P. (1950). Gas-liquid partition chromatography; the separation and micro-estimation of volatile fatty acids from formic acid to dodecanoic acid., *Biochem. J.*, **50**, pp. 679–690.
3. Gehrke, C. W., Wixom, R. L., and Bayer, E. (2001). *Chromatography: A Century of Discovery* 1900–2000: The bridge to the sciences/technology (Elsevier, Amsterdam).

4. Ruthven, D. M. (1984). *Principles of Adsorption and Adsorption Processes* (Wiley-Interscience, New York).
5. Hamende, M., and Cavoy, E., (2000). Chiral chromatography: From analytical to production scale, *Chim. Nouvelle*, **18**, pp. 3124–3126.
6. Shaimi, M. (2001). Direct sample injection technique for preparative supercritical fluid chromatography: Solventless injection, *PREP 2001 Conference*, Washington, DC.
7. Stringham, R. W., Krueger, B. R., and Marshall, J. (2007). Use of elevated flow rates in preparative subcritical fluid chromatography. *J. Chromatogr A.*, **1175**, pp. 112–116.
8. Van der Vorst, G., Van Langenhove, H., De Paep, F., Aelterman, W., Dngenen. J., and Dewulf, J. (2009). Exergetic life cycle analysis for the selection of chromatographic separation processes in the pharmaceutical industry: Preparative HPLC *versus* preparative SFC, *Green Chem.*, **11**, pp. 1007–1012.
9. Tarafder, A., and Guiochon, G. (2011). Use of isopycnic plots in designing operations of supercritical fluid chromatography: I. The critical role of density in determining the characteristics of the mobile phase in supercritical fluid chromatography. *J. Chromatogr A*, **1218**, pp. 4569–4575.
10. Dingenen, J. (1998). Columns and packing methods, *Analusis*, **26**, pp. 18–32.
11. Colin, H., Hilaireau, P., and de Tournemire, J. (1991). Dynamic axial compression columns for preparatrive high performance liquid chromatography, *LCGC*, **8**, pp. 4–9.
12. Jusforgues, P. (1995). Instrumental design and separation in large-scale industrial supercritical fluid chromatography, in *Process Scale Liquid Chromatography* (Subramanian, G., ed.), Chapter 7, pp. 153–162 (VCH, Weinheim).
13. Johnson Matthey Pharma Services (2012). Web site: http:// www. jmpharmaservices.com / capacity-separations.htm.
14. Han, N. M., May, C. Y., and Hawari, Y. (2008). Pilot scale supercritical fluid chromatography: An experience, *SFC Conference*, Zurich, Switzerland.
15. Nieto, S., Cordoba, A. M., Sanhueza, J., and Valenzuela, A. (1997). *Grasas y Aceites*, **48**, pp. 197–199.
16. Krumbholz, R., Lembke, P., and Schirra, N. (2011). Chromatography process for recovering a substance or group of substances from a mixture., *US Patent*, 8 063 235.

17. Grill, C. M., and Hampton, T. (1997) Discontinuous countercurrent chromatographic process and apparatus, *US Patent,* 5630943.
18. Valery, E., and Ludemann-Hombourger, O., Method and device for separating fractions of a mixture., *US Patent Appl.,* 20110168632.
19. Clavier, J. Y., Nicoud, R. M., and Perrut, M. (1996). A new efficient fractionation process: The simulated moving bed with supercritical eluent. *Process Technol. Proc.,* **12 (C)**, pp. 429–434.
20. Depta, A., Giese, T., Johanssen, M., and Brunner, G. (1999). Separation of stereoisomers in a simulated moving bed-supercritical fluid chromatography plant. *J. Chromatogr. A,* **865**, pp. 175–186.
21. Ludemann-Hombourger, O., Nicoud, R. M., and Bailly, M. (2000). The VARICOL process: A new multicolumn continuous chromatographic process. *Sep. Sci. Technol.,* **35**, pp. 1827–1860.
22. Peper, S., Johanssen, M., and Brunner, G. (2007). Preparative chromatography with supercritical fluids. Comparison of simulated moving bed and batch processes. *J. Chromatogr. A,* **1176**, pp. 246–253.
23. Di Giovanni, O., Mazzotti, M., Morbidell, M., Denet, F., Hauck, W., Nicoud, R. M. (2001). Supercritical fluid simulated moving bed chromatography II. Langmuir isotherm. *J. Chromatogr A.,* **919**, pp. 1–12.
24. Rubio-Rodriguez, N., Beltran, S., Jaime, I., de Diego, S. M., Sanz, M. T., and Carballido, J. R. (2010). Production of omega-3 polyunsaturated fatty acid concentrates: A review, *Innovative Food Sci. Emerg. Technol.,* **11**, pp. 1–12.
25. http://en.wikipedia.org/wiki/Omega-3_fatty_acid and references therein.
26. Nieto, S., Cordoba, A. M., Sanhueza, J., and Valenzuela, A. (1997). *Grasas y Aceites,* **48**, pp. 197–199.
27. Higashidate, S., Yamauchi, Y., and Saito, M., (1990). Enrichment of eicosapentaenoic acid and docosahexaenoic acid esters from esterified fish oil by programmed extraction-elution with supercritical carbon dioxide. *J. Chromatogr. A,* **515**, pp. 295–303.
28. Berger, C., Jusforgues, P., and Perrut, M. (1988). Purification of unsaturated fatty acid esters by preparative supercritical fluid chromatography, in *Proceedings of the 1st International Symposium on Supercritical Fluids,* Nice, pp. 397–404.
29. Brunner, G., and Reichmann, F. (1998). Process for recovering unsaturated fatty acids and derivatives thereof, *US Patent,* 5777141.
30. Montanes, F., Catchpole, O. J., Tallon, S., Mitchell, K., and Lagutin, K. (2012). Semi-preparative supercritical scale plant for polyunsaturated

fatty acids purification. *10th International Symposium on Supercritical Fluids*, San Francisco.

31. Aaltonen, O., Alkio, M., Lundell, J., Ruohonen, S., Parvinen, L., and Suoninen, V. (1998). *Pharmacuetical Technology Europe*, **10,** pp. A42–A54.
32. FDA ruling on Menhaden oil (2005). *70 FR 14530.*

Index